D0630790

Problem Books in Mathematics

Edited by P. R. Halmos

Problem Books in Mathematics

Series Editor: P.R. Halmos

Unsolved Problems in Intuitive Mathematics, Volume I: Unsolved Problems in Number Theory
by *Richard K. Guy*
1981. xviii, 161 pages. 17 illus.

Theorems and Problems in Functional Analysis
by *A.A. Kirillov* and *A.D. Gvishiani* (trans. *Harold H. McFaden*)
1982. ix, 347 pages. 6 illus.

Problems in Analysis
by *Bernard Gelbaum*
1982. vii, 228 pages. 9 illus.

Bernard Gelbaum

Problems in Analysis

With 9 Illustrations

Springer-Verlag
New York Heidelberg Berlin

Bernard R. Gelbaum
Department of Mathematics
106 Diefendorf Hall
State University of New York
Buffalo, NY 14214
U.S.A.

Library of Congress Cataloging in Publication Data
Gelbaum, Bernard R.
Problems in analysis.
(Problem books in mathematics)
Bibliography: p.
Includes index.
1. Mathematical analysis—Problems, exercises, etc.
I. Title. II. Series.
QA301.G44 1982 515 82-10465

Typeset by J. W. Arrowsmith Ltd., Bristol, England.
Printed and bound by R. R. Donnelley & Sons, Harrisonburg, VA.

Printed in the United States of America.

9 8 7 6 5 4 3 2 1

ISBN 0-387-90692-4 Springer-Verlag New York Heidelberg Berlin
ISBN 3-540-90692-4 Springer-Verlag Berlin Heidelberg New York

Preface

These problems and solutions are offered to students of mathematics who have learned real analysis, measure theory, elementary topology and some theory of topological vector spaces. The current widely used texts in these subjects provide the background for the understanding of the problems and the finding of their solutions. In the bibliography the reader will find listed a number of books from which the necessary working vocabulary and techniques can be acquired.

Thus it is assumed that terms such as *topological space*, *σ-ring*, *metric*, *measurable*, *homeomorphism*, etc., and groups of symbols such as $A \cap B$, $x \in X$, $f:\mathbb{R} \ni x \mapsto x^2 - 1$, etc., are familiar to the reader. They are used without introductory definition or explanation. Nevertheless, the index provides definitions of some terms and symbols that might prove puzzling.

Most terms and symbols peculiar to the book are explained in the various introductory paragraphs titled Conventions. Occasionally definitions and symbols are introduced and explained within statements of problems or solutions.

Although some solutions are complete, others are designed to be sketchy and thereby to give their readers an opportunity to exercise their skill and imagination.

Numbers written in boldface inside square brackets refer to the bibliography.

I should like to thank Professor P. R. Halmos for the opportunity to discuss with him a variety of technical, stylistic, and mathematical questions that arose in the writing of this book.

Buffalo, NY
August 1982

B.R.G.

Contents

Problems

1. Set Algebra

Conventions

The set of positive integers is $\mathbb{N}$; the set of real numbers is $\mathbb{R}$. The set of all subsets of a set X is 2^X. If $\mathsf{E} \subset 2^X$, then $\mathsf{R}(\mathsf{E})$, ($\sigma\mathsf{R}(\mathsf{E})$, $\mathsf{A}(\mathsf{E})$, $\sigma\mathsf{A}(\mathsf{E})$) is the intersection of the (nonempty) set of rings (σ-rings, algebras, σ-algebras) containing E and contained in 2^X. It is the ring (σ-ring, algebra, σ-algebra) generated by E. The set of x in X such that $\cdots$ is $\{x: \cdots\}$. If $A \subset X$ then $A' = \{x: x \notin A\}$ and if $B \subset X$ then $A \backslash B = A \cap B'$. The cardinality of X is $\text{card}(X)$. If X is a topological space then $\mathsf{O}(X)$ ($\mathsf{F}(X)$, $\mathsf{K}(X)$) is the set of open (closed, compact) subsets of X. A subset M of 2^X is monotone if it is closed with respect to the formation of countable unions and intersections of monotone sequences in M, i.e., if $\{A_n: n = 1, 2, \ldots\}$ is a sequence in M and $A_n \subset A_{n+1}$ ($A_n \supset A_{n+1}$) for n in $\mathbb{N}$ then $\bigcup_n A_n (\bigcap_n A_n)$ is in M. The set $\mathsf{M}(\mathsf{E})$ is the monotone subset of 2^X generated by E.

1. Show that if M is monotone and closed with respect to the formation of finite unions and intersections, it is closed with respect to the formation of countable unions and intersections.

2. Show that if M is monotone, R is a ring and $\mathsf{M} \supset \mathsf{R}$, then $\mathsf{M} \supset \sigma\mathsf{R}(\mathsf{R})$.

3. Show that if $2^{\mathbb{R}} \supset \mathsf{M} \supset \mathsf{O}(\mathbb{R})$ and M is monotone then $\mathsf{M} \supset \mathsf{F}(\mathbb{R})$. Repeat, with $\mathbb{R}$ in the preceding sentence replaced by X, a metric space.

4. Show that if $2^{\mathbb{R}} \supset \mathsf{M} \supset \mathsf{O}(\mathbb{R})$ and M is monotone then $\mathsf{M} \supset \sigma\mathsf{R}(\mathsf{O}(\mathbb{R}))$ and $\sigma\mathsf{R}(\mathsf{O}(\mathbb{R})) = \sigma\mathsf{R}(\mathsf{F}(\mathbb{R})) = \sigma\mathsf{R}(\mathsf{K}(\mathbb{R}))$.

5. Show that if S is a σ-ring then $\text{card}(\mathsf{S}) \neq \text{card}(\mathbb{N})$.

6. Show that if $A \in \sigma\mathsf{R}(\mathsf{E})$ then there is in E a finite or countable subset E_0 such that $A \in \sigma\mathsf{R}(\mathsf{E}_0)$.

7. Assume that for each sequence $\{p, q, r, \ldots\}$ of positive integers there is a sequence $\{A_p, B_{pq}, A_{pqr}, \ldots\}$ contained in 2^X. Let the following conditions obtain: i) $A_p = \bigcap_q B_{pq}$, $B_{pq} = \bigcup_r A_{pqr}, \ldots$; ii) for each sequence S of sets $A_p, B_{pq}, A_{pqr}, \ldots,$ there is in $\mathbb{N}$ an $m(S)$ such that each member of S with more than $m(S)$ indices is in E. Let A be the set of all countable unions of sets A_p. Show that A is closed with respect to the formation of countable unions and intersections of its members. Assume additionally: iii) if $E \in \mathsf{E}$ then $E' \in \mathsf{E}$. Show that $\mathsf{A} = \sigma\mathsf{A}(\mathsf{E})$.

2. Topology

Conventions

The set of rational numbers is $\mathbb{Q}$. If A and B are subsets of $\mathbb{R}$ then $A+B=\{x: x=a+b, a\in A, b\in B\}$. Similar conventions apply to AB and in general to "products" of subsets of algebraic structures. The set of complex numbers is $\mathbb{C}$ and $\mathbb{T}=\{z: z\in\mathbb{C}, |z|=1\}$; the latter is regarded as a group under ordinary multiplication. The set of Borel sets of $\mathbb{R}$ is $\sigma\mathsf{R}(\mathsf{O}(\mathbb{R}))$ (see Problem 4).

If Γ and X are sets, X^Γ is the set of all maps of Γ into X. Equivalently, if $\Gamma=\{\gamma\}$ and if for all γ, $X_\gamma=X$ then X^Γ is the Cartesian product $\prod_\gamma X_\gamma$. (To reconcile these notations with 2^X as defined earlier in Set Algebra, Conventions, regard 2 as the set $\{0, 1\}$.) In particular, X^2, and more generally X^n, n in $\mathbb{N}$, is in some sense isomorphic to the n-factor Cartesian product of X with itself.

If X is a topological space containing A, A^0 is the interior of A (the union of all open sets contained in A), $\bar{A}$ is the closure of A (the intersection of all closed sets containing A). The set A is dense in X iff $\bar{A}=X$. If X and Y are topological spaces, $C(X, Y)$ is the set of continuous maps $f: X\mapsto Y$. A map $f: X^\Gamma\mapsto Y$ is finitely (countably) determined iff there is a finite (countable) set $\{\gamma_k\}$ in Γ so that if $x=\{x_\gamma\}$, $y=\{y_\gamma\}$, and $x_{\gamma_k}=y_{\gamma_k}$, $k=1, 2, \ldots$, then $f(x)=f(y)$. The expression $x_n\uparrow x$ ($x_n\downarrow x$) means that the set $\{x_n\}_{n=1}^\infty$ is a monotone increasing (decreasing) sequence of real numbers converging to x. For a topological space X metrized by d, if $E\subset X$ then $\text{diam}(E)=\sup\{d(x, y): x, y\in E\}$. The ball $\{x: d(x, y)\leqq r\}$ is $B(y, r)$ and for positive r, $B(y, r)^0$ is the open ball (both centered at y). If $x\in\mathbb{R}^n$ and $x=(x_1, x_2, \ldots, x_n)$ then $\|x\|=\sqrt{\sum_{k=1}^n x_k^2}$. This norm endows $\mathbb{R}^n$ with the standard Euclidean metric: $d:(x, y) \to \|x-y\|$. If $T:X \mapsto X$ is a map and n is in $\mathbb{N}$, T^n is the nth iterate of T, i.e., $T^n(x) = T(T^{n-1}(x))$, $n = 2, 3, \ldots$.

If $\{X_\gamma\}_{\gamma\in\Gamma}$ is a set of topological spaces and $X = \prod_\gamma X_\gamma$, a basic neighborhood is defined by finitely many open sets U_{γ_i}, each contained in X_{γ_i}, $i = 1, 2, \ldots, n$, and engendering the subset $U_{\gamma_1} \times U_{\gamma_2} \times \cdots \times U_{\gamma_n} \times \prod_{\gamma \notin \{\gamma_1, \gamma_2, \ldots, \gamma_n\}} X_\gamma$.

8. Prove or disprove: there is a continuous map of $[0, 1)$ onto $\mathbb{R}$.

9. Find in $[0, 1]$ an uncountable subset E such that the interior of $E-E$ is empty.

10. Metrize $[0, 1)$ so that it is complete and homeomorphic to $[0, 1)$ in its usual topology.

11. Show that if f maps $[0, 1]$ homeomorphically onto $A \times B$ (Cartesian product) then A or B consists of precisely one element.

12. Let *the* Cantor set C be $\{a: a = \sum_{k=1}^{\infty} \alpha_k 3^{-k}, \alpha_k = 0 \text{ or } 2, k = 1, 2, \ldots\}$. Let D_n be $\{0, 1\}$ in the discrete topology, $n = 1, 2, \ldots$. Show C is homeomorphic to $\prod_{n=1}^{\infty} D_n$ and that $f_k: C \ni a \mapsto (-1)^{\alpha_k/2}$ is continuous, $k = 1, 2, \ldots$.

13. Assume $\mathsf{F}(\mathbb{R}) \ni F \subset \bigcup_{\gamma\in\Gamma}(a_\gamma, b_\gamma]$. Show there is in Γ a countable subset $\{\gamma_n\}$ such that $F \subset \bigcup_n (a_{\gamma_n}, b_{\gamma_n}]$.

14. Metrize $\mathbb{R}$ by $d: \mathbb{R} \times \mathbb{R} \ni (x, y) \mapsto |\arctan x - \arctan y|$. Prove or disprove that $\mathbb{R}$ is complete in the metric d.

15. Assume $S \subset [0, \infty)$, $u = \sup S$, $u < 1$ and that if x and y are in S and $x < y$ then $x/y \in S$. Show that $u \in S$.

16. Prove or disprove: $\mathbb{R}\backslash\mathbb{Q}$ and $(\mathbb{R}\backslash\mathbb{Q}) \cap (0, 1)$ are homeomorphic.

17. Let E be a compact, countable, and nonempty subset of $\mathbb{R}^2$. Show that there is an isolated point in E.

18. Show that if $\mathbb{R}^2$ is the countable union of closed sets F_n then the union of their interiors is dense in $\mathbb{R}^2$.

19. Show that if A is an uncountable subset of $\mathbb{R}^2$ then there is in A an x such that for every neighborhood $U(x)$, $U(x) \cap A$ is uncountable.

20. Construct a compact metrizable space X and a self-homeomorphism T of X (onto) X such that for no metric d compatible with the topology of X is it true that $d(Tx, Ty) = d(x, y)$ for all x and y in X.

21. Let X be a compact metric space and let $\{T_\gamma\}_{\gamma\in\Gamma}$ be an equicontinuous subset of $C(X, X)$. Show that for every f in $C(X, X)$ the set $\overline{\{f \circ T_\gamma\}}_{\gamma\in\Gamma}$ is compact.

22. Show that if X and Y are compact metric spaces and $f \in Y^X$, then f is continuous iff the graph $\{(x, f(x)) : x \in X\}$ of f is closed in $X \times Y$.

23. Let X be a compact metric space with metric d and let f be in $C(X, X)$ and such that $d(f(a), f(b)) \geqq d(a, b)$ for all a and b in X. Show that $d(f(a), f(b)) = d(a, b)$ for all a, b in X.

24. Show that if X is a separable metric space then $\text{card}(\mathsf{F}(X)) \leqq \text{card}(\mathbb{R})$.

25. Let A be $C([0, 1], [0, 1])$ metrized according to $d: A \times A \ni (f, g) \mapsto \sup\{|f(x) - g(x)| : x \in [0, 1]\}$. Let A_i be the set of injective and A_s be the set of surjective elements of A and let A_{is} be $A_i \cap A_s$. Prove or disprove: i) A_i is closed; ii) A_s is closed; iii) A_{is} is closed; iv) A is connected; v) A is compact.

26. Let X be a complete metric space having no isolated points. Show that if U is a nonempty open set of X show then $\text{card}(U) \geqq \text{card}(\mathbb{R})$.

27. Let X and Y be compact Hausdorff spaces and $f: X \mapsto Y$ a continuous surjection such that for all y in Y, $f^{-1}(y)$ is connected. Show that for every connected subset C of Y, $f^{-1}(C)$ is connected. Give a counterexample to the conclusion if the hypothesis *Y is Hausdorff* is dropped.

28. Let $\{X_\gamma\}_{\gamma \in \Gamma}$ be a set of compact Hausdorff spaces. Show that if $X = \prod_\gamma X_\gamma$, $f \in C(X, \mathbb{R})$, and $\varepsilon > 0$ then there is a finitely determined $g: X \mapsto \mathbb{R}$ such that $\sup_x |f(x) - g(x)| < \varepsilon$. Show f is countably determined.

29. Let X be a compact space and let $f: X \mapsto \mathbb{R}$ be such that for all x, $f^{-1}([x, \infty))$ is closed. Show that for some M in $\mathbb{R}$ and for all x in X, $f(x) \leqq M$ and that for some x_0 in X, $f(x_0) = \sup_x f(x)$.

30. Let K be compact and a subset of the union of two open sets U and V in a Hausdorff space X. Show there are compact sets K_U and K_V contained respectively in U and V and such that $K = K_U \cup K_V$.

31. For all γ in Γ let I_γ be $[0, 1]$ and let X be $\prod_\gamma I_\gamma$. Show that if $\text{card}(\Gamma) = \text{card}(\mathbb{R})$ then there is a countable dense subset in X.

32. Show that $[0, 1]^{[0,1]}$ is not metrizable.

33. Show that if $f \in C([0, 1], \mathbb{R})$ and the subset A of $[0, 1]$ is the countable union of closed sets, i.e., A is an F_σ, then $f(A)$ is an F_σ.

34. Show that if $f \in C(\mathbb{T}, \mathbb{R})$ then there is in $\mathbb{T}$ a z such that $f(z) = f(ze^{i\pi})$. ("For some x in $\mathbb{R}$, $f^{-1}(x)$ contains two antipodal points.")

35. Show that if $f \in C(\mathbb{R}, \mathbb{R})$ and V is open then $f(V)$ is a Borel set.

36. If Y is a topological space such that for all n in $\mathbb{N}$, Y^n and Y are homeomorphic, need Y and $Y^{\mathbb{N}}$ be homeomorphic?

37. Let $\{F_n\}_{n=1}^{\infty}$ be a subset of $\mathsf{F}([0, 1])$. Show that if all F_n are nonempty and they are pairwise disjoint, then $[0, 1] \neq \bigcup_n F_n$.

3. Limits

Conventions

The series $\sum_{n=1}^{\infty} a_n$ may or may not converge. When it does its sum is $\sum_{n=1}^{\infty} a_n$. If p is a polynomial its degree is $\deg(p)$. The characteristic function of a set E is χ_E. If $n \geqq 2$ and if E is a Borel set in $\mathbb{R}^n$ the Lebesque measure of E is $\lambda_n(E)$. If $n=1$, the corresponding number is $\lambda(E)$. If ambiguity is unlikely λ_n will be written λ.

38. Show there are real constants C and D such that if $n \geqq 2$ then $C \log n \leqq \sum_{k=1}^{\infty} (1-(1-2^{-k})^n) \leqq D \log n$.

39. Show that if $0 < a_n \leqq \sum_{k=n+1}^{\infty} a_k$, $n=1, 2, \ldots$, and $\sum_{k=1}^{\infty} a_k = 1$, then for every x in $(0, 1)$ there is a subseries $\sum_{p=1}^{\infty} a_{k_p}$ whose sum is x.

40. Show that if $a_n, b_n \in \mathbb{R}$, $(a_n + b_n)b_n \neq 0$, $n = 1, 2, \ldots$, and both $\sum_{n=1}^{\infty} a_n/b_n$ and $\sum_{n=1}^{\infty} (a_n/b_n)^2$ converge, then $\sum_{n=1}^{\infty} a_n/(a_n+b_n)$ converges.

41. Find $\{a_n\}_{n=1}^{\infty}$ in $[0, \infty)$ so that $na_n \to 1$ as $n \to \infty$ and yet $\sum_{n=1}^{\infty}(-1)^n a_n$ diverges.

42. Show that if $b_n \downarrow 0$ and $\sum_{n=1}^{\infty} b_n = \infty$ then there is in $\mathbb{R}$ a sequence $\{a_n\}_{n=1}^{\infty}$ such that $a_n/b_n \to 1$ as $n \to \infty$ and $\sum_{n=1}^{\infty}(-1)^n a_n$ diverges.

43. Prove or disprove: If $\{p_n\}_{n=1}^{\infty}$ is a sequence of polynomials for which $\deg(p_n) \leqq M < \infty$, n in $\mathbb{N}$, and $p_n \to f$ uniformly on $[0, 1]$ as $n \to \infty$ then f is a polynomial. Is pointwise convergence enough?

44. Show that $\int_x^{\infty} e^{-t^2/2}\, dt\, e^{x^2/2}$ is a monotone decreasing function of x on $[0, \infty)$ and that its limit as $x \to \infty$ is 0.

45. Show that $\lim_{n\to\infty} n \sin(2\pi e n!) = 2\pi$ (whence $e \notin \mathbb{Q}$).

46. Show that if $\{r_n\}_{n=1}^{\infty} \subset \mathbb{R}$ then $\lim_{n\to\infty} \int_0^\infty e^{-x}[\sin(x + r_n\pi/n)]^n \, dx = 0$.

47. Show $\lim_{\varepsilon\to 0} \int_0^\infty (1 - e^{(\varepsilon x)^2}) \, e^{-x^3} \sin^4 x \, dx = 0$.

48. Evaluate: $\lim_{n\to\infty} \int_{-\infty}^{\infty} (1 - e^{-t^2/n}) \, e^{-|t|} \sin^3 t \, dt$.

49. Let f be

$$[0, 1] \ni x \mapsto \begin{cases} (x \log x)/(x-1) \text{ if } 0 < x < 1 \\ 0 \text{ if } x = 0. \\ 1 \text{ if } x = 1 \end{cases}$$

Show that $\int_0^1 f(x) \, dx = 1 - \sum_{n=2}^{\infty} 1/n^2(n-1)$.

4. Continuous Functions

Conventions

If A is a subset of a vector space V over a field $\mathbb{K}$ (usually $\mathbb{K}=\mathbb{R}$ or $\mathbb{K}=\mathbb{C}$) the linear span of A is the set $\{\sum_{k=1}^{n} \alpha_k a_k : \alpha_k \in \mathbb{K},\ a_k \in A,\ n \in \mathbb{N}\}$ and the convex hull of A is the set $\{\sum_{k=1}^{n} \alpha_k a_k : \alpha_k \in \mathbb{R},\ 0 \leqq \alpha_k,\ \sum_{k=1}^{n} \alpha_k = 1,\ a_k \in A,\ n \in \mathbb{N}\}$. If X is a topological space and $f \in \mathbb{R}^X$ then for x in X, $\limsup_{y=x} f(y)$ ($\liminf_{y=x} f(y)$) is $\inf\{\sup_{y\in U} f(y): U$ a neighborhood of $x\}$ ($\sup\{\inf_{y\in U} f(y): U$ a neighborhood of $x\}$); f is upper (lower) semicontinuous, *usc* (*lsc*) iff $f(x)=\limsup_{y=x} f(y)$ ($\liminf_{y=x} f(y)$). The set $C_b(X, \mathbb{C})$ is the set of bounded continuous functions on X; its norm is given by $\|\cdot\cdot\cdot\|_\infty : f \mapsto \sup_x |f(x)|$.

The set $C^k(A, \mathbb{C})$, k in $\mathbb{N}$, consists of all functions having a kth derivative continuous on A; $C^\infty(A, \mathbb{C}) = \bigcap_{k=1}^{\infty} C^k(A, \mathbb{C})$; for k in $\mathbb{N}$ the norm $\|\cdot\cdot\cdot\|^{(k)}$ for $C^k([0,1], \mathbb{C})$ maps f into $\sum_{j=0}^{k} \|f^{(j)}\|_\infty$. Similar definitions apply when $\mathbb{C}$ is replaced by $\mathbb{R}$. The unit ball of a Banach space is the set of elements with norm not greater than one.

To emphasize that an integration is carried out in the sense of Lebesgue rather than of Riemann the notation $\int_E f(x)\, d\lambda(x)$ will be used occasionally for the Lebesgue integral of f whereas $\int_E f(x)\, dx$ will be the only notation for the Riemann integral of f (in both cases over the set E). If X is a set, S is a σ-ring of sets in X, and μ is a measure defined on S then (X, S, μ) denotes the situation just described. If p is positive $L^p(X, \mu)$ is the set of (equivalence classes of) measurable functions f such that $\|f\|_p^p$ given by $\int_X |f(x)|^p\, d\mu(x)$ is finite; $L^\infty(X, \mu)$ is the set of (equivalence classes of) essentially bounded measurable functions f and $\|f\|_\infty$ is the essential supremum of $|f|$.

If X is a locally compact Hausdorff space, $C_0(X, \mathbb{C})$ is the set of continuous functions "vanishing at infinity", i.e., functions f such that for positive ε there is a compact set $K(\varepsilon, f)$ off of which $|f|$ is not more than ε; $C_{00}(X, \mathbb{C})$ is the set of continuous functions having compact support (supp(f), the support of f, is the closure of the set where f is not zero).

If f, g are $\mathbb{R}$-valued functions on a set X, $f^+ = (|f|+f)/2$, $f^- = (|f|-f)/2$; thus $g+(f-g)^+ = \max(f, g)$ and $g-(f-g)^- = \min(f, g)$, $f \vee g = \max(f, g)$, $f \wedge g = \min(f, g)$.

If E is a topological vector space over $\mathbb{K}$, E^* is the vector space of continuous linear maps of E onto $\mathbb{K}$; E^* is the conjugate or dual space of E.

50. Show that if $f, g \in C([0, 1], \mathbb{R})$ and $g(y_1) = g(y_2)$ whenever $f(y_1) = f(y_2)$ then there is a sequence $\{p_n\}_{n=1}^\infty$ of polynomials such that $p_n(f) \to g$ uniformly on $[0, 1]$ as $n \to \infty$.

51. Let f be defined as follows:

$$f(x) = \begin{cases} (3x+1)/2, & -1 \leqq x \leqq -1/3 \\ 0, & -1/3 < x \leqq 1/3 \\ (3x-1)/2, & 1/3 < x \leqq 1. \end{cases}$$

Let L belong to $C([-1, 1], \mathbb{C})^*$ and assume that if $n = 1, 2, \ldots$, $L(\underbrace{f \circ f \circ f \circ \cdots \circ f}_{n}) = 0$. Show that if $g \in C([-1, 1], \mathbb{C})$ and

$$g([-1/3, 1/3]) = \{0\}$$

then $L(g) = 0$.

52. For the maps $f_n: [0, 1] \ni x \mapsto e^{nx}$, n in $\mathbb{N}$, let S_N be $\{f_n : n \geqq N\}$, N in $\mathbb{N}$. Show that for all N in $\mathbb{N}$ the linear span of S_N is dense in $C([0, 1], \mathbb{C})$.

53. Show that if $f \in C([0, 1], \mathbb{R})$ and $\int_0^1 x^n f(x)\,dx = 0$ or $\int_0^1 e^{\pm 2\pi inx} f(x)\,dx = 0$ for all n in $\mathbb{N} \cup \{0\}$ then $f = 0$.

54. Construct a sequence $\{a_n\}_{n=1}^\infty$ in $\mathbb{C}$ so that for any f in $C([0, 1], \mathbb{C})$ and for which $f(0) = 0$ there is in $\mathbb{N}$ a sequence $\{n_k\}_{k=1}^\infty$ (dependent on f) such that $\sum_{n=1}^{n_k} a_n x^n \mapsto f$ uniformly on $[0, 1]$ as $k \to \infty$.

55. Show that if G is an open unbounded subset of $[0, \infty)$ and if $D = \{x : x \in (0, \infty),\ nx \in G$ for infinitely many n in $\mathbb{N}\}$ then D is dense in $[0, \infty)$.

56. Show that if $f \in C((0, \infty), \mathbb{R})$, $0 < a < b < \infty$, and for all h in (a, b), $f(nh) \to 0$ as $n \to \infty$ then $f(x) \to 0$ as $x \to \infty$.

57. Let f_0 be in $C([0, 1], \mathbb{R})$ and let $f_n(x)$ be $\int_0^x f_{n-1}(t)\,dt$ for n in $\mathbb{N}$ and x in $[0, 1]$. Show that if for each x in $[0, 1]$ there is in $\mathbb{N}$ an n (dependent on x) such that $f_n(x) = 0$, then the following are true: i) there is in $[0, 1]$ a nonempty open set on which f_0 is 0; ii) for every n in $\mathbb{N}$ and every b in $(0, 1]$, f_n has infinitely many zeros in $(0, b)$.

58. Let g be in $C([0,1],[0,1])$ and such that $g(0)=1-g(1)=0$. Let (*per* convention) g^n denote the n-fold composition $\underbrace{g\circ g\circ\cdots\circ g}_{n}$ and assume there is an m such that $g^m(x)=x$ for all x in $[0,1]$. Show $g(x)=x$ for all x in $[0,1]$.

59. Prove or disprove: i) if f is left-continuous on $[0,1]$ then f is bounded; ii) if f is *usc* on $[0,1]$ then f is bounded above.

60. Show that if $f\in C([0,1],\mathbb{R})$ and $f(0)=0$ then the sequence $\{\underbrace{f.f.\cdots.f}_{n}\}_{n=1}^{\infty}$ is equicontinuous iff $\|f\|_\infty<1$.

61. For f in $C([0,1],\mathbb{R})$ and n in $\mathbb{N}$ let a_n be $(\int_0^1 x^n f(x)\,dx)/(\int_0^1 x^n\,dx)$. Show that $\lim_{n\to\infty} a_n$ exists.

62. Let f be in $C([0,1],\mathbb{R})$ and assume that for some c in $(0,1)$ $\lim_{\substack{h\in\mathbb{Q},h\neq 0\\ h\to 0}}[f(c+h)-f(c)/h]$ exists and is L. Show f is differentiable at c.

63. Show that if $f,g\in C(\mathbb{R},\mathbb{R})$ and if, for all compactly supported h in $C^\infty(\mathbb{R},\mathbb{R})$, $\int_{-\infty}^{\infty} f(x)h(x)\,dx=-\int_{-\infty}^{\infty} g(x)h'(x)\,dx$, then g is differentiable and $g'=f$.

64. Let A be $\{f: f\in C^3([0,1],\mathbb{R}), \|f\|_\infty, \|f'''\|_\infty\leqq 1\}$. Show there is a constant K such that for all f in A, $\|f'\|_\infty, \|f''\|_\infty\leqq K$.

65. Assume $\{f_n\}_{n=1}^{\infty}\subset C([0,1],\mathbb{R})$, that each f_n is differentiable and that $\|f_n'\|_\infty\leqq 1$. Show that if $\int_0^1 f_n(x)g(x)\,dx\to 0$ as $n\to\infty$ for all g in $C([0,1],\mathbb{R})$ then $\|f_n\|_\infty\to 0$ as $n\to\infty$.

66. Prove or disprove: if $f\in C([1,\infty),\mathbb{R})$ there is a sequence $\{p_n\}_{n=1}^{\infty}$ of polynomials such that $p_n\to f$ uniformly on $[1,\infty)$ as $n\to\infty$.

67. Show that if $\{a_n\}_{n=1}^{\infty}\subset\mathbb{C}$ then there is in $C^\infty(\mathbb{R},\mathbb{C})$ an f such that for n in $\mathbb{N}$, $f^{(n)}(0)=a_n$.

68. Show that if $f\in C(\mathbb{R},\mathbb{R})$ and $|f|$ is improperly Riemann integrable then f is improperly Riemann integrable and Lebesgue integrable and that $\int_{\mathbb{R}} f(x)\,d\lambda(x)=\lim_{\substack{t\to\infty\\ s\to-\infty}}\int_s^t f(x)\,dx$.

69. Let f be in $C^1(\mathbb{R},\mathbb{C})$ and assume f' is real analytic (for each a in $\mathbb{R}$ there is a sequence $\{b_n(a)\}_{n=1}^{\infty}$ and a positive $r(a)$ such that for all x in $(a-r(a),a+r(a))f'(x)=\sum_{n=0}^{\infty} b_n(a)(x-a)^n$). Show f is real analytic.

70. Show that if $g\in C(\mathbb{R},\mathbb{R})$ and $\lim_{T\to\infty}\int_{-T}^{T} g(x)h(x)\,dx$ exists for all h in $L^2(\mathbb{R},\lambda)$, then $g\in L^2(\mathbb{R},\lambda)$.

71. Show that if $0<a, f\in C(\mathbb{R}^n,\mathbb{R}^n)$, and for all x, y in $\mathbb{R}^n$, $|f(x)-f(y)|\geqq a|x-y|$, then $f(\mathbb{R}^n)=\mathbb{R}^n$ (f is surjective).

72. Give a useful necessary and sufficient condition that a set F closed in $[0,\infty)$ be such that every f in $C([0,\infty),\mathbb{R})$ is uniformly approximable on

F by "polynomials in x^2", i.e., by compositions of polynomials p and the function $x \mapsto x^2$.

73. Let U be $\{x : x \in \mathbb{R}^2, \|x\| < 1\}$. Show that if $f \in C(\bar{U}, \mathbb{R})$ and for all x in U and all r in $(0, 1 - \|x\|)$ there obtains:

$$f(x) = \frac{\oint_{\|y-x\|=r} f(y)\, dy}{2\pi r} \text{ (line integral),}$$

then f is constant iff $f(0) = \sup_{x \in \bar{U}} |f(x)|$.

74. Let (a, b) be a finite interval in $\mathbb{R}$. Prove or disprove: if f is uniformly continuous on (a, b) then f is bounded on (a, b).

75. In $C([0, 2\pi], \mathbb{C})$ let A be the subset consisting of those functions f corresponding to Fourier series $\sum_{n=-\infty}^{\infty} a_n e^{inx}$ for which $\sum_{n=-\infty}^{\infty} |a_n|(1 + |n|) \leqq 1$. (Note: $a_n = \int_0^{2\pi} f(x) e^{-inx}\, dx/2\pi$, $n = 0, \pm 1, \pm 2, \ldots$.) Show that $\bar{A}$ is compact in the topology induced by $\|\cdot \cdot \cdot\|_\infty$.

76. Show that if $f \in C^1([0, 2\pi], \mathbb{C})$, $f(0) = f(2\pi)$ and $\int_0^{2\pi} f(x)\, dx = 0$, then $\|f\|_2 \leqq \|f'\|_2$ and equality obtains iff for some a, b, $f(x) = a \cos x + b \sin x$ for all x.

77. Show that if $f, g \in C(\mathbb{R}, \mathbb{C})$ and for all x, $f(x+1) = f(x)$, $g(x+1) = g(x)$, then $\lim_{n \to \infty} \int_0^1 f(x) g(nx)\, dx = (\int_0^1 f(x)\, dx)(\int_0^1 g(x)\, dx)$.

78. For f in $C(\mathbb{T}, \mathbb{R})$ and x in $\mathbb{T}$ the translate $f_{(x)}$ is the map $z \mapsto f(xz)$. Show that in the closure of the convex hull of the set A of translates of f there is precisely one constant function and find its value (in terms of f).

79. Let X be a complete metric space and let $\mathcal{F}$ be a subset of $C(X, \mathbb{C})$. Assume that for each x in X, $\sup_{f \in \mathcal{F}} |f(x)| \leqq M_x < \infty$. Show there is in X a nonempty open subset Ω and there is a positive M such that $\sup_{f \in \mathcal{F}} |f(x)| \leqq M$ for all x in Ω.

80. Let X be a compact Hausdorff space and let $\{f_n\}_{n=1}^{\infty}$ be a subset of $C(X, \mathbb{C})$. Show that if, for all n in $\mathbb{N}$, $\|f_n\| \leqq M$ and if, for all x in X, there exists $\lim_{n \to \infty} f_n(x)$, then for every L in $C(X, \mathbb{C})^*$ there exists $\lim_{n \to \infty} L(f_n)$.

81. Show that if X and Y are compact Hausdorff spaces and A is the set $\{\sum_{i=1}^{n} f_i g_i : (f_i, g_i) \in C(X, \mathbb{C}) \times C(Y, \mathbb{C}),\ n \text{ in } \mathbb{N}\}$ then A is dense in $C(X \times Y, \mathbb{C})$.

82. Let X be a metric space and let f be in $\mathbb{C}^X$. Show that the set C of points where f is continuous is the countable intersection of open sets (C is a G_δ).

83. If X is a locally compact and not compact Hausdorff space, let B_1 be the unit ball of $C_0(X, \mathbb{R})$. Find the set of extreme points of B_1. (An extreme point P of a convex set S in a vector space is one that cannot be described as $\alpha A + \beta B$ for A, B in S, one of $A, B \neq P$, α, β positive and $\alpha + \beta = 1$.)

84. Let A be closed in $[0, 1]$. Assume that for each compact metric space K there is a continuous map $f_K: A \mapsto K$ such that $f_K(A) = K$ (f_K is a continuous surjection of A (onto K)). Show: i) the cardinality of the set $\mathscr{C}$ of components of A is card($\mathbb{R}$); ii) A is the union of three disjoint sets P, D, U such that P is homeomorphic to *the* Cantor set (see 12.), D is countable, and contains no nonempty subset dense in itself and U is open; iii) if $B = A \backslash A^0$, then, for every compact metric space K, there is a continuous surjection $g_K: B \mapsto K$.

85. Prove or disprove: if $\{f_n\}_{n=0}^{\infty} \subset C([0, 1], \mathbb{C})$ and for each L in $C([0, 1], \mathbb{C})^*$ $\lim_{n \geqq 1, n \to \infty} L(f_n) = L(f_0)$, then whenever $x_n \to x$ as $n \to \infty$, $\lim_{n \geqq 1, n \to \infty} f_n(x_n) = f_0(x)$.

86. Assume $L \in C([0, 1], \mathbb{C})^*$ and $\sup\{|L(f)|: \|f\|_\infty \leqq 1\} = 1$. Prove or disprove there is in $C([0, 1], \mathbb{C})$ an f such that $\|f\|_\infty \leqq 1$ and $L(f) = 1$.

87. Let A be $\{f: f \in C([0, 1], \mathbb{C}),\ f(\frac{1}{2}) = 0\}$. Prove or disprove: i) A is a principal ideal in $C([0, 1], \mathbb{C})$, i.e., there is in A an f_0 such that $f_0 . C([0, 1], \mathbb{C}) = A$; ii) there is in A a g_0 such that $\overline{g_0 . C([0, 1], \mathbb{C})} = A$.

88. Show that the inclusion map $T: C^1([0, 1], \mathbb{C}) \ni f \hookrightarrow f \in C([0, 1], \mathbb{C})$ is compact, i.e., for every bounded set B, $\overline{T(B)}$ is compact.

89. On $C^1([0, 1], \mathbb{C})$ let two norms be given, *viz.*, $\|\cdot \cdot \cdot\|': f \mapsto (\int_0^1 |f(x)|^2\, dx)^{1/2}$ and $\|\cdot \cdot \cdot\|'': f \mapsto (\int_0^1 (|f(x)|^2 + |f'(x)|^2) dx)^{1/2}$. Let E_1 and E_2 be the respective completions *re* these norms of $C^1([0, 1], \mathbb{C})$. Let D be the operator of differentiation. Show it has a continuous extension $\tilde{D}: E_2 \to E_1$ and that $\tilde{D}^{-1}(0) = \{f: f \text{ is constant}\}$.

90. Let X be a subspace of $C^1([0, 1], \mathbb{C})$. Show that if X is also a closed subspace of $C([0, 1], \mathbb{C})$, then i) X is a closed subspace of $C^1([0, 1], \mathbb{C})$; ii) there are positive constants k, K such that for all f in X, $k\|f\|^{(1)} \leqq \|f\|_\infty \leqq K\|f\|^{(1)}$; iii) X is finite-dimensional.

91. Let K be a compact subset of $[0, 1]$ and let A_K be $\{f: f \in C([0, 1], \mathbb{C}),\ f(K) = 0\}$ and B_K be $\{f: f \in C([0, 1], \mathbb{C}), f(V) = 0$ for some open V containing $K\}$. Show: i) $\bar{B}_K = A_K$; ii) if $\varnothing \neq K$ then A_K is not a principal ideal.

92. In $C([0, 1], \mathbb{R})$ let A be $\{f: f(\mathbb{Q}) \subset \mathbb{Q}\}$. Describe A^0; show A is a dense Borel set, i.e., show $\bar{A} = C([0, 1], \mathbb{R})$ and $A \in \sigma\mathsf{R}(\mathsf{O}(C([0, 1], \mathbb{R})))$.

93. Let A be a closed subspace of $C([0, 1], \mathbb{R})$ and let g in $\mathbb{R}^{[0,1]}$ be such that for all f in A, $gf \in A$. Show that the map $M_g: A \ni f \mapsto gf$ is linear and continuous.

94. Let f be in $C([0, 1], \mathbb{R})$. Prove or disprove: for each positive ε there is a finite set of rectangles covering the graph of f (see Problem 22) and having a total area less than ε (the Jordan content of the graph of f is zero).

5. Functions from $\mathbb{R}^n$ to $\mathbb{R}^m$

Conventions

If $\varphi: E \mapsto \mathbb{R}$ is a map into $\mathbb{R}$ of a convex subset E of a vector space V, φ is convex iff whenever $x, y \in E$, $0 \leqq \alpha, \beta$, and $\alpha + \beta = 1$ then $\varphi(\alpha x + \beta y) \leqq \alpha\varphi(x) + \beta\varphi(y)$. If V and W are (topological) vector spaces $\text{Hom}(V, W)$ is the vector space of (continuous) linear maps from V to W. If $V = W$, $\text{Hom}(V, W)$ is denoted $\text{End}(V)$ and the identity map is *id.*

The differential (called by some the derivative) at x_0 of the map $f: \mathbb{R}^n \mapsto \mathbb{R}^m$ is, if it exists, a continuous linear map $df(x_0): \mathbb{R}^n \mapsto \mathbb{R}^m$ such that

$$\lim_{\substack{h \neq 0 \\ \|h\| \to 0}} \frac{\|f(x_0 + h) - f(x_0) - df(x_0)(h)\|}{\|h\|} = 0.$$

By induction higher differentials $d^k f$, $k = 2, 3, \ldots$, may be defined similarly. The domains and ranges deserve special attention. Thus, e.g., df is a map from $\mathbb{R}^n$ to $\text{Hom}(\mathbb{R}^n, \mathbb{R}^m)$, whence $d^2 f$ is a map from $\mathbb{R}^n$ to $\text{Hom}(\mathbb{R}^n, \text{Hom}(\mathbb{R}^n, \mathbb{R}^m))$. For consistency $d^0 f = f$. The symbol $C^k(\mathbb{R}^n, \mathbb{R}^m)$ stands for the set of functions f having k continuous differentials; $C^\infty(\mathbb{R}^n, \mathbb{R}^m) = \bigcap_{k=1}^\infty C^k(\mathbb{R}^n, \mathbb{R}^m)$.

The boundary ∂S of a set S in a topological space X is $\{x$: for all open sets U containing x, both $U \cap S$ and $U \cap (X \backslash S)$ are nonempty$\}$.

If $f \in \mathbb{R}^{[a,b]}$, f is in $\text{Lip}(\alpha)$ iff for some constant K and all x, y in $[a, b]$, $|f(x) - f(y)| \leqq K|x - y|^\alpha$.

If ∞ is regarded as adjoined to $\mathbb{N}$ or $\mathbb{R}$ or $\mathbb{Z}$ (the set of all integers) and if the neighborhoods of ∞ are the complements of bounded closed sets, then $\limsup_{x=\infty}$ and $\liminf_{x=\infty}$ may be regarded as defined by previous conventions; similar remarks apply to $\limsup_{x=-\infty}$ and $\liminf_{x=-\infty}$ for $\mathbb{R}$

and $\mathbb{Z}$. More directly these may be defined, e.g., as follows: $\limsup_{x=\infty} f(x) = \inf_r\{\sup_{x>r} f(x)\}$.

In particular if $\{E_n\}_{n=1}^{\infty}$ is a sequence of sets in a set X, then for each x, $\chi_{E_n}(x)$ may be regarded as an element of $\{0, 1\}^{\mathbb{N}}$, i.e., $2^{\mathbb{N}}$; $\limsup_{n=\infty}$ and $\liminf_{n=\infty}$ applied to the resulting function define the characteristic functions of two sets denoted $\limsup_{n=\infty} E_n$ resp. $\liminf_{n=\infty} E_n$. More directly these may be defined, e.g., as follows: $\limsup_{n=\infty} E_n = \bigcap_{n=1}^{\infty} \bigcup_{m=n}^{\infty} E_m$. Finally, if $f \in \mathbb{R}^{\mathbb{R}}$ then

$$\begin{matrix} \limsup_{x\downarrow a} f(x) & & \limsup_{x\uparrow a} f(x) \\ & \text{and} & \\ \liminf_{x\downarrow a} f(x) & & \liminf_{x\uparrow a} f(x) \end{matrix}$$

(four items) are to be interpreted in an obvious fashion. In all instances, if "lim sup = lim inf" then "lim" is said to exist and to be the common "value" of "lim sup" and "lim inf". (If $\mathbb{R}$-valued functions are the arguments, "value" is an element of $\mathbb{R} \cup \{\infty\} \cup \{-\infty\}$; if sets are the arguments, "value" is set in X.)

For z in $\mathbb{C}$ the map

$$\text{sgn: } z \mapsto \left\{ \begin{matrix} 0, \text{ if } z = 0 \\ |z|/z, \text{ if } z \neq 0 \end{matrix} \right\}$$

defines "sign(z)" or "signum(z)". Hence $z \cdot \text{sgn}(z) = |z|$.

As noted parenthetically earlier, the set of all integers, $0, \pm 1, \pm 2, \ldots$ is $\mathbb{Z}$.

The determinant of a square matrix M is $\det(M)$.

If X and Y are normed linear spaces and $T \in \text{Hom}(X, Y)$, then $\|T\| = \sup_{\|x\|=1} \|T(x)\|$.

If A is a subset of the metric space (X, d) then $\text{diam}(A) = \sup\{d(x, y) : x, y \in A\}$.

95. Let f be in $\mathbb{R}^{\mathbb{R}}$ and assume for each x and some positive δ depending on x that whenever $x - \delta < a < x < b < x + \delta$, $f(a) \leqq f(x) \leqq f(b)$. Show that if $p < q$ then $f(p) \leqq f(q)$. ("If f is locally monotone (increasing) then f is monotone (increasing).")

96. Prove or disprove: If $\varphi, \psi \in \mathbb{R}^{\mathbb{R}}$ and both are convex so is $\varphi \circ \psi$. Repeat for the case that ψ is monotone increasing.

97. Give an example of a convex positive function φ such that $\log(\varphi)$ is not convex.

98. Show that if $\varphi \in \mathbb{R}^{\mathbb{R}}$, $\varphi > 0$, and $\log(\varphi)$ is convex than φ is convex.

99. Show that if φ is convex and $x \leqq x' < y \leqq y'$ then $(\varphi(y) - \varphi(x))/(y - x) \leqq (\varphi(y') - \varphi(x'))/(y' - x')$. Give a geometrical interpretation to the result.

100. (Extension of Problem 99.) Show that if φ is as in Problem 99 then it is in Lip(1) and right and left differentiable everywhere.

101. Show that if φ is as in Problem 99, then φ is monotone increasing, monotone decreasing, or there is a (possibly degenerate) closed interval $[p, q]$ such that on $(-\infty, p)$, φ is monotone decreasing, on $[p, q]$, φ is constant, and on (q, ∞), φ is monotone increasing. Give examples of each kind of convex function.

102. (Jensen's Inequality.) Show that if φ is as in Problem 99, and if $f \in L^1([0, 1], \lambda)$ then $\varphi(\int_0^1 f(t)\, dt) \leqq \int_0^1 \varphi(f(t))\, dt$.

103. Show that if $g \in \mathbb{R}^{(0,1)}$, $g \geqq 0$, and $g(x) \to \infty$ as $x \to 0$ then there is in $\mathbb{R}^{(0,1)}$ a convex function φ such that $\varphi \leqq g$ and $\varphi(x) \to \infty$ as $x \to 0$.

104. Prove or disprove: if $g \in \mathbb{R}^{(0,\infty)}$, $g \geqq 0$ and $g(x) \to \infty$ as $x \to \infty$ there is a convex function φ such that $\varphi \leqq g$ and $\varphi(x) \to \infty$ as $x \to \infty$.

105. Show: i) if φ'' exists and is positive on (a, b) then φ is convex on (a, b); ii) if φ is convex on (a, b) and φ'' exists then $\varphi'' \geqq 0$ on (a, b).

106. Let f be a measurable function positive a.e. on $[0, 1]$. Show that if $\{E_n\}_{n=1}^{\infty}$ is a sequence of measurable subsets of $[0, 1]$ and $\int_{E_n} f(x)\, ds \to 0$ as $n \to \infty$ then $\lambda(E_n) \to 0$ as $n \to \infty$.

107. Let a, b, c, y be nonnegative functions in $C([0, \infty), \mathbb{R})$ and assume that for all t in $[0, \infty)$, $y(t) \leqq \int_0^t [a(s)y(s) + b(s)]\, ds + c(t)$. Show $y(t) \leqq [\int_0^t b(s)\, ds + \max_{0 \leqq s \leqq t} c(s)]\, e^{\int_0^t a(s) ds}$.

108. Show that if $q \in L^1([0, \infty), \lambda)$ and $y'' + y = -qy$, $y(0) = 0$, $y'(0) = 1$, then for some M and for all x in $[0, \infty)$, $|y(x)| \leqq M$.

109. Assume $f \in C^2([0, 1], \mathbb{R})$ and that $\lambda(f^{-1}(0)) = 0$. Show: i) $|f|''$ exists a.e. and is a bounded measurable function; ii) if g is in $C^2([0, 1], \mathbb{R})$, $g \geqq 0$, and $g(x) = 0$ for all x in some open set containing 0 and 1 then $\int_0^1 g(x)|f|''(x)\, dx \leqq \int_0^1 |f|(x) g''(x)\, dx$.

110. Let $\{a_n\}_{n=1}^{N}$ be a finite subset of $\mathbb{R}$ and let s_m be $\sum_{n=1}^{m} a_n$, $m = 1, 2, \ldots, N$. Call an index n distinguished if there is an n' greater than n and such that $s_{n'} > s_{n-1}$ $(s_0 = 0)$. If the set of distinguished elements is nonempty, call it D, $D = \{n_k\}_{k=1}^{K}$, $n_1 < n_2 < \cdots < n_K$. Call a maximal chain of distinguished indices that are consecutive integers a block. Show that if n is in a block, if $n^{\#} > n$, and $n^{\#}$ is the last element in the block then $s_n^{\#} > s_{n-1}$. (In particular $\sum_{n \in D} a_n > 0$.)

111. (An analog of 110.) Let f be a bounded element of $\mathbb{R}^{(0,1)}$. Call x in $(0, 1)$ distinguished if there is in $(x, 1)$ an x' such that $\limsup_{y=x} f(y) < f(x')$. Show that the set S of distinguished elements of $(0, 1)$ is open. If $S \neq \varnothing$ let $\{(a_n, b_n): 1 \leqq n < M \leqq \infty\}$ be the unique sequence of pairwise disjoint open intervals the union of which is S. Show that if $x \in (a_n, b_n)$ then $f(x) \leqq \limsup_{y=b_n} f(y)$.

112. Let $\{a_r\}_{r=1}^R$ and $\{b_s\}_{s=1}^S$ be two finite sequences of real numbers. Show that over $\mathbb{R}$ there is a polynomial p such that: $\deg(p) \leqq R+S-1$, $p^{(r-1)}(1) = a_r$, $1 \leqq r \leqq R$, $p^{(s-1)}(2) = b_s$, $1 \leqq s \leqq S$.

113. Let f be in $(0, 1)^{(0,1)}$. Prove or disprove: i) if f is continuous and $\{a_n\}_{n=1}^\infty$ is a Cauchy sequence in $(0, 1)$ then $\{f(a_n)\}_{n=1}^\infty$ is a Cauchy sequence; ii) if $\{f(a_n)\}_{n=1}^\infty$ is a Cauchy sequence whenever $\{a_n\}_{n=1}^\infty$ is then f is continuous.

114. If $f, g \in \mathbb{R}^{(0,\infty)}$, $\lim_{x\to 0} g(x) = L$, and for all a, b in $(0, \infty)$, $|f(b)-f(a)| \leqq |g(b)-g(a)|$, show $\lim_{x\to 0} f(x)$ exists.

115. Let $\{a_n\}_{n=1}^\infty$ be a sequence in $(0, \infty)$ and let $\{b_n\}_{n=1}^\infty$ be a sequence in $\mathbb{R}$. Assume $\sum_{n=1}^\infty a_n$ converges. Show there is a monotone increasing function f, continuous on $\mathbb{R}\backslash\{b_n\}_{n=1}^\infty$ and such that for all n, $f(b_n+0)-f(b_n-0) = a_n$.

116. Let $\{a_n\}_{n=-\infty}^\infty$ be a (bilateral) sequence in $\mathbb{C}$. Assume that for some positive K, all N in $\mathbb{N}$, and all sequences $\{c_n\}_{n=-\infty}^\infty$ in $\mathbb{C}$, $|\sum_{n=-N}^N a_n c_n| \leqq K \sup_t |\sum_{n=-N}^N c_n e^{-int}|$. Show there is on $[0, 2\pi]$ a complex Borel measure μ such that $|\mu| \leqq K$ and for all n, $a_n = \int_0^{2\pi} e^{int}\, d\mu(t)$.

117. Let $\{c_n\}_{n=-\infty}^\infty$ be a sequence in $\mathbb{C}$. Show that there is on $[0, 2\pi]$ a complex Borel measure μ such that $c_n = \int_0^{2\pi} e^{-int}\, d\mu(t)$ iff for some finite M, $\|\sum_{n=-N}^N c_n(1-|n|/(N+1))\, e^{int}\|_1 \leqq M$.

118. Let X be a set and let $\mathcal{F}$ be a subset of $\mathbb{C}^X$. Assume that for all f in $\mathcal{F}$ the set S_f defined to be $\{x : f(x) \neq 0\}$ is finite and for all g in $\mathbb{C}^X$, $\sup_{f\in\mathcal{F}} |\sum_{x\in X} f(x)g(x)| < \infty$. Show there is in X a finite subset X_0 such that all f in $\mathcal{F}$ are 0 off X_0.

119. Find in $\mathbb{R}^{[0,4\pi]}$ a g such that g is monotone decreasing and for all real r, $\lambda\{x : x \in [0, 4\pi], \sin x > r\} = \lambda\{x : g(x) > r\}$.

120. Prove or disprove: if $\{f_n\}_{n=1}^\infty$ is a sequence in $C^2(\mathbb{R}, \mathbb{R})$, for some finite M'' and all n, $\|f_n''\|_\infty \leqq M''$, and $f_n \to 0$ uniformly as $n \to \infty$, then for some M' and all n, $\|f_n'\|_\infty \leqq M'$.

121. Let $\{f_n\}_{n=1}^\infty$ be a sequence of monotone functions in $\mathbb{R}^{[0,1]}$ and assume that for all x in $[0, 1]$, $\lim_{n\to\infty} f_n(x) = 1$. Show $\liminf_{n=\infty} f_n'(x) = 0$ a.e. Give an example of a sequence as described and for which $\lim_{n\to\infty} f_n'(x) = \infty$ for some x.

122. Let $\mathcal{A}$ be the set of real algebraic numbers (real zeros of polynomials over $\mathbb{Z}$). Does there exist in $\mathbb{R}^\mathbb{R}$ an f such that the set of points of continuity of f is $\mathcal{A}$?; $\mathbb{R}\backslash\mathcal{A}$?

123. Let f be in $\mathbb{R}^\mathbb{R}$ and assume f is continuous a.e. Show f is Lebesgue measurable.

124. Show that there is in $\mathbb{R}^\mathbb{R}$ no function f differentiable on an open set U containing 0, equal to 0 at 0, and such that for all x in U, $f'(x) = (\chi_{(-\infty,0]} \circ f)(x)$.

125. Let f be in $C(\mathbb{R}, \mathbb{R})$ and assume that for all x,

$$\limsup_{h\downarrow 0} (f(x+h)-f(x))/h \geqq 0.$$

Show f is monotone increasing.

126. Let f be in $C(\mathbb{R}, \mathbb{R})$ and assume $\limsup_{h\downarrow 0} (f(x+h)-f(x))/h \geqq 0$ a.e. Show f is monotone increasing.

127. Show that if $f \in C((0, 1), \mathbb{R})$ there is a remetrization of $(0, 1)$, say with a new metric d, so that f is uniformly continuous relative to d and the topology induced by d is the standard topology of $(0, 1)$.

128. Let f in $\mathbb{R}^{\mathbb{R}^n}$ be such that for all x in some open ball $B(0, r)^0$ in $\mathbb{R}^n$, $f(x) = f(x_1, x_2, \ldots, x_n) = \sum_{(k_1,k_2,\ldots,k_n)=(1,1,\ldots,1)}^{(\infty,\infty,\ldots,\infty)} a_{k_1,k_2,\ldots,k_n} x_1^{k_1} x_2^{k_2} \ldots x_n^{k_n}$. Show: either $f = 0$ in $B(0, r)^0$ or $\lambda_n(f^{-1}(0) \cap B(0, r)^0) = 0$.

129. Let S be $\{f: f \in C(\mathbb{R}^3, \mathbb{R})$, for some M_f, k_f, and all x, $|f(x)| \leqq M_f(1+\|x\|)^{k_f}\}$. Let F be a linear map of S into $\mathbb{R}$ and assume: i) if $f \geqq 0$, $F(f) \geqq 0$ (F is positive); ii) for some Borel measure μ on $\mathbb{R}^3$, for all f in $C_0(\mathbb{R}^3, \mathbb{R})$, $F(f) = \int_{\mathbb{R}^3} f(x)\, d\mu(x)$. Show that for all f in S, $F(f) = \int_{\mathbb{R}^3} f(x)\, d\mu(x)$.

130. Let Σ be $\{x: x \in \mathbb{R}^n, \|x\| = 1\}$ (the surface of the unit sphere in $\mathbb{R}^n$). Assume $f \in \mathbb{R}^{(\Sigma\times\mathbb{R})}$ and that f, $\partial f/\partial t \in C(\Sigma \times \mathbb{R}, \mathbb{R})$, $f^2 + (\partial f/\partial t)^2 > 0$. Show that for each t_0 in $\mathbb{R}$ there is an open set $U(t_0)$ containing t_0 and such that for all x in Σ there is in $U(t_0)$ at most one t such that $f(x, t) = 0$.

131. Let f, g belong to $C(\mathbb{R}^n, \mathbb{R})$ and assume they are both homogeneous of positive degree m ($f(tx) = t^m f(x)$, $g(tx) = t^m g(x)$). Show that if $f \geqq 0$ and if $g(x) > 0$ whenever $x \neq 0$ and $f(x) = 0$, then for some constants C, D, and all x, $Cf(x) + Dg(x) \geqq \|x\|^m$.

132. For $B(0, 1)$ in $\mathbb{R}^n$ let f be in $C^1(B(0, 1), \mathbb{R}^n)$. Show there is a positive δ such that if $\sup_x \|df(x) - id\| < \delta$ then f is one-one (on $B(0, 1)$).

133. Let f be in $C^2(\mathbb{R}^n, \mathbb{R})$ and assume that $df(x_0) = 0$ and that $(d^2 f(x_0))^{-1}$ exists. Show there is an open set U containing x_0 and such that $df(y) \neq 0$ for all y in $U \backslash \{x_0\}$.

6. Measure and Topology

Conventions

A measure μ is nonnegative unless otherwise qualified. If μ is complex then $|\mu|$ is the measure defined by the equation:

$$|\mu|(E) = \sup\left\{ \sum_{n=1}^{\infty} |\mu(E_n)| : \{E_n\}_{n=1}^{\infty} \text{ a sequence of pairwise disjoint measurable sets, } E = \bigcup_{n=1}^{\infty} E_n \right\}.$$

If X is a topological space, a Borel set E is inner (outer) regular iff $\mu(E) = \sup\{\mu(K): K \text{ compact}, K \subset E\}$ ($\mu(E) = \inf\{\mu(U): U \text{ open}, U \supset E\}$); E is regular iff E is both inner and outer regular; μ is inner (outer) regular iff every Borel set E is inner (outer) regular; if μ is complex, μ is inner (outer) regular iff $|\mu|$ is inner (outer) regular. A discrete measure μ is one for which there exists a map $f: X \ni x \mapsto f(x) \in [0, \infty)$ such that for any set E, $\mu(E) = \sum_{x \in E} f(x)$.

In the situation (X, S, μ), if $E \in \mathsf{S}$, E is an atom iff $\mu(E) > 0$ and for every measurable subset A of E either $\mu(A) = \mu(E)$ or $\mu(A) = 0$; μ is nonatomic iff there are no atoms.

A basis for a topology is a set $\{U_\lambda\}_{\lambda \in \Lambda}$ of open sets such that every open set is the union of (some of) the U_λ.

A partially ordered set is a pair $(\Gamma = \{\gamma\}, <)$, or more simply Γ, in which the order $<$ is transitive and the relation $\gamma < \gamma'$ (also written $\gamma' > \gamma$) obtains for a (possibly empty) set of pairs (γ, γ'). The partially ordered set is directed iff for every pair (γ, γ') there is a γ'' such that $\gamma'' > \gamma$ and $\gamma'' > \gamma'$. If Y is a set, a net is a map $\gamma \mapsto y_\gamma$ from Γ to Y. If $(\Gamma, <)$ is directed and Y is topologized, y_γ converges to y iff for every open set U containing y there is a γ_U such that $y_\gamma \in U$ whenever $\gamma > \gamma_U$.

The support of a measure μ in a topological space X is supp(μ). It is the complement of the union of all open sets of measure zero (supp(μ)$=$ $X \setminus \cup\{U\colon U$ open, $\mu(U)=0\}$).

134. For A a subset of $\mathbb{R}^n$, p, ε positive, let $\rho_\varepsilon^p(A)$ be

$$\inf\left\{\sum_{k=1}^{\infty} (\operatorname{diam}(U_k))^p : \{U_k\}_{k=1}^{\infty}, \text{ a sequence of bounded open sets,}\right.$$

$$\left.\bigcup_{k=1}^{\infty} U_k \supset A, \operatorname{diam}(U_k) < \varepsilon, k \in \mathbb{N}\right\}$$

and let $\rho^p(A)$ be $\sup_{\varepsilon>0} \rho_\varepsilon^p(A)$. (Similar definitions apply if $\mathbb{R}^n$ is replaced by an arbitrary metric space.) Show i) $\rho^p(A) = \lim_{\varepsilon\to 0} \rho_\varepsilon^p(A)$; ii) ρ^p is an outer measure on $2^{\mathbb{R}^n}$; iii) if γ is a simple rectifiable curve in $\mathbb{R}^n$ ($\gamma\colon [0,1] \mapsto \mathbb{R}^n$) then length($\gamma$)$=\rho^1(\gamma([0,1])$. The function ρ^p is usually called p-dimensional Hausdorff measure.

135. Show that if $\rho^p(A) < \infty$ and $q > p$, then $\rho^q(A) = 0$.

136. Let λ_p^* denote p-dimensional Lebesgue outer measure. Show that if $p \in \mathbb{N}$ there is a positive constant c_p such that $c_p\rho^p(A) \leqq \lambda_p^*(A) \leqq \rho^p(A)$ for all subsets A of $\mathbb{R}^p$.

137. Show that if A, B are subsets of a metric space (X, d) and if A and B are a positive distance apart, i.e., $\inf\{d(a,b)\colon a \in A, b \in B\} = \delta > 0$, then $\rho^p(A \cup B) = \rho^p(A) + \rho^p(B)$.

138. The set of (Caratheodory-) ρ^p-measurable sets in X is $\{A\colon$ for all S in 2^X, $\rho^p(S) = \rho^p(S \cap A) + \rho^p(S \setminus A)\}$. Show that every closed set is ρ^p-measurable.

139. Show that there is a constant K_p such that $\lambda_p^* = K_p\rho^p$.

140. Let the situation (X, S, μ) be such that X is a metric space, $\mathsf{S} = \sigma\mathsf{R}(\mathsf{K}(X))$ and μ is finite. Show μ is regular.

141. Let the situation (X, S, μ) be such that X is a separable, complete, metric space, $\mathsf{S} = \sigma\mathsf{R}(\mathsf{O}(X))$, and μ is finite. Show μ is regular.

142. Let μ be a finite Borel measure on a compact metric space X. Assume that for each x in X, $\mu(x) = 0$. Show that if $\varepsilon > 0$ there is a positive $\delta(\varepsilon)$ such that whenever the diameter of a Borel set E is less than $\delta(\varepsilon)$ then $\mu(E) < \varepsilon$.

143. Show that if a Borel measure μ on $\mathbb{R}^n$ is such that $\mu(B(0,r)) < \infty$ for all r in $[0, \infty)$ then μ is regular.

144. Let μ be a regular, finite Borel measure on a compact Hausdorff space X. Show there is in X a minimal closed subset F such that $\mu(X\backslash F)=0$, i.e., if F_1 is closed and $\mu(X\backslash F_1)=0$, then $F\subset F_1$. Show that if $f\in C(X,\mathbb{C})$ then $f=0$ a.e. iff $f^{-1}(0)\supset F$.

145. Let ν be a finite, finitely additive, and positive set function defined on the Borel sets of a compact Hausdorff space X. Using the obvious interpretation of "regular" as applied to ν, show that if ν is regular then ν is countably additive.

146. Let μ be a finite Borel measure on a compact Hausdorff space X. Define a partially ordered set $\Gamma=\{\gamma\}$ and a net $\gamma\mapsto\mu_\gamma$ such that each μ_γ is discrete and such that for every f in $C(X,\mathbb{C})$, $\int_X f(x)\,d\mu_\gamma$ converges to $\int_X f(x)\,d\mu(x)$.

147. Show that if μ_1 and μ_2 are regular complex measures then $\mu_1-\mu_2$ is also a regular complex measure.

148. Let μ be a finite, nonatomic, Borel measure on a compact metric space (X,d). Show there is for the topology of X a (countable) basis $\{U_n\}_{n=1}^\infty$ such that for all n, $\mu(\partial U_n)=0$.

149. Let X be a compact metric space, let S be $\sigma\mathsf{R}(\mathsf{O}(X))$, and let $\{(X,\mathsf{S},\mu_n): n=0,1,\ldots\}$ be a sequence of measure situations such that for all n, $\mu_n(X)<\infty$. Assume that for each f in $C(X,\mathbb{C})$, $\int_X f(x)\,d\mu_n(x)\to\int_X f(x)\,d\mu_0(x)$ as $n\to\infty$. Show that if U is open and $\mu_0(\partial U)=0$, then $\mu_n(U)\to\mu_0(U)$ as $n\to\infty$.

150. (Converse of Problem 149.) In the notation of Problem 149 assume that for all open sets U for which $\mu_0(\partial U)=0$, $\mu_n(U)\to\mu_0(U)$ as $n\to\infty$. Show that if $f\in C(X,\mathbb{C})$, then $\int_X f(x)\,d\mu_n(x)\to\int_X f(x)\,d\mu_0(x)$ as $n\to\infty$.

151. Repeat Problem 149 but without the assumption that $\mu_0(X)$ is finite.

7. General Measure Theory

Conventions

If (X, S, μ) is a measure situation and $A \in \mathsf{S}$, a partition of A is a sequence $\{A_n\}_{n=1}^{\infty}$ of a pairwise disjoint sets in S and such that $A = \bigcup_{n=1}^{\infty} A_n$. The measure μ is signed iff the range of μ is a subset of $(-\infty, \infty]$ or $[-\infty, \infty)$. If μ is a signed measure and $A \in \mathsf{S}$, $|\mu|(A)$ is defined to be $\sup\{\sum_{n=1}^{\infty} |\mu(A_n)| \colon \{A_n\}_{n=1}^{\infty}$ a partition of $A\}$. If μ is signed there are two positive measures $\mu^{\pm}$ such that $\mu = \mu^{+} - \mu^{-}$. Furthermore there are in S disjoint sets $P^{\pm}$ such that for all E in S, $\mu^{\pm}(E) = \mu(P^{\pm} \cap E)$. The sets $P^{\pm}$ constitute a Hahn decomposition. Finally, $|\mu|(E) = \mu^{+}(E) + \mu^{-}(E)$. The measure μ is complex iff its range is a subset of $\mathbb{C}$; $|\mu|$ is defined by the formula used above for signed measures. Note that if μ is complex then $|\mu|$ is bounded (finite).

If (X, S, μ_i), $i = 1, 2$, are two measure situations, μ_1 is absolutely continuous with respect to μ_2 ($\mu_1 \ll \mu_2$) iff $\mu_1(E) = 0$ whenever $\mu_2(E) = 0$. If there is a measurable function f such that for all E in S, $\mu_1(E) = \int_E f(x)\, d\mu_2(x)$, then $d\mu_1/d\mu_2$ is the symbol for f (which is defined uniquely modulo μ_2-null sets, i.e., sets having μ_2-measure zero). The measures are mutually singular ($\mu_1 \perp \mu_2$) iff there are disjoint sets A_1, A_2 such that for all E in S, $\mu_i(E) = \mu_i(E \cap A_i)$, $i = 1, 2$; μ_i is then said to live or to be concentrated on A_i, $i = 1, 2$. The Lebesgue–Radon–Nikodým theorem gives conditions under which μ_1 may be decomposed into a sum $\mu_{1a} + \mu_{1s}$ such that $\mu_{1a} \ll \mu_2$, $\mu_{1s} \perp \mu_2$, $\mu_{1s} \perp \mu_{1a}$, and $d\mu_{1a}/d\mu_2$ exists.

If A and B are sets, $A \,\Delta B = (A \backslash B) \cup (B \backslash A)$ is the symmetric difference of A and B; $A \doteq B$ iff $A \,\Delta B$ is a null set (with respect to the measure in question). A set A is σ-finite iff it is the countable union of measurable sets of finite measure.

If $\{(X_\gamma, S_\gamma, \mu_\gamma): \gamma \in \Gamma\}$ is a set of measure situations, $(X, S, \mu) = \prod_{\mu \in \Gamma} (X_\gamma, S_\gamma, \mu_\gamma)$ iff $X = \prod_\gamma X_\gamma$; S is the σ-ring generated by the set of all "finite rectangles", i.e., sets of the form $A_{\gamma 1} \times \cdots \times A_{\gamma_n} \times \prod_{\gamma \neq \gamma_1, \ldots, \gamma_n} X_\gamma$; for all γ, $\mu_\gamma(X_\gamma) = 1$; and μ (finite rectangle) $= \prod_i \mu_{\gamma_i}(A_{\gamma_i})$ (perforce $A_{\gamma_1} \in S_{\gamma_i}$, $X_\gamma \in S_\gamma$). By abuse of language μ may be denoted $\prod_\gamma \mu_\gamma$; S may be denoted $\prod_\gamma S_\gamma$. If f is S-measurable, $f^{x_{\gamma'}}$ maps $\prod_{\gamma \neq \gamma'} X_\gamma$ according to the rule: $\{x_\gamma\}_{\gamma \neq \gamma'} \mapsto f(\{x_\gamma\})$.

152. Let $\sum_{n=1}^{\infty} a_n$ be a convergent series of complex numbers. For every finite subset E of $\mathbb{N}$ let $\mu(E)$ be $\sum_{n \in E} a_n$. Show that μ is a finitely but not necessarily countably additive set function.

153. Let R be a ring of subsets of a set X and assume that $(X, \sigma\text{R}(\text{R}), \mu_i)$, $i = 1, 2$, are measure situations such that for all E in R, $0 \leqq \mu_1(E) = \mu_2(E) < \infty$. Show that $\mu_1 = \mu_2$ on $\sigma\text{R}(\text{R})$. Give a counterexample to the conclusion if "$<\infty$" is replaced by "$\leqq\infty$".

154. Let $\mathcal{U}$ be a subalgebra of $\mathbb{C}^X$ (if $f, g \in \mathcal{U}$ and if $\alpha, \beta \in \mathbb{C}$, then $\alpha f + \beta g$ and $f \cdot g$ are in $\mathcal{U}$). Show that if $\mathcal{U}$ contains $x \mapsto 1$ and if $\mathcal{U}$ is closed with respect to the formation of sequential limits (if $\{f_n\}_{n=1}^{\infty} \subset \mathcal{U}$ and $\lim_{n\to\infty} f_n$ exists then it is in $\mathcal{U}$) then $\{A: \chi_A \in \mathcal{U}\}$, denoted $\mathcal{F}$, is a σ-algebra.

155. Show that if in Problem 154 X is a locally compact metric space and $\mathcal{U} \supset C_0(X, \mathbb{C})$, then $\sigma\text{R}(\text{K}(X)) \subset \mathcal{F}$.

156. Let (X, S, μ) be a measure situation and assume μ is signed. Let $\mathcal{M}$ be $\{\nu: (X, S, \nu)$ a measure situation, for all A in S, $-\mu(A), \mu(A) \leqq \nu(A)\}$. Show that if $A \in S$ then $|\mu|(A) = \inf_{\nu \in \mathcal{M}} \nu(A)$.

157. Let (X, S, μ) be a measure situation in which μ is complex. For E in S let $M(E)$ be $\sup\{|\mu(A)|: A \in S, A \subset E\}$. Show that $M(E) \leqq |\mu|(E) \leqq 4M(E)$.

158. (Borel–Cantelli lemma). Let (X, S, μ) be a measure situation. Show that if $\{E_n\}_{n=1}^{\infty} \subset S$ and $\sum_{n=1}^{\infty} \mu(E_n) < \infty$, then $\mu(\limsup_{n=\infty} E_n) = 0$.

159. Let (X, S, μ) be a measure situation such that $X \in S$ and $\mu(X) < \infty$. Let $f: X \mapsto X$ be a map such that for every A in S, $f^{-1}(A) \in S$ and for every null set N, $f^{-1}(N)$ is also a null set. Show that there is in $L^1(X, \mu)$ an h such that for every g in $L^\infty(X, \mu)$, $\int_X g \circ f(x)\, d\mu(x) = \int_X g(x) h(x)\, d\mu(x)$.

√ **160.** Let (X, S, μ) be a measure situation and assume $\{f_n\}_{n=1}^{\infty}$ is a sequence of nonnegative measurable functions such that $f_n \to f_0$ a.e. as $n \to \infty$ and $\int_X f_n(x)\, d\mu(x) \to 0$ as $n \to \infty$. Show $\int_X f_0(x)\, d\mu(x) = 0$.

√ **161.** Let (X, S, μ) be a measure situation in which X is σ-finite. Assume $\{f_n\}_{n=0}^{\infty}$ is a sequence of nonnegative measurable functions such that $f_n \to f_0$ a.e. as $n \to \infty$ and $\int_X f_n(x)\, d\mu(x) \to \int_X f_0(x)\, d\mu(x)$ as $n \to \infty$. Show that for each E in S, $\int_E f_n(x)\, d\mu(x) \to \int_E f_0(x)\, d\mu(x)$ as $n \to \infty$.

162. Let (X, S, μ) be a measure situation such that $X \in \mathsf{S}$, $\mu(X) < \infty$. Let $\{f_n\}_{n=1}^{\infty}$ be a sequence of $\mathbb{R}$-valued measurable functions. Show that $f_n \to 0$ in measure as $n \to \infty$ iff $\int_X (|f_n(x)|/(1+|f_n(x)|))\, d\mu(x) \to 0$ as $n \to \infty$.

163. Let (X, S, μ) be a measure situation in which $X \in \mathsf{S}$, $\mu(X) < \infty$ and let $\{f_n\}_{n=1}^{\infty}$ be an orthonormal sequence in $L^2(X, \mu)$. Show that if there is an M such that $|f_n(x)| < M$ a.e. for all n and $\sum_{n=1}^{\infty} a_n f_n$ converges a.e., then $a_n \to 0$ as $n \to \infty$.

164. Let (X, S, μ) be a measure situation such that X is σ-finite. If $\{f_n\}_{n=1}^{\infty}$ is an orthonormal sequence in $L^2(X, \mu)$, let E be $\{x: \lim_{n\to\infty} f_n(x) \text{ exists}\}$. Show that the map

$$f: x \mapsto \begin{cases} \lim_{n\to\infty} f_n(x), & \text{if } x \in E \\ 0, & \text{if } x \notin E \end{cases}$$

is zero a.e.

165. Let (X, S, μ) be a measure situation such that $X \in \mathsf{S}$ and $\mu(X) < \infty$. Show that if $f \in L^{\infty}(X, \mu)$ and $\|f\|_{\infty} > 0$ then $\|f\|_{\infty} = \lim_{n\to\infty} \|f\|_{n+1}^{n+1}/\|f\|_n^n$.

166. Let (X, S, μ) be a measure situation such that $L^1(X, \mu) = L^{\infty}(X, \mu)$. Show $L^1(X, \mu)$ is finite-dimensional.

167. Show that the converse of Problem 166 is true.

168. Let (X, S, μ) be a measure situation such that $X \in \mathsf{S}$ and $\mu(X) = 1$. Assume $\{f_n\}_{n=1}^{\infty} \subset L^{\infty}(X, \mu)$, f_n are $\mathbb{R}$-valued, $|f_n(x)| \leqq 1$ for all n and x, and $\int_X f_n(x)\, d\mu(x) \to 0$ as $n \to \infty$. Prove or disprove: there is a subsequence $\{f_{n_k}\}_{k=1}^{\infty}$ that converges to zero a.e.

169. Let (X, S, μ) be a measure situation. For f in $L^{\infty}(X, \mu)$ let T_f be the map $L^2(X, \mu) \ni g \mapsto f.g$. Show that the operator norm $\|T_f\|$ is $\|f\|_{\infty}$ for all f iff X contains no infinite atoms.

170. In the notation of Problem 169, when and only when is T_f surjective?

171. Let (X, S, μ) be a measure situation. Show that if $f \in L^1(X, \mu)$ and for some constant a and all measurable sets E of finite measure, $\int_E f(x)\, d\mu(x) \leqq a$, then $\int_X f(x)\, d\mu(x) \leqq a$.

172. Construct a counterexample to the conclusion in Problem 171 if the hypothesis that $f \in L^1(X, \mu)$ is dropped.

173. Give an example of a measure situation (X, S, μ) and, in $L^1(X, \mu)$, a sequence $\{f_n\}_{n=1}^{\infty}$ such than $f_n \to 0$ uniformly as $n \to \infty$ and $\int_X f_n(x)\, d\mu(x) \to \infty$ as $n \to \infty$.

174. Let (X, S, μ) be a measure situation. Show that if $\{f_n\}_{n=1}^{\infty} \subset L^1(X, \mu)$, $0 \leqq f_n \leqq 1$, for all x, $\lim_{n\to\infty} f_n(x) = 1$, and for all n and some measurable set E of finite measure, $f_n = 1$ off E, then $\lim_{n\to\infty} \int_X (1 - f_n(x))\, d\mu(x) = 0$.

175. Give an example of a measure situation (X, S, μ), a sequence $\{E_n\}_{n=1}^{\infty}$ of measurable sets of finite measure, and a sequence $\{f_n\}_{n=1}^{\infty}$ of functions such that f_n and $1-f_n$ are integrable, $0 \leqq f_n \leqq 1$, $f_n = 1$ on E_n, $\lim_{n\to\infty} f_n(x) = 1$ a.e., and $\int_X (1-f_n(x))\, d\mu(x) \not\to 0$ as $n \to \infty$.

176. Let (X, S, μ) be a measure situation. Show that if $f \in L^1(X, \mu)$ and $E_n = \{x : |f(x)| \geqq n\}$ then $n\mu(E_n) \to 0$ as $n \to \infty$.

177. Let (X, S, μ) be a measure situation. Show that if $0 < p < \infty$, $\varepsilon > 0$, and f is a measurable function then $\mu\{x : |f(x)| \geqq \varepsilon\} \leqq \int_X |f(x)|^p\, d\mu(x)/\varepsilon^p$.

178. Let (X, S, μ_i) be measure situations such that X is measurable, $i = 1, 2$, and $\mu_1(X) < \infty$. Show that if $1 \leqq p < \infty$ there is in $L^p(X, \mu_1)$ an f satisfying: $\mu_2(E) = \int_E f(x)\, d\mu_1(x)$ for all E in S iff the following is true: for some real a and all partitions $\{E_n\}_{n=1}^{\infty}$ of X, $\sum_{n=1}^{\infty} (\mu_2(E_n))^p/(\mu_1(E_n))^{p-1} \leqq a$. (The standard convention: $0 \cdot \infty = 0$ is to be observed.)

179. If (X, S, μ) is a measure situation such that $X \in \mathsf{S}$ and X is σ-finite, let p be in $(1, \infty)$ and let f be in $L^p(X, \mu)$. Show that if $F_t = \{x : |f(x)| \geqq t\}$ then $\|f\|_p^p = p \int_0^{\infty} t^{p-1} \mu(F_t)\, dt$.

180. Let (X, S, μ_i), $i = 1, 2, 3$, be measure situations such that X is measurable and $\mu_1(X) + \mu_2(X) + \mu_3(X) < \infty$. Show that if $\mu_j = \mu_{ja} + \mu_{js}$ are such that $\mu_{ja} \ll \mu_3$, $\mu_{js} \perp \mu_3$, $j = 1, 2$, then $\mu_{1s} \perp \mu_{2a}$.

181. Let (X, S, μ_i), $i = 1, 2$, be measure situations such that X is measurable and $\mu_1(X) + \mu_2(X) < \infty$. Show: i) there is in S an E such that, $\mu_{iE}(A)$ denoting $\mu_i(A \cap E)$, $i = 1, 2$, $\mu_{iE} \ll \mu_{jE}$ and $\mu_{i(X\setminus E)} \perp \mu_{j(X\setminus E)}$, $i \neq j$; ii) if, for all F in S, $\mu_{iF} \ll \mu_{jF}$ and $\mu_{i(X\setminus F)} \perp \mu_{j(X\setminus F)}$, $i \neq j$, then $\mu_1(E \,\Delta F) + \mu_2(E \,\Delta F) = 0$.

182. Let (X, S, μ_i), $i = 1, 2$, be measure situations where the μ_i are complex. Show that $\mu_1 \perp \mu_2$ iff for all a_1, a_2 in $\mathbb{C}$, $|a_1| \cdot |\mu_1| + |a_2| \cdot |\mu_2| = |a_1\mu_1 + a_2\mu_2|$.

183. Let (X, S, μ) be a measure situation such that X is measurable and of finite measure. Let $\mathscr{F}$ be a set $\{f_\gamma\}_{\gamma\in\Gamma}$ of measurable functions such that if $f_{\gamma_1}, f_{\gamma_2} \in \mathscr{F}$ then $f_{\gamma_1} \vee f_{\gamma_2}$, denoted $f_{\gamma_1 \vee \gamma_2}$, $\in \mathscr{F}$. Show: i) there is a measurable function g such that for all γ, $f_\gamma \leqq g$ a.e.; ii) if, for all γ, $f_\gamma \leqq h$ a.e., and h is measurable, then $g \leqq h$ a.e. (The function g may be regarded as a minimal measurable cover for $\mathscr{F}$.)

184. Let (X, S, μ) be a measure situation such that X is σ-finite. Show that if $f \in L^1(X, \mu)$ and $f \geqq 0$, then $\int_X f(x)\, d\mu(x) = (\mu \times \lambda)(\{(x, y) : 0 \leqq y \leqq f(x)\})$.

185. Let (X, S, μ) be a measure situation such that X is measurable and $\mu(X) = 1$. Show that if $f \in L^2(X, \mu)$ and f is $\mathbb{R}$-valued, then $\int_X (f(x) - \int_X f(y)\, d\mu(y))^2\, d\mu(x)$, denoted $\operatorname{var}(f)$ (the variance of f), is $\int_X (f(x))^2\, d\mu(x) - (\int_X f(x)\, d\mu(x))^2$.

186. (An extension of Problem 184.) In the notation of Problem 184, let F_n be the map $X^n \ni (x_1, x_2, \ldots, x_n) \mapsto \sum_{k=1}^n f(x_k)/n$. Show that if X^n is endowed with product measure derived from μ, then $\operatorname{var}(F_n) = \operatorname{var}(f)/n$.

187. (Corollary to Problem 185.) Show that if $0 \leqq x \leqq 1$ then

$$\sum_{k=0}^n (x - k/n)^2 \binom{n}{k} x^k (1-x)^{n-k} \leqq 1/4n, \qquad n \text{ in } \mathbb{N}.$$

8. Measures in $\mathbb{R}^n$

Conventions

Hereafter I denotes $[0, 1]$, $\mathsf{S}_\beta(I)$ is $\sigma\mathsf{R}(\mathsf{K}(I))$, $\mathsf{S}_\lambda(I)$ is the set of Lebesgue measurable sets in I. If $A \subset \mathbb{R}^n$, $\mathsf{S}_\beta(A)$ and $\mathsf{S}_\lambda(A)$ have analogous meanings (even if A is neither Borel nor Lebesgue measurable). When the context permits little ambiguity, the symbols S_β and S_λ will be used by themselves.

If (X, S, μ) is a measure situation and if $\mathcal{F} \subset L^1(X, \mu)$, for each f in $\mathcal{F}$ let $G(f, k)$ be $\{x: |f(x)| \geqq k\}$, k in $\mathbb{N}$. The set $\mathcal{F}$ is uniformly integrable HS (Hewitt–Stromberg) iff for each positive ε there is in $\mathbb{N}$ a $K(\varepsilon)$ such that for all f in $\mathcal{F}$, $\int_{G(f,k)} |f(x)|\, d\mu(x) < \varepsilon$ if $k > K(\varepsilon)$.

If $x \in \mathbb{R}$, $[x] = \max\{n: n \in \mathbb{Z}, n \leqq x\}$ = "the greatest integer in x".

A Borel measure μ in $\mathbb{R}^n$ is, by definition, finite on compact sets (if K is compact, $\mu(K) < \infty$). As needed for clarity λ_n denotes Lebesgue measure in $\mathbb{R}^n$, $n \geqq 2$; $\int_{\mathbb{R}^n} f(x)\, d\lambda_n(x)$ and $\int_{\mathbb{R}^n} f(x)\, dx$ are used interchangeably.

188. Find: i) $\operatorname{card}(\mathsf{S}_\beta(\mathbb{R}^n))$; ii) $\operatorname{card}\{(\mathbb{R}^n, \mathsf{S}_\beta, \mu): \mu(\{x\}) = 0$ for all x in $\mathbb{R}^n$, $\mathbb{R}^n$ is σ-finite $(\mu)\}$.

189. Construct $(I, \mathsf{S}_\beta, \mu)$ so that μ is nonatomic and $\operatorname{supp}(\mu) = \{t: t = \sum_{n=1}^\infty \varepsilon_n 10^{-n}, \varepsilon_n = 0 \text{ or } 7\}$, denoted E (see Measure and Topology, Conventions).

190. Does there exist a measure situation $(I, \mathsf{S}_\beta, \mu)$ such that the range $\mu(\mathsf{S}_\beta)$ is *the* Cantor set? (See Problem 12.)

191. Show that if $(I, \mathsf{S}_\beta, \mu)$ is a measure situation, $\mu(I) = 1$, and f is $\mathbb{R}$-valued and in $L^1(I, \mu)$, then $\exp(\int_I f(x)\, d\mu(x)) = \int_I e^{f(x)}\, d\mu(x)$ iff f is constant a.e. (μ).

192. For the measure situation $(I, \mathsf{S}_\beta, \mu)$ assume that for all f in $C(I, [0, \infty))$, $f(0) \geqq \int_I f(x)\, d\mu(x)$. Show there is in I a c such that for all f in $C(I, \mathbb{C})$, $\int_I f(x)\, d\mu(x) = cf(0)$.

193. For the measure situation $(I, \mathsf{S}_\beta, \mu)$ in which μ is signed (see General Measure Theory, Conventions) assume $\int_I \sin^k \pi x\, d\mu(x) = 0$ for all k in $\mathbb{N}$. Show that if $E \in \mathsf{S}_\beta([0, \frac{1}{2}])$ then $\mu(E) = -\mu(1 - E)$.

194. In the situation described in Problem 193 show that if $\int_I \cos^k \pi x\, d\mu(x) = 0$ for all k in $\mathbb{N}$ then for all E in $\mathsf{S}_\beta(I \backslash \{\frac{1}{2}\})$, $\mu(E) = 0$.

195. Show that if $\{a_n\}_{n=1}^N \subset \mathbb{R}$ there is a measure situation $(I, \mathsf{S}_\beta, \mu)$ in which μ is signed and $\int_I \cos^n \pi x\, d\mu(x) = a_n$, $n = 1, 2, \ldots, N$.

196. For the measure situation $(I, \mathsf{S}_\lambda, \lambda)$ assume $\mathscr{F}$ is a family of uniformly integrable HS functions and that $\mathscr{F} \subset L^1(I, \lambda)$. Show that if $\{f_n\}_{n=1}^\infty \subset \mathscr{F}$ there is a subsequence $\{f_{n_k}\}_{k=1}^\infty$ and in $C(I, \mathbb{C})$ a g such that $\int_0^y f_{n_k}(x)\, dx \to g(y)$ uniformly as $k \to \infty$.

197. Let P be $\{\mu : (I, \mathsf{S}_\beta, \mu)$ is a measure situation and $\mu(I) = 1\}$. Find the extreme points of P regarded as a subset of M, the Banach space of all signed Borel measures on I.

198. Let $\{t_n\}_{n=0}^\infty$ be a sequence in $\mathbb{R}$ and assume that for all N in $\mathbb{N}$, $\sum_{n=0}^N a_n t_n \geqq 0$ whenever $\sum_{n=0}^N a_n e^{nx} \geqq 0$ for all x in I. Show there is a unique measure situation $(I, \mathsf{S}_\beta, \mu)$ such that for all n in $\mathbb{N}$, $t_n = \int_I e^{nx}\, d\mu(x)$.

199. Let X be $\{(x, y) : 0 \leqq x \leqq y \leqq 1\}$ and let S be the set of Borel subsets of X. Do there exist measures μ_1 and μ_2 defined on S and such that whenever $0 \leqq a \leqq b \leqq 1$ then $\mu_1([0, a] \times [a, b]) = ab + a^2 b^2$ and $\mu_2([0, a] \times [a, b]) = ab - a^3 b^3$?

200. For the measure situation $(\mathbb{R}, \mathsf{S}_\beta, \mu)$ in which μ is complex assume $f \in L^1(\mathbb{R}, \mu)$ and that for all h in $C_{00}^\infty(\mathbb{R}, \mathbb{C})$, $\int_{\mathbb{R}} h'(x)\, d\mu(x) = -\int_{\mathbb{R}} f(x) h(x)\, dx$. Show that $\mu \ll \lambda$ and that $d\mu/d\lambda = (x \mapsto \int_{-\infty}^x f(t)\, dt)$.

201. For the measure situation $(\mathbb{R}, \mathsf{S}_\lambda, \lambda)$ assume $E \in \mathsf{S}_\lambda$ and that for all x in $\mathbb{R}$, $\lambda(E\, \Delta(x + E)) = 0$. Show: i) $\lambda(\mathbb{R} \backslash E) \Delta(x + (\mathbb{R} \backslash E)) = 0$ for all x in $\mathbb{R}$; ii) $\lambda(E) \cdot \lambda(\mathbb{R} \backslash E) = 0$.

202. For the measure situation $(\mathbb{R}, \mathsf{S}_\beta, \mu)$ in which μ is complex, let $\hat{\mu}$ be the map $\mathbb{R} \ni x \mapsto \int_{\mathbb{R}} e^{-itx}\, d\mu(t)$. i) Show that if $\hat{\mu}(n) = 0$ for n in $\mathbb{Z}$ then for every Borel set E, $\mu(\bigcup_{n \in \mathbb{Z}} (E + 2\pi n)) = 0$. ii) Find a μ such that $\hat{\mu}(n) = 0$ for n in $\mathbb{Z}$ and an E such that $\sum_{n \in \mathbb{Z}} \mu(E + 2\pi n) \neq 0$.

203. i) Let $(I, \mathsf{S}_\beta, \mu)$ be a measure situation such that $\mu \perp \lambda$ and $\mu(I) < \infty$. Show that if $f \in L^1(I, \lambda)$ then $\int_I f(x - y)\, d\mu(y)$ exists and is finite a.e. (λ) on $\mathbb{R}$. ii) Show that if $(I, \mathsf{S}_\beta, \mu)$ is a measure situation and $\mu(I) < \infty$, then for every f in $L^1(I, \lambda)$, $\int_I f(x - y)\, d\mu(y)$ exists and is finite a.e. (λ) on $\mathbb{R}$.

204. Let $\{\mathbb{R}, \mathsf{S}_\beta, \mu_n)\}_{n=1}^\infty$ be a sequence of measure situations in each of which μ_n is complex and nonzero. i) Construct a Borel measure ν such that for all n, $\mu_n \ll \nu$. ii) If all μ_n are positive, is there a measure ν such that $0 \neq \nu \ll \mu_n$ for all n?

205. Let $(\mathbb{R}, \mathsf{S}_\beta, \mu)$ be a measure situation such that $\mu(\mathbb{R}) < \infty$ and let E be a Borel set. Show there are Borel measures ν_1 and ν_2 such that $\mu = \nu_1 + \nu_2$, for all x in $\mathbb{R}$, $\nu_1(x+E) = 0$, and such that there is in $\mathbb{R}$ a sequence $\{x_n\}_{n=1}^\infty$ for which $\nu_2(\mathbb{R} \setminus \bigcup_n (x_n + E)) = 0$.

206. Show that if F is a closed subset of $\mathbb{R}$ there is a Borel measure μ such that $\operatorname{supp}(\mu) = F$.

207. Let μ be a finite nonatomic Borel measure on $\mathbb{R}$. Show that if $a \in [0, \mu(\mathbb{R})]$ there is a Borel set E such that $\mu(E) = a$.

208. Let $Q(a_1, a_2)$ be $\{(x_1, x_2): |x_1 - a_1|, |x_2 - a_2| \leqq \frac{1}{2}\}$. For the measure situation $(\mathbb{R}^2, \mathsf{S}_\beta, \mu)$ assume that for any horizontal or vertical line L, $\mu(L) = 0$. Show that the map $f: (x_1, x_2) \mapsto \mu(Q(x_1\ x_2))$ is continuous.

209. Let μ be a finite Borel measure on $\mathbb{R}^2$ and assume that for any line L, $\mu(L) = 0$. Show that if E is a bounded Borel set and $0 < a < \mu(E)$ there is a Borel set F contained in E and such that $\mu(F) = a$.

210. Let $\{(\mathbb{R}^n, \mathsf{S}_\beta, \mu_m)\}_{m=0}^\infty$ be a sequence of measure situations. Show that if for all f in $C_{00}(\mathbb{R}^n, \mathbb{C})$, $\lim_{m\to\infty} \int_{\mathbb{R}^n} f(x)\, d\mu_m(x) = \int_{\mathbb{R}^n} f(x)\, d\mu_0(x)$ then: i) for any open set U in $\mathbb{R}^n$, $\liminf_{m=\infty} \mu_m(U) \geqq \mu_0(U)$; and ii) for any Borel set E such that $\mu_0(\partial E) = 0$, $\lim_{m\to\infty} \mu_m(E) = \mu_0(E)$. (See Problem 149.)

211. Let K be a compact perfect subset of $\mathbb{R}^n$ (K is compact and contains no isolated points). Show that there is on $\mathbb{R}^n$ a nonatomic Borel measure μ such that $\operatorname{supp}(\mu) = K$.

212. Show that $\mathsf{S}_\beta(\mathbb{R}^n) \times \mathsf{S}_\beta(\mathbb{R}^m) = \mathsf{S}_\beta(\mathbb{R}^{n+m})$.

213. Let μ be a Borel measure on $\mathbb{R}^n$. Show that if $\operatorname{supp}(\mu)$, denoted K, is compact, and for some $\mathbb{R}$-valued polynomial p and all $\mathbb{R}$-valued polynomials q, $\int_{\mathbb{R}^n} p(x) q^2(x)\, d\mu(x) \geqq 0$ then $p \geqq 0$ on K.

214. Let μ be a complex Borel measure on $[0, \infty)$. Show that if $\int_0^\infty e^{-nx}\, d\mu(x) = 0$ for all n in $\mathbb{N}$ then $\mu = 0$.

215. Let μ be a Borel measure on $[0, \infty)$ and assume $\mu([0, \infty)) = 1$. Show that $\int_0^\infty (1 - \mu([0, x))\, dx = \int_0^\infty x\, d\mu(x)$.

216. Let f be a nonnegative bounded Borel measurable function on $[0, \infty)$. Show that if μ is a Borel measure on $[0, \infty)$, then $f \in L^1([0, \infty), \mu)$ iff $\int_0^\infty \mu(\{y: x < f(y)\}\, dy < \infty$. (Compare Problems 215 and 216 with Problem 179.)

9. Lebesgue Measure in $\mathbb{R}^n$

Conventions

If $(\prod_\gamma X_\gamma, \prod_\gamma \mathsf{S}_\gamma, \prod_\gamma \mu_\gamma)$ is a (product-) measure situation and if $E \in \prod_\gamma \mathsf{S}_\gamma$, then $E_{x_{\gamma'}}$ is the subset of $\prod_{\gamma \neq \gamma'} X_\gamma$ that consists of all points $\{x_\gamma\}_{\gamma \neq \gamma'}$ such that $\{x_\gamma\}_\gamma \in E$.

In a metric space (X, d) a set E is of the first category iff E is the countable union of nowhere dense sets (sets A such that the closures $\bar{A}$ have empty interiors: $(\bar{A})^0 = \varnothing$); a set E is of the second category if E is not of the first category.

If (X, S, μ) is a measure situation and $E \subset X$ then $\mu_*(E) = \sup\{\mu(B): B \in \mathsf{S}, B \subset E\}$ and $\mu^*(E) = \inf\{\mu(C): C \in \mathsf{S} . C \supset E\}$; $\mu_*(\mu^*)$ is inner (outer) measure corresponding to μ.

If $f, g \in L^1(\mathbb{R}^n, \lambda_n)$ then (see Problem 201)

$$f * g : x \mapsto \int_{\mathbb{R}^n} f(x-y) g(y)\, d\lambda_n(y)$$

exists and is in $L^1(\mathbb{R}^n, \lambda_n)$; $f * g$ is called the convolution of f and g (see Problem 203). The notation $f_{(t)}$ is for the function $x \mapsto f(x+t)$.

217. Show that if $\{[a_n, b_n]\}_{n=1}^N$ is a finite set of closed intervals in $\mathbb{R}$ and if $\bigcup_{n=1}^N [a_n, b_n] \supset I$ then $\sum_{n=1}^N (b_n - a_n) \geqq 1$.

218. Show that $\mathbb{R}^2$ is not the countable union of lines.

219. Let $\{r_n\}_{n=1}^\infty$ be an enumeration of $\mathbb{Q}$. Show that

$$\mathbb{R} \backslash \bigcup_n (r_n - 1/n^2, r_n + 1/n^2) \neq \varnothing.$$

Does there exist an enumeration $\{s_n\}_{m=1}^{\infty}$ of $\mathbb{Q}$ such that

$$\mathbb{R}\backslash\bigcup_m (s_m - 1/m, s_m + 1/m) \neq \varnothing?$$

219. Let $\{r_n\}_{n=1}^{\infty}$ be an enumeration of $\mathbb{Q}$. Show that $\mathbb{R}\backslash\bigcup_n (r_n - 1/n^2, r_n + 1/n^2) \neq \varnothing$. Does there exist an enumeration $\{s_n\}_{m=1}^{\infty}$ of $\mathbb{Q}$ such that $\mathbb{R}\backslash\bigcup_m (s_m - 1/m, s_m + 1/m) \neq \varnothing$?

220. Prove or disprove: if E is a Lebesgue measurable subset of I^2 and if $\lambda(E_{x_1}) \leqq \frac{1}{2}$ a.e. in I then $\lambda(\{x_2 \colon \lambda(E_{x_2}) = 1\}) \leqq \frac{1}{2}$.

221. Show that $\mathbb{R}^2\backslash\{(x_1, x_2) \colon x_1 - x_2 \in \mathbb{Q}\}$, denoted E, contains no measurable "rectangle" $A_1 \times A_2$ of positive measure $(\lambda_2(A_1 \times A_2) > 0)$.

222. Show that if $f \in \mathbb{R}^{\mathbb{R}^2}$, f^{x_1} is Borel measurable for all x_1, and $f^{x_2} \in C(\mathbb{R}, \mathbb{R})$ for all x_2 then f is Borel measurable.

223. Let E be dense in $\mathbb{R}$ and let f be in $\mathbb{R}^{\mathbb{R}^2}$. Show that if f^{x_1} is Lebesgue measurable for all x_1 in E and $f^{x_2} \in C(\mathbb{R}, \mathbb{R})$ a.e. then f is Lebegue measurable.

224. Show that if $f \in \mathbb{R}^{\mathbb{R}^2}$, f^{x_1} is Lebesgue measurable for all x_1, and $f^{x_2} \in C(\mathbb{R}, \mathbb{R})$ for all x_2 then for every Lebesgue measurable function g in $\mathbb{R}^{\mathbb{R}}$, $h \colon x_2 \mapsto f(g(x_2), x_2)$ is Lebesgue measurable.

225. Is $f \colon (x_1, x_2) \mapsto x_1x_2/(x_1^2 + x_2^2)$ in $L^1([-1, 1]^2, \lambda_2)$?

226. For the map

$$f \colon (x_1, x_2) \mapsto \begin{cases} (x_1^2 - x_2^2)/(x_1^2 + x_2^2), & \text{if } x_1^2 + x_2^2 > 0 \\ 0, & \text{if } x_1 = x_2 = 0 \end{cases}$$

how are $\int_I (\int_I f(x_1, x_2)\, dx_1)\, dx_2$ and $\int_I (\int_I f(x_1, x_2)\, dx_2)\, dx_1$ related?

227. In $C((0, 1)^2, \mathbb{R})$ find a nonnegative f such that

$$\int_{I^2} f(x_1, x_2)\, d\lambda_2(x_1, x_2) < \infty$$

and yet for some x_1 in $(0, 1)$, $\int_I f(x_1, x_2)\, dx_2 = \infty$.

228. Prove or disprove: if (X, S, μ_i), $i = 1, 2$, are measure situations such that $\mu_1(X) + \mu_2(X) < \infty$, $\mu_i \ll \mu_j$, $i \neq j$, and $d\mu_1/d\mu_2 \in L^{\infty}(X, \mu_2)$ then $d\mu_2/d\mu_1 \in L^{\infty}(X, \mu_1)$.

229. In the measure situation $(I, \mathsf{S}_{\beta}, \mu)$ where $\mu \ll \lambda$ and $\mu(I) < \infty$, show that $\lim_{a>0, a\to 0} \mu(I \cap (x-a, x+a))/\lambda(I \cap (x-a, x+a))$ exists a.e.

230. Prove or disprove: there is in $\mathsf{S}_{\lambda}(I)$ an E such that $\lambda(E \cap [a, b]) = (b-a)/2$ for all a, b in $I (a < b)$.

231. Let $\mathcal{J}$ be a set of half-open intervals (if $J \in \mathcal{J}$ then $J = [a, b)$ or $J = (a, b]$, $(a < b)$). Prove that $\bigcup_{J \in \mathcal{J}} J$ is Lebesgue measurable.

232. Let $\{J_n\}_{n=1}^{\infty}$ be a sequence of open intervals in $\mathbb{R}$ and let $\{C_n\}_{n=1}^{\infty}$ be the set of components of $\bigcup_n J_n$. Show that $\sum_n \lambda(C_n) \leqq \sum_n \lambda(J_n)$.

233. Let $\{J_n\}_{n=1}^{N}$ be a (finite) set of open intervals in $\mathbb{R}$. Prove there is a subset $\{J_{n_k}\}_{k=1}^{K}$ of pairwise disjoint intervals such that $\lambda(\bigcup_{n=1}^{N} J_n) \leqq 2\lambda(\bigcup_{k=1}^{K} J_{n_k})$.

234. Prove or disprove: there is a measure situation $(I, \mathsf{S}_\lambda, \mu)$ such that μ is signed, not identically zero, $\mu \ll \lambda$, and for all a in I, $\mu([0, a]) = 0$.

235. Show that λ_n is rotation invariant, i.e., if T is a rotation of $\mathbb{R}^n$ about some point x in $\mathbb{R}^n$ and if $E \in \mathsf{S}_\lambda(\mathbb{R}^n)$ then $T(E) \in \mathsf{S}_\lambda(\mathbb{R}^n)$ and $\lambda_n(T(E)) = \lambda_n(E)$.

236. Find in I a subset A of the second category, Lebesgue measurable, and such that $\lambda(A) = 0$.

237. Let E be a Lebesgue measurable set in $\mathbb{R}$ and assume $0 < \lambda(E) < \infty$. Show that if f is Lebesgue measurable and nonnegative then $g \mapsto \int_E f(x-t)\,dt$ is in $L^1(\mathbb{R}, \lambda)$ iff $f \in L^1(\mathbb{R}, \lambda)$.

238. Show that if $E \in \mathsf{S}_\lambda(\mathbb{R})$ and $\lambda(E) > 0$ then for some positive a, if $|x| \leqq a$ then $(x + E) \cap E \neq \varnothing$.

239. Show that if $E \in \mathsf{S}_\lambda(\mathbb{R}^n)$ and $0 < a < \lambda_n(E)$, there is in E a compact subset K such that $\lambda_n(K) = a$.

240. Let $(\mathbb{R}, \mathsf{S}_\beta, \mu)$ be a measure situation such that $\mu(\mathbb{R}) = 1$. Show that if E is a Borel set then $\lambda(E) = \int_{\mathbb{R}} \mu(x + E)\,dx$, i.e., "$\mu * \lambda = \lambda$".

241. Let $T: \mathbb{R}^n \mapsto \mathbb{R}^m$ be a map satisfying the following conditions: i) $T(x+y) = T(x) + T(y)$; ii) if U is open in $\mathbb{R}^m$ then $T^{-1}(U) \in \mathsf{S}_\beta(\mathbb{R}^n)$. Show T is continuous (and hence linear).

242. If $x \in [0, 1)$ represent x as $\sum_{n=1}^{\infty} a_n 10^{-n}$, a_n in $\mathbb{N}$. If x has two such representations choose the one such that $a_n = 0$ for all but finitely many n. Let k be different from 0 and 1 and let f be the map.

$$x \mapsto \begin{cases} k, & \text{if each } a_n \neq 0 \\ 1, & \text{if the first nonzero } a_n \text{ has an even subscript.} \\ 0, & \text{otherwise} \end{cases}$$

Show f is Lebesgue measurable and calculate $\int_I f(x)\,dx$.

243. Let A be $\{x: x \in I, x = \sum_{n=1}^{\infty} a_n 10^{-n}, a_n = 2 \text{ or } 7\}$. Prove or disprove: i) A is closed; ii) A is open; iii) A is countable; iv) A is dense in I; v) A is Borel measurable. If A is Lebesgue measurable, find $\lambda(A)$.

244. For x in I let $A_n(x)$ be the number of 7's among the first n digits in the decimal representation of x. (If x has two decimal representations use the one involving only finitely many nonzero digits.) Let E be

$\{x: \lim_{n\to\infty} A_n(x)/n$ does not exist$\}$. Show that E is nonempty, Lebesgue measurable and neither open nor closed.

245. Construct an E in $\mathsf{S}_\lambda(\mathbb{R})$ so that $\lambda(E)<\infty$ and such that for all intervals $[a, b]$, $0<\lambda(E\cap[a, b])<b-a$ $(a<b)$.

246. In $\mathbb{R}$ construct a set E such that for all intervals $[a, b]$ $(a<b)$, $\lambda(E\cap[a, b]) . \lambda([a, b]\backslash E)>0$.

247. Assume $\{E_n\}_{n=1}^\infty\subset\mathsf{S}_\lambda(I)$ and that for some positive a and all n, $\lambda(E_n)\geqq a$. Prove or disprove: There is a subsequence $\{E_{n_k}\}_{k=1}^\infty$ such that $\lambda(\bigcap_{k=1}^\infty E_{n_k})>0$.

248. If $\{A_n\}_{n=1}^\infty$ is a sequence of Lebesgue measurable sets no two of which are the same and if $\sum_n \lambda(A_n)<\infty$, let G_k be $\{x: x\in A_n$ for exactly k different values of n, i.e., for k different values of n and no others, $x\in A_n\}$. Show that for all k, G_k is Lebesgue measurable and that $\sum_{k=1}^\infty k\lambda(G_k)=\sum_n \lambda(A_n)$.

10. Lebesgue Measurable Functions

Conventions

If $f \in L^1([-\pi, \pi], \lambda)$ and for n in $\mathbb{Z}$, $c_n = (2\pi)^{-1} \int_{-\pi}^{\pi} f(x) e^{-inx} dx$, then $\sigma_N(f): x \mapsto (N+1)^{-1} \sum_{k=0}^{N} (\sum_{n=-k}^{k} c_n e^{inx})$ is the average of the first $N+1$ partial sums S_k, $k = 0, 1, \ldots, N$, of the Fourier series for f. If $f \in L^1(\mathbb{R}, \lambda)$, then $\hat{f}: t \mapsto \int_{\mathbb{R}} f(x) e^{-itx} dx$ is the Fourier transform of f. Furthermore, $\hat{f} \in C_0(\mathbb{R}, \mathbb{C})$. $\hat{f} = 0$ iff $f = 0$ a.e., and if $f, g \in L^1(\mathbb{R}, \lambda)$ then $(f * g)\hat{} = \hat{f} . \hat{g}$. If $\mathbb{T}$ is regarded as an abelian group and arc length is the foundation of measure λ on $\mathbb{T}$ then $L^p((-\pi, \pi], \lambda)$ and $L^p(\mathbb{T}, \lambda)$ are isomorphic. Furthermore $\hat{f}$ is the map $\mathbb{Z} \ni n \mapsto (2\pi)^{-1} \int_{\mathbb{T}} f(t) e^{-int} dt$.

A map $f: X \mapsto Y$ is injective iff f is one-one; is surjective iff $f(X) = Y$; is bijective iff f is injective and surjective.

A curve is a continuous map $f: I \mapsto X$ into a topological space X. If there is a metric d compatible with the topology of X then the length of f is defined as follows: if P is a partition of $I \backslash \{1\}$ into disjoint intervals $[t_k, t_{k+1})$, then $l(P) = \sum_{k=0}^{n-1} d(f(t_{k+1}), f(t_k))$ and the length l_f of the curve S is $\sup_P l(P)$.

A countably subadditive nonnegative set function μ^* ($\mu^*(\bigcup_{n=1}^{\infty} E_n) \leqq \sum_{n=1}^{\infty} \mu^*(E_n)$) defined on 2^X and such that $\mu^*(\emptyset) = 0$ is an outer measure. The set of (Caratheodory) μ^*-measurable sets is $\{E$: for all A in 2^X, $\mu^*(A \cap E) + \mu^*(A \backslash E) = \mu^*(A)\}$. The μ^*-measurable sets constitute a σ-algebra denoted S_μ; μ^* confined to S_μ is a measure, the measure induced by μ^* and denoted μ.

For h in $\mathbb{C} \backslash \{0\}$ and f a map from $\mathbb{C}$ to a vector space X, $\Delta_h f$ is the map $x \mapsto (f(x+h) - f(x))/h$.

249. Assume f is Lebesgue measurable and that for every Lebesgue measurable set J such that $\lambda(J)=1$ or $2^{1/2}$, $\int_J f(x)\,dx=0$. Show $f=0$ a.e.

250. Assume $f\in L^\infty_{\mathbb{R}}([-\pi,\pi],\lambda)$, $\|f\|_\infty=1$, and for some N and some x_0, $|\sigma_N(f)(x_0)|f=1$. Show there is a constant c such that $f=c$ a.e. on $[-\pi,\pi]$.

251. Construct a map $f:I\mapsto I$ so that: i) f is bijective; ii) E is Lebesgue measurable iff $f(E)$ is Lebesgue measurable; iii) if E is Lebesgue measurable then $\lambda(E)=\lambda(f(E))$; iv) $f((1/4, 3/4))=[1/4, 3/4]$.

252. Construct a nonnegative Lebesgue measurable function f, finite everywhere and such that if $a<b$ then $\int_a^b f(x)\,dx=\infty$.

253. Show that if f is Lebesgue measurable and $\mathbb{R}$-valued there is a sequence $\{f_n\}_{n=1}^\infty$ of $\mathbb{Q}$-valued functions converging uniformly and monotonely to f as $n\to\infty$.

254. Assume $f, g\in\mathbb{R}^{\mathbb{R}}$, f is continuous, g is Lebesgue measurable, and for every null set (λ) N, $f^{-1}(N)$ is Lebesgue measurable. Show $g\circ f$ is Lebesgue measurable.

255. Assume $\lambda^*(A)>0$ and $\theta\in(0,1)$. Show there is an interval J, say (a,b), such that $\lambda^*(A\cap J)>\theta\,.\,\lambda^*(J)$.

256. Assume $f\in\mathbb{C}^{I^2}$, for all x_1, $f^{x_1}\in C(I,\mathbb{C})$, for all x_2, $f^{x_2}\in L^1(I,\lambda)$, and that there is in $L^1(I,\lambda)$ a g such that for all x_1, x_2, $|f(x_1,x_2)|\leqq g(x_1)$. Show that $h:x_2\mapsto\int_I f(x_1,x_2)\,dx_1$ is continuous on I.

257. Show that if f is Lebesgue measurable there is a g, also Lebesgue measurable and such that $\sup_x|g(x)|=\|f\|_\infty$ and $f=g$ a.e.

258. Show that if $f\in\mathbb{R}^I$ and is Lebesgue measurable there is a unique a_0 such that $\lambda\{x: f(x)\geqq a_0\}\geqq\frac{1}{2}$ and for all a in (a_0,∞), $\lambda\{x: f(x)\geqq a\}<\frac{1}{2}$.

259. Let g_C be the Cantor function for *the* Cantor set (see Solution 189) and consider the curve $f:I\ni t\mapsto(t, g_C(t))$. Find: i) $\int_I g_C(t)\,dt$; ii) the length l_{g_C} of g_C.

260. Assume that f' exists for all x in I and that if $x\in(0,1)$, $|f'(x)|\leqq M<\infty$. Show that for any Lebesgue measurable set E, $f(E)$ is Lebesgue measurable and $\lambda(f(E))\leqq M\lambda(E)$.

261. Assume $f, g\in L^\infty((0,\infty),\lambda)$ and $\int_{(0,\infty)}(|f(x)|+|g(x)|)/x\,dx<\infty$. Show $\int_{(0,\infty)}(|f(xy)g(1/y)|)/y\,dy<\infty$ a.e.

262. Show that if $f\in L^\infty(I,\lambda)$ and for all n in $\mathbb{N}$, $\int_I t^n f(t)\,dt=0$ then $f=0$ a.e.

263. Show that if $\{f_n\}_{n=0}^\infty\subset L^\infty(\mathbb{R},\lambda)$, $f_n\to f_0$ in measure as $n\to\infty$, and for all n in $\mathbb{N}$, $|f_n(x)|\leqq e^{-x^2}$, then $f_0\in L^1(\mathbb{R},\lambda)$ and $\int_{\mathbb{R}} f_n(x)\,dx\to\int_{\mathbb{R}} f_0(x)\,dx$ as $n\to\infty$.

264. Show that if $\lambda(E)>0$ and whenever $x, y \in E$ so also $\frac{1}{2}(x+y)\in E$ then there is a nonempty open set contained in E.

265. Show card $(\mathsf{S}_\lambda)=2^{\text{card}\,(\mathbb{R})}$.

266. Let E be a subset of $\mathbb{R}^n$ and assume that for each a in $\mathbb{R}^n$ there is a positive r_a such that $B(a, r_a)\cap E\in\mathsf{S}_\lambda$. Show $E\in\mathsf{S}_\lambda$.

267. Let E be a null set (λ) and assume card$(E)>$card$(\mathbb{N})$. Show there is in E a subset F such that $F\in\mathsf{S}_\lambda\backslash\mathsf{S}_\beta$.

268. Assume μ^* is an outer measure on I, $\mathsf{S}_\mu\supset\mathsf{S}_\beta$, and $\mu\ll\lambda$. Show $\mathsf{S}_\lambda\subset\mathsf{S}_\mu$.

269. Prove or disprove: if $\{f_n\}_{n=0}^\infty\subset\mathbb{R}^I$, each f_n is monotone increasing, and $f_n\to f_0$ in measure as $n\to\infty$, then $f_n\to f_0$ at each point of continuity of f_0.

270. Let $\theta:[0,1)\to[0,1)/(\mathbb{Q}/\mathbb{Z})$ be the canonical map of $[0,1)$, regarded as the abelian group $\mathbb{R}/\mathbb{Z}$ onto its quotient by its subgroup $\mathbb{Q}\cap[0,1)$. Assume that for some subset S of $\mathbb{R}$, $\theta(S/\mathbb{Z})=[0,1)/\mathbb{Q}\cap[0,1)$ and that $\theta|_{S/\mathbb{Z}}$ is bijective. (In other words $S/\mathbb{Z}$ consists of a complete system of coset representatives, one from each coset.) Let $\{r_k\}_{k=1}^\infty$ be an enumeration of $\mathbb{Q}$ and let S_k be $(r_k+S)/\mathbb{Z}$. If, for t in $(0,1]$, k_t is the unique element of $\mathbb{N}$ for which $2^{-(k_t+1)}\leqq t<2^{-k_t}$ let f_t be the map

$$x\mapsto\begin{cases}1, & \text{if } x\in S_{k_t} \text{ and } x=2^{k_t+1}t-1\\ 0, & \text{otherwise.}\end{cases}$$

Show $\lim_{t\to 0}f_t(x)=0$ for all x and that for some positive a,

$$\lambda^*\{x: f_t(x)>\tfrac{1}{2}\}>a.$$

271. Construct a sequence $\{f_n\}_{n=1}^\infty$ of Lebesgue measurable functions defined on I and such that the sequence converges everywhere on I and yet for every Lebesgue measurable set E for which $\lambda(E)=1$ the sequence fails to converge uniformly on E. (This constitutes sharpening of Egorov's theorem.)

272. Let E be a null (λ) subset of I. Find a monotone increasing function f defined on I and such that for all x in E, $\Delta_h f(x)\to\infty$ as $h\to 0$.

273. Let f be a nonnegative and Lebesgue measurable function defined on I. Show that if $A_f=\{g: g\in L^1(I,\lambda),\ |g|\leqq f \text{ a.e.}\}$ then i) A_f is closed in $L^1(I,\lambda)$ and ii) if A_f is compact then $f\in L^1(I,\lambda)$.

274. Find a Lebesgue measurable set E such that $E+E$ is not measurable. (See Problem 457.)

11. $L^1(X, \mu)$

Conventions

The map

$$\mu: 2^X \ni B \mapsto \begin{cases} \text{card}(B), & \text{if } \text{card}(B) < \text{card}(\mathbb{N}) \\ \infty, & \text{otherwise} \end{cases}$$

is counting measure. If $w: X \mapsto [0, \infty)$ is a map, the map $\mu: 2^X \ni B \mapsto \sum_{x \in B} w(x)$ is discrete measure. Counting measure and discrete measure are the same iff $w = 1$. If μ is counting measure on $\mathbb{N}$ or $\mathbb{Z}$, $L^p(\mathbb{N}, \mu)$ resp. $L^p(\mathbb{Z}, \mu)$ are occasionally denoted $l^p(\mathbb{N})$ resp. $l^p(\mathbb{Z})$, $1 \leqq p \leqq \infty$.

If E is a Banach space and if $\{T_\gamma\}_{\gamma \in \Gamma}$ is a set of continuous linear maps $T_\gamma: E \mapsto F$ into a Banach space F, then $\sup_\gamma \|T_\gamma\| < \infty$ iff $\sup_\gamma \|T_\gamma(x)\| < \infty$ for all x in some subset S of the second category in E. (This result is often called the uniform boundedness principle.)

If E is a Banach space, the weak topology $\sigma(E, E^*)$ has as a basis for its open sets $\{U(x; f_1, \ldots, f_n, a) = \{y: |f_k(y) - f_k(x)| < a,\ k = 1, \ldots, n\}: x \in E,\ f_1, \ldots, f_n \in E^*,\ n \in \mathbb{N},\ a > 0\}$. Correspondingly, the weak* topology $\sigma(E^*, E)$ has a basis for its open sets $\{U(f; x_1, \ldots, x_n) = \{g: |g(x_k) - f(x_k)| < a,\ k = 1, \ldots, n\}: f \in E^*,\ x_1, \ldots, x_n \in E,\ n \in \mathbb{N},\ a > 0\}$. Alaoglu's theorem asserts that $B(0, 1)$ in E^* is weak*-compact.

The map $T: E \ni x \mapsto F_x \in E^{**}$ according to the formula $f(x) = F_x(f)$ for all f in E^* is an isometric linear (hence injective) transformation of the Banach space E into its second dual E^{**}. By definition E is reflexive iff T is surjective ($T(E) = E^{**}$). Eberlein's theorem states that E is reflexive iff the unit ball $B(0, 1)$ of E is weakly sequentially compact in $\sigma(E, E^*)$, i.e., if $\{x_n\}_{n=1}^\infty \subset B(0, 1)$ there is a subsequence $\{x_{n_k}\}_{k=1}^\infty$ converging with respect to the topology $\sigma(E, E^*)$ to some x_0 in $B(0, 1)$. Alternatively, E is reflexive iff $\sigma(E^*, E) = \sigma(E^*, E^{**})$.

275. Let $\{\{x_{nm}\}_{m=1}^{\infty}\}_{n=0}^{\infty}$ be a sequence in $l^1(\mathbb{N})$ and assume that for each $\{y_m\}_{m=1}^{\infty}$ in $l^{\infty}(\mathbb{N})$. $\lim_{n\to\infty} \sum_{m=1}^{\infty} x_{nm}y_m = \sum_{m=1}^{\infty} x_{0m}y_m$. Show that

$$\|\{x_{nm}\}_{m=1}^{\infty} - \{x_{0m}\}_{m=1}^{\infty}\|_1 \to 0$$

as $n \to \infty$. ("If a sequence in $l^1(\mathbb{N})$ conveges weakly it converges strongly." Nevertheless the weak and strong topologies of $l^1(\mathbb{N})$ are different, since, e.g., no weak neighborhood of zero is contained in any strong neighbourhood (ball).)

276. Find in $C^{\infty}([0, \infty), \mathbb{C}) \cap L^1([0, \infty), \lambda)$ a nonnegative function f such that $\sum_{n=0}^{\infty} f(n) = \infty$.

277. Show $x \mapsto (\sin x)/x \notin L^1((0, \infty), \lambda)$.

278. Let $\{(X, \mathsf{S}, \mu_n)\}_{n=0}^{\infty}$ be a sequence of measure situations such that $X \in \mathsf{S}$, $\mu_0(X) = 1 = \lim_{n\to\infty} \mu_n(X)$ and $\mu_n \leqq \mu_0$. Show there is a subsequence $\{\mu_{n_k}\}_{k=1}^{\infty}$ such that for all g in $L^1(X, \mu_0)$, $\int_X g(x)\, d\mu_{n_k}(x) \to \int_X g(x)\, d\mu_0(x)$ as $k \to \infty$.

279. Let (X, S, μ) be a measure situation such that there is a countable sequence $\{A_n\}_{n=1}^{\infty}$ contained in S and consisting of pairwise disjoint sets of finite positive measure. Define the map $T: L^1(X, \mu) \ni f \mapsto Tf \in (L^{\infty}(X, \mu))^*$ by the rule: for g in $L^{\infty}(X, \mu)$, $(Tf)(g) = \int_X f(x)g(x)\, d\mu(x)$. Show that $T(L^1(X, \mu)) \subsetneqq (L^{\infty}(X, \mu))^*$.

280. Show that if f is a nonnegative Lebesgue measurable function on I and if $E_n = \{x : n-1 \leqq f(x) < n\}$ then $f \in L^1(1, \lambda)$ iff $\sum_{n=1}^{\infty} n\lambda(E_n) < \infty$.

281. Show that if f is

$$x \mapsto \begin{cases} x^2 \sin(1/x^2), & \text{if } x \neq 0 \\ 0, & \text{otherwise} \end{cases}$$

then f' exists everywhere and $f' \notin L^1(\mathbb{R}, \lambda)$.

282. Prove or disprove: If $f \in L^1([0, n], \lambda)$ for n in $\mathbb{N}$ and if $x^{-1} \int_0^x f(t)\, dt \geqq f(x)$ a.e. on $(0, \infty)$ then f is monotone decreasing.

283. Show that if $f \in L^1(I, \lambda)$ and if $\int_I f(x)\, dx \neq 0$ there is a Lebesgue measurable function g such that $\int_I |f(x)g(x)|\, dx < \infty$ and $\int_I |f(x)| \cdot |g(x)|^2\, dx = \infty$.

284. Show that if f is a nonnegative element in $L^1(I, \lambda)$ and if, for all n in $\mathbb{N}$, $\int_I (f(x))^n\, dx$ is a constant c (independent of n) then for some Lebesgue measurable set E, $f = \chi_E$ a.e. How is the conclusion altered if f is not restricted to be nonnegative?

285. Assume $f \in L^1(\mathbb{R}, \lambda)$ and that for every open set U such that $\lambda(U) = 1$, $\int_U f(x)\,dx = 0$. Show that $f = 0$ a.e.

286. Prove or disprove: if $f \in L^1(\mathbb{R}, \lambda)$ and if for all n in $\mathbb{N}$, $\int_{\mathbb{R}} |x|^n |f(x)|\,dx \leqq 1$ then $f = 0$ a.e. on $\{x: |x \geqq 1\}$.

287. Assume $f \in L^1([a, b], \lambda)$, that f is $\mathbb{R}$-valued, and for all c, d satisfying: $a < c < d < b$, $\lim_{|h| \downarrow 0} h^{-1} \int_c^d (f(x+h) - f(x))\,dx = 0$. Show there is a constant k such that $f = k$ a.e.

288. Show that if $f \in L^1(\mathbb{R}, \lambda)$ and $\|f_{(t)} - f\|_1 \leqq |t|^2$ then $f = 0$ a.e.

289. Show that if $q \in L^1(\mathbb{R}, \lambda)$ there is a constant C_q such that for all f in $C_{00}^{\infty}(\mathbb{R}, \mathbb{C})$, $|\int_{\mathbb{R}} q(x)(f(x))^2\,dx| \leqq C_q \int_{\mathbb{R}} |(f(x))^2 + (f'(x))^2|\,dx$.

290. Show that if $f \in L^1(\mathbb{R}, \lambda)$ then $g: x \mapsto f(x - 1/x) \in L^1(\mathbb{R}, \lambda)$ and $\int_{\mathbb{R}} f(x)\,dx = \int_{\mathbb{R}} g(x)\,dx$.

291. Show that if $f, g \in L^1(I, \lambda)$ then $f = g$ a.e. iff for all h in $C^{\infty}(I, \mathbb{R})$, $\int_I f(x)h(x)\,dx = \int_I g(x)\,h(x)\,dx$.

292. Assume $f \in L^1([0, \infty), \lambda)$ and that g is Lebesgue measurable. Show that if for all t in $[1, \infty)$, $|g(t)/t| \leqq M < \infty$ then $\lim_{t \to \infty} t^{-1} \int_1^t f(s)g(s)\,ds = 0$.

293. Assume $f \in L^1(\mathbb{R}, \lambda)$ and that for every open set U, $\int_U f(x)\,dx = \int_{\bar{U}} f(x)\,dx$. Show $f = 0$ a.e.

294. Assume $f \in L^1(I, \lambda)$ and that f is continuous at zero. Show that for all n in $\mathbb{N}$, $f_n: x \mapsto f(x^n)$ is in $L^1(I, \lambda)$.

295. Show that if $f \in L^1(\mathbb{R}, \lambda)$ and if $f_n = f_{(n)} \cdot \chi_I$, n in $\mathbb{N}$, then $\{\sum_{n=1}^N f_n\}_{N=1}^{\infty}$ is a Cauchy sequence in $L^1(I, \lambda)$.

296. Assume $\{f\}_{n=1}^{\infty} \subset L^{\infty}(G, \lambda)$, G is open in $\mathbb{R}^n$, $\lambda(G) < \infty$, $\|f_n\|_{\infty} \leqq M < \infty$, $\{f_n\}_{n=1}^{\infty}$ is a Cauchy sequence in $L^1(G, \lambda)$, and $f_n \to f$, $g_n \to g$ in $L^1(G, \lambda)$ as $n \to \infty$. Show $f_n g_n \to fg$ in $L^1(G, \lambda)$ as $n \to \infty$.

297. Show that if $\{f_n\}_{n=1}^{\infty} \subset L^1(I, \lambda)$, $f_n \geqq 0$, and $\int_I f_n(x)\,dx \to 0$ as $n \to \infty$, then $\int_I (1 - e^{-f_n(x)})\,dx \to 0$ as $n \to \infty$.

298. Show that if $\{f_n\}_{n=1}^{\infty} \subset L^1(\mathbb{R}, \lambda)$ and $f_n \to f$ a.e. as $n \to \infty$ then $f \in L^1(\mathbb{R}, \lambda)$ and $f_n \to f$ in $L^1(\mathbb{R}, \lambda)$ as $n \to \infty$ iff i) for each positive a there is a Lebesgue measurable set A_a such that $\lambda(A_a) < \infty$ and $\sup_n \int_{\mathbb{R} \setminus A_a} |f_n(x)|\,dx < a$, and ii) $\lim_{\lambda(B) \to 0} \sup_n \int_B |f_n(x)|\,dx = 0$.

299. Show that the partial sums S_n of the Maclaurin series $\sum_{n=0}^{\infty} 1 . 3 \ldots (2n-1)x^n/n!2^n$ representing $f: x \mapsto (1-x)^{-1/2}$ in $(-1, 1)$ converge in $L^1((-1, 1), \lambda)$ to f.

300. Assume $f \geqq 0$, f is monotone increasing on $\mathbb{R}$, $x/f(x) \to 0$ as $x \to \infty$, $\{g_n\}_{n=1}^{\infty}$ is a sequence of Lebesgue measurable functions, $g_n \to g$ a.e. as

$n \to \infty$, and $\int_I f(|g_n(x)|)\,dx \leqq M < \infty$, n in $\mathbb{N}$. Show $\int_I |g_n(x) - g(x)|\,dx \to 0$ as $n \to \infty$.

301. Let A be $\{f : f \in L^1(I, \lambda),\ |f| \geqq 1 \text{ a.e.}\}$. Prove or disprove that A is closed in the norm-induced topology of $L^1(I, \lambda)$.

302. For the set described in Problem 301, prove or disprove that A is closed in the weak topology $(\sigma(L^1(I, \lambda), L^\infty(I, \lambda)))$ of $L^1(I, \lambda)$.

303. Let f be in $L^1(I, \lambda)$ and let S be $\{x : x \in I,\ f(x) \in \mathbb{Z}\}$. Show that $\lim_{n\to\infty} \int_I |\cos \pi f(x)|^n\,dx = \lambda(S)$.

304. Assume that $g \in L^1(\mathbb{T}, \lambda)$. Show that $T: C(\mathbb{T}, \mathbb{C}) \ni f \mapsto (x \mapsto \int_{\mathbb{T}} g(x-y) f(y)\,dy) \in C(\mathbb{T}, \mathbb{C})$ is a compact (linear) transformation, i.e., that T is linear and maps sets bounded with respect to the norm $\|\cdot\cdot\cdot\|_\infty$ into sets having compact closure in the norm-induced topology. In particular, show that if f_n is $x \mapsto \cos nx$ show then $T(f_n)$ converges uniformly to zero.

305. Show that if $f \in L^1(\mathbb{R}, \lambda)$ and for some n in $\mathbb{N}$, $\int_{\mathbb{R}\setminus[-n,n]} |f(x)|\,dx = 0$, then $\hat{f} \in C_0^\infty(\mathbb{R}, \mathbb{C})$.

306. Show that if $f \in L^1(\mathbb{R}, \lambda)$ and $f = f * f$ then $f = 0$ a.e.

307. Show that if $\{z_n\}_{n=1}^N$ is a set of N different points in $\mathbb{T}$ and if $\{a_n\}_{n=1}^N$ are such that $a_n \geqq 0$, $\sum_n a_n = 1$, then $|\sum_n a_n z_n| \leqq 1$ and equality obtains iff only one a_n is not zero.

308. (See Problem 307.) Let E be a Lebesgue measurable set in $\mathbb{R}$ and assume $\lambda(E) > 0$. Show that if $f \in L^1(\mathbb{R}, \lambda)$, $f \geqq 0$, and $\int_E f(x)\,dx = 1$, then $|\int_E f(x) e^{ix}\,dx| < 1$.

309. (See Problem 308.) Show that if $f \in L^1(\mathbb{R}, \lambda)$, $f \geqq 0$, and $\|f\|_1 > 0$ then for all nonzero t, $|\hat{f}(t)| < \|f\|_1$.

310. Assume $f \in L^1(\mathbb{R}, \lambda)$ and that $\text{supp}(\hat{f})$ is compact. Show there is in $L^1(\mathbb{R}, \lambda) \cap C(\mathbb{R}, \mathbb{C})$ a g such that g is not identically zero and $f * g = 0$ a.e.

311. Assume $f \in L^1(\mathbb{R}, \lambda)$, for some n in $\mathbb{N}$, $\int_{\mathbb{R}\setminus[-n,n]} |f(x)|\,dx = 0$, and $\text{supp}\,(\hat{f})$ is compact. Show $f = 0$ a.e.

312. Partially order the open sets containing zero in $\mathbb{R}^n$ according to the rule: $U \leqq V$ iff $V \subset U$. Let $\{g_U\}$ be a net with values in $L^1(\mathbb{R}^n, \lambda)$ and assume: i) $g_U \geqq 0$; ii) $\|g_U\|_1 = 1$; iii) $g_U = 0$ off U. Show that if $f \in L^1(\mathbb{R}^n, \lambda)$ then $\lim_U \|g_U * f - f\|_1 = 0$. (If Γ is a partially ordered set and if A is a Banach algebra, a map $\Gamma \ni \gamma \mapsto a_\gamma \in A$ is called an approximate (left) identity if, for all b in A, $\lim_\gamma \|a_\gamma b - b\| = 0$. Thus the problem is to prove that $U \mapsto g_U$ is an approximate left identity.)

313. (See Problem 312.) Show that if $\gamma \mapsto g_\gamma$ is an approximate identity for $L^1(\mathbb{R}, \lambda)$ then $\lim_\gamma (\hat{g}_\gamma - 1)_\infty = 0$.

314. Show that if $\gamma \mapsto g_\gamma$ is an approximate identity for the Banach algebra A then so is $\gamma \mapsto (g_\gamma)^n$, n in $\mathbb{N}$, an approximate identity for A.

315. For t in $(0, \infty)$ let g_t be $x \mapsto k_t e^{-x2/t}$. Determine the value of the constant k_t so that $\|g_t\|_1 = 1$. Show $\lim_{t\to 0} \|g_t * f - f\|_1 = 0$ for all f in $L^1(\mathbb{R}, \lambda)$.

316. Let k be in $L^\infty(\mathbb{R}, \lambda)$ and assume $\int_{\mathbb{R}} e^{-(x-y)2}\, k(y)\, dy =$ for all x in $\mathbb{R}$. Show $\mathrm{k} = 0$ a.e.

317. Show that T as defined in Problem 304, but with domain $L^2(\mathbb{R}, \lambda)$ and g in $L^1(\mathbb{R}, \lambda)$, is not compact if $\|g\|_1 \neq 0$.

318. Let the map $w: [1, \infty) \to [1, \infty)$ be the identity and let μ be the corresponding discrete measure on $[1, \infty)$. Describe $(L^1([1, \infty), \mu))^*$.

319. Refer to the solution of Problem 250. Show that if $f \in C(\mathbb{T}, \mathbb{C})$ then $\lim_{N\to\infty} \|\sigma_N(f) - f\|_\infty = 0$.

320. Repeat Problem 319 with the hypothesis: $f \in C(\mathbb{T}, \mathbb{C})$ replaced by $f \in L^1(\mathbb{T}, \lambda)$ and the conclusion replaced by $\lim_{N\to\infty} \|\sigma_N(f) - f\|_1 = 0$.

321. Determine a nonempty open interval J such that if $p \in J$ and if $f \in L^1(\mathbb{R}, \lambda)$ then $x \mapsto \int_{\mathbb{R}} f(x-y)|\sin y^{-1}| \,.\, |y|^{-1/2}\, dy$ is in $L^p(\mathbb{R}, \lambda)$.

12. $L^2(X, \mu)$ or $\mathfrak{H}$ (Hilbert Space)

Conventions

The gradient map ∇ is $C^1(\mathbb{R}^n, \mathbb{C}) \ni f \mapsto (\partial f/\partial x_1, \ldots, \partial f/\partial x_n)$. The Laplacian map Δ is $C^2(\mathbb{R}^n, \mathbb{C}) \ni f \mapsto \sum_{i=1}^n \partial^2 f/\partial x_i^2$.

If $f, g \in \mathfrak{H}$ (Hilbert space), (f, g) is the inner product of f and g; $f \perp g$ iff $(f, g) = 0$. If $S \subset \mathfrak{H}$, $S^\perp = \{f\colon (f, g) = 0$ for all g in $S\}$. If $T \in \operatorname{End}(\mathfrak{H})$ then T^* in $\operatorname{End}(\mathfrak{H})$ is the (unique) element such that for all f, g in $\mathfrak{H}$, $(Tf, g) = (f, T^*g)$. If $U \in \operatorname{Hom}(\mathfrak{H}_1, \mathfrak{H}_2)$, then U is unitary iff U is bijective and for all f, g in $\mathfrak{H}_1$, $(Uf, Ug) = (f, g)$.

If X and Y are Banach spaces and if $T\colon X \mapsto Y$ is a map, T is closed iff its graph, $\{(x, Tx)\colon x \in X\}$ is closed in the product topology of $X \times Y$.

The variation of a function $f\colon I \mapsto \mathbb{C}$ is $\sup\{\sum_{i=1}^n |f(t_{i+1}) - f(t_i)|\colon 0 \leqq t_1 < t_2 < \cdots < t_{n+1} \leqq 1, n$ in $\mathbb{N}\}$. The function f is of bounded variation iff its variation is finite.

If G is an open connected set (a region) in $\mathbb{C}$, $H(G)$ is the set of functions holomorphic in G.

322. Assume $\{a_n\}_{n=1}^\infty \in l^2(\mathbb{N})$ and that for all t in $(-\frac{1}{2}, \frac{1}{2})$, $\sum_{n=1}^\infty a_n/(n-t) = 0$. Show that for all n, $a_n = 0$.

323. Let F_N be $l^2(\mathbb{N}) \ni \{a_n\}_{n=1}^\infty \mapsto (\sum_{n \geqq N} |a_n|^2)^{1/2}$. Show that $F_N \to 0$ uniformly on compact sets (i.e., compact with respect to the norm-induced topology) as $N \to \infty$.

324. Assume $f, g \in C^2(\mathbb{R}^2, \mathbb{C}) \cap L^2(\mathbb{R}^2, \lambda)$, $\Delta f, \Delta g \in L^2(\mathbb{R}^2, \lambda)$, and that $\nabla f, \nabla g \in L^2(\mathbb{R}^2, \lambda)$ (i.e., each component of each vector is in $L^2(\mathbb{R}^2, \lambda)$). Show that $\int_{\mathbb{R}^2} f(x) \Delta g(x)\, dx = \int_{\mathbb{R}^2} g(x) \Delta f(x)\, dx$.

325. Find a curve $\gamma:[0, 1] \mapsto \mathfrak{H}$ such that whenever $0 \leqq t_1 < t_2 \leqq t_3 < t_4 \leqq 1$, $(\gamma(t_2) - \gamma(t_1)) \perp (\gamma(t_4) - \gamma(t_3))$.

326. Assume $\int_{[0,\infty)} |f_n'(x)|^2 \, dx \leqq M^2 < \infty$ and $|f_n(x)| \leqq x^{-1}$ on $(0, \infty)$ for all n in $\mathbb{N}$. Prove or disprove: i) the sequence contains a (pointwise) convergent subsequence; ii) the sequence contains a subsequence converging uniformly on $[0, \infty)$; iii) the sequence contains a subsequence converging in the norm-induced topology of $L^2([0, \infty), \lambda)$.

327. Let E be $\{(x, y): 0 \leqq |x| \leqq y \leqq 1\}$. Show that if $f \in L^2(E, \lambda)$ then $\liminf_{y=0} \int_{-y}^{y} |f(x, y)| \, dx = 0$.

328. Show that if $\{f_n\}_{n=1}^{\infty} \subset L^2(I, \lambda)$, $f_n \to 0$ in measure as $n \to \infty$, and $\|f_n\|_2 \leqq 1$, n in $\mathbb{N}$, then $\|f_n\|_1 \to 0$ as $n \to \infty$.

329. Show that if $f, g \in L^2(\mathbb{R}, \lambda)$ and $\lim_{h \to 0} \int_{\mathbb{R}} |\Delta_h f(x) - g(x)|^2 \, dx = 0$, then there is a constant c such that $f(x) = \int_0^x g(t) \, dt + c$ a.e.

330. Assume T is $L^2(I, \lambda) \ni f \mapsto \int_0^x f(t) \, dt$. Show: i) $\|T\| \leqq 2^{-1/2}$; ii) if $P = T^* + T$ then $P^2 = P$ and $P(L^2(I, \lambda)) = \mathbb{C}$; iii) T is compact.

331. Assume $S \subset C(I, \mathbb{C}) \cap L^2(I, \lambda)$ and that S is a closed subspace of $L^2(I, \lambda)$. Show: i) S is closed in $C(I, \mathbb{C})$; ii) for all f in S and some M in $(0, \infty)$, $\|f\|_2 \leqq \|f\|_\infty \leqq M\|f\|_2$; iii) for all y in I there is in $L^2(I, \lambda)$ a k_y such that for all f in S, $f(y) = \int_I k_y(x) f(x) \, dx$.

332. Assume that $f \in L^2([-\pi, \pi], \lambda)$ and that $\int_{-\pi}^{\pi} |h \Delta_h f(x)|^2 \, dx \leqq C|h|^{1+a}$ for some positive constants C and a. Show that the Fourier series for f is absolutely convergent.

333. Let A be $\{f: f \in C^\infty(I, \mathbb{C}), f(0) = 0, \int_{(0,1]} f(x)/x \, dx = 0\}$. Show that with respect to the norm-induced topology of $L^2(I, \lambda)$, A is a dense subset of $L^2(I, \lambda)$.

334. Show that if $f \in L^2(I, \lambda)$ and for all n in $\mathbb{N}$, $\int_0^1 t^n f(t) \, dt = 1/(n+2)$ then $f(t) = t$ a.e. on I.

335. Assume that $\{f_n\}_{n=1}^{\infty} \subset L^2(X, \mu)$ and that for all n in $\mathbb{N}$, $\|f_n - f_{n+1}\|_2 \leqq 2^{-n}$. Show there is in $L^2(X, \mu)$ an f such that $f_n \to f$ a.e. and in the norm-induced topology of $L^2(X, \mu)$ as $n \to \infty$.

336. Let (X, S, μ) be a measure situation and let $\{f_n\}_{n=1}^{\infty}$ be an orthonormal set in $L^2(X, \mu)$. Show that if $E = \{x: \lim_{n \to \infty} f_n(x) = f(x) \text{ exists}\}$ then $f = 0$ a.e. on E.

337. Show that if $\{f_n\}_{n=1}^{\infty}$ is an orthonormal set in $L^2(I, \lambda)$ then $\sup_n(\text{variation of } f_n) = \infty$.

338. Assume $\{f_n\}_{n=1}^{\infty}$ is an orthonormal set in $L^2(X, \mu)$, that $\mu(X) < \infty$, and that for some M in $(0, \infty)$, $|f_n(x)| \leqq M$ for all n and all x. Show that $\sum_{n=1}^{\infty} n^{-1} f_n(x)$ converges a.e.

339. Assume $K \in L^2(I^2, \lambda)$ and let T be $L^2(I, \lambda) \ni f \mapsto \int_I K(x, y) f(y)\, dy$. Show: i) $T(L^2(I, \lambda)) \subset L^2(I, \lambda)$; ii) for each positive a there is in $\text{End}(L^2(I, \lambda))$ a T_a such that the range of T_a is finite-dimensional and $\|T_a - T\| < a$.

340. Assume that T is a self-map of I and that T preserves the measurability and measure of every set. Show that if $f \geqq 0$, $f \in L^2(I, \lambda)$, and $\int_I f(T^n(x)) f(x)\, dx = 0$ for all n in $\mathbb{N}$ then $f = 0$.

341. Let U be in $\text{End}(\mathfrak{H})$ and unitary and let f be in $\mathfrak{H}$. Show there is in $\mathfrak{H}$ a g such that $\|(N+1)^{-1} \sum_{n=0}^N U^n(f) - g\| \to 0$ as $N \to \infty$. (This result is the mean ergodic theorem.)

342. Let $\mathfrak{H}$ be an infinite-dimensional separable Hilbert space and let $\mathsf{O}(\mathfrak{H})$ be the set of open sets in the norm-induced topology of $\mathfrak{H}$. Show that if $(\mathfrak{H}, \sigma\mathsf{R}(\mathsf{O}(\mathfrak{H})), \mu)$ is a measure situation in which $\mu \neq 0$ and $\mu(B(0, 1)) < \infty$ then μ is not translation-invariant, i.e., it is not true that for each measurable set A and each f in $\mathfrak{H}$, $f + A$ is measurable and $\mu(f + A) = \mu(A)$.

343. Show that if $\mathfrak{H}$ is an infinite-dimensional separable Hilbert space then for all p in $[0, \infty)$ $\rho^p(A) = \infty$ for all nonempty open subsets A of $\mathfrak{H}$. (See 134.)

344. Give an example of a Hilbert space $\mathfrak{H}$ endowed with two norms, $\|\cdots\|$ and $\|\cdots\|'$, the first derived from the inner product and the second satisfying the inequality $\|f\| \geqq \|f\|'$ for all f in $\mathfrak{H}$, and such that the topology induced by $\|\cdots\|'$ is not stronger than $\sigma(\mathfrak{H}, \mathfrak{H}^*) = \sigma(\mathfrak{H}, \mathfrak{H})$. ("It is possible to find a weaker norm that induces a topology not stronger than the weak topology.")

345. Assume that M is a closed subspace of $l^2(\mathbb{N})$ and that D is $\{\{a_n\}_{n=1}^\infty : \{a_n\}_{n=1}^\infty$ and $\{na_n\}_{n=1}^\infty \in l^2(\mathbb{N})\}$. Show that if $M \cap D = \{0\}$ then A given by $\{\{a_n\}_{n=1}^\infty : \{a_n\}_{n=1}^\infty \in D$ and $\{na_n\}_{n=1}^\infty \in M^\perp\}$ is norm-dense in $l^2(\mathbb{N})$.

346. Assume $\{x_n\}_{n=1}^\infty$ is a sequence of elements of $l^2(\mathbb{N})$ and that $x_n \to 0$ in the weak topology as $n \to \infty$. Show there is a sequence $\{x_{n_k}\}_{k=1}^\infty$ such that $\|K^{-1} \sum_{k=1}^K x_{n_k}\|_2 \to 0$ as $K \to \infty$.

347. Show that if $S = \{f : f(x) = \sum_{n=1}^\infty a_n \sin 2n\pi x,\ x \text{ in } I,\ \sum_{n=1}^\infty |na_n| \leqq 1\}$ then S is norm-compact in $L^2(I, \lambda)$.

348. Let S be $\{f : f \in L^2(I, \lambda)$ and for some g in $L^2(I, \lambda)$ and some c in $\mathbb{C}$, $f(x) = \int_0^x g(t)\, dt + c\}$. Show that the map $T : f \mapsto g$ is well-defined and that its graph is norm-closed (i.e., its graph is closed in the product topology of $\mathfrak{H} \times \mathfrak{H}$ when both factors are given the norm-induced topology).

349. Assume that M is a subspace of $L^2(I, \lambda)$ and that for some C and all f in M, $|f(x)| \leqq C\|f\|_2$ a.e. on I. Show that the (linear) dimension of M is not more than C^2.

350. For the measure situation (X, S, μ) assume $\mu(X) < \infty$. Show that if $\{f_n\}_{n=0}^{\infty} \subset L^2(X, \mu)$, $f_n \to f_0$ a.e., and $\|f_n\|_2 \to \|f_0\|_2$ as $n \to \infty$, then $\|f_n - f_0\|_2 \to 0$ as $n \to \infty$.

351. i) Describe in different and simple terms the norm-closure M in $L^2([-1, 1], \lambda)$ of $\{f: f \in C([-1, 1], \mathbb{C}), f(x) = f(-x)\}$.

ii) Give an orthonormal basis for M and an orthonormal basis for $M^{\perp}$.

iii) Show that M has an orthonormal basis consisting of polynomials.

352. Show that if $\{x_n\}_{n=1}^{\infty}$ is an orthonormal sequence then $x_n \to 0$ weakly in $\mathfrak{H}$ as $n \to \infty$.

353. Show that if $\{x_n\}_{n=0}^{\infty} \subset \mathfrak{H}$, if $\|x_n\| = 1$ for all n, and if $x_n \to x_0$ weakly as $n \to \infty$, then $\|x_n - x_0\| \to 0$ as $n \to \infty$.

354. Assume that $\{x_n\}_{n=0}^{\infty} \subset \mathfrak{H}$, that for all y in $\mathfrak{H}$, $(x_n, y) \to (x_0, y)$ as $n \to \infty$, and $\|x_n\| \to \|x_0\|$ as $n \to \infty$. Show $\|x_n - x_0\| \to 0$ as $n \to \infty$.

355. Assume $\{x_n\}_{n=1}^{\infty}$ and $\{y_n\}_{n=1}^{\infty}$ are subsets of $B(0, 1)$, the unit ball of $\mathfrak{H}$, and that $(x_n, y_n) \to 1$ as $n \to \infty$. Show $\|x_n - y_n\| \to 0$ as $n \to \infty$.

356. Let $x, y \mapsto b(x, y)$ be a conjugate bilinear form on $\mathfrak{H} \times \mathfrak{H}$, i.e., b is linear in x and conjugate linear in y. Show that if $|b(x, y)| \leqq C\|x\| \cdot \|y\|$ for some C then there is in End($\mathfrak{H}$) an A such that $\|A\| \leqq C$ and $b(x, y) = (x, Ay)$.

357. Show that $\mathfrak{H}$ in its weak topology is a Hausdorff space.

358. Show that $\mathfrak{H}$ in the weak topology is of the second category iff $\mathfrak{H}$ is finite-dimensional.

359. Assume $\{x_n\}_{n=1}^{\infty}$ is an orthonormal system in $\mathfrak{H}$ and let E be $\{x_m + mx_n : n > m,\ m, n \text{ in } \mathbb{N}\}$. Show: i) the weak closure of E contains zero; ii) if F is a norm-bounded subset of E then zero is not in the weak closure of F; iii) no subsequence of E converges weakly to zero.

360. Show that if M is a closed subspace of $\mathfrak{H}$ and if $x_0 \in \mathfrak{H}$ then $a = \inf\{\|x - x_0\| : x \in M\} = \sup\{|(x_0, y)| : y \in M^{\perp} \cap \partial B(0, 1)\} = b$.

361. Assume $f \in L^2(\mathbb{T}, \lambda)$ and that $\hat{f}(n) \neq 0$ for all n in $\mathbb{Z}$. Show that the linear span of $\{f_{(t)}\}_{t \in \mathbb{T}}$ is norm-dense in $L^2(\mathbb{T}, \lambda)$.

13. $L^p(X, \mu)$, $1 \leqq p \leqq \infty$

Conventions

If $1<p$ then q is the conjugate of p, $1/p+1/q=1$. If $p=1$ its conjugate is ∞.

If A is a subset of a Banach space X, $A^{\perp}=\{x^*: x^* \in X^* \text{ and } x^*(A)=0\}$. For convenience as required, a Banach space X is regarded as isometrically, isomorphically, and canonically embedded in X^{**}, $X \ni x \mapsto x^{**} \in X^{**}$ according to the formula: $x^{**}(x^*)=x^*(x)$ for all x^* in X^*. The Banach space X is reflexive iff the canonical image of X in X^{**} is X^{**}.

A step-function on $\mathbb{R}$ is a (finite) linear combination of characteristic functions of intervals. An analogous definition applies for a step-function on $\mathbb{R}^n$, $n>1$.

If X is a set, $N \subset X$ and $f \in \mathbb{C}^X$ then $\operatorname{osc}_N(f)=\sup_{a,b \in N}|f(a)-f(b)|=$ oscillation of f on N.

A partition of unity $\{\varphi_\gamma\}_{\gamma \in \Gamma}$ subordinate to an open cover $\{U_\gamma\}_{\gamma \in \Gamma}$ of a topological space X is a set of nonnegative continuous functions such that $\varphi_\gamma=0$ off U_γ, for each x only finitely many of the $\varphi_\gamma(x)$ are nonzero, and furthermore $\sum_\gamma \varphi_\gamma(x)=1$.

362. Assume E is an equivalence class (of functions) corresponding to an element in $L^p(I, \lambda)$, $1 \leqq p \leqq \infty$. Show: i) there is at most one continuous function in E; ii) there is some equivalence class E containing no continuous function.

363. Assume $1<p<\infty$ and $\{f_n\}_{n=0}^{\infty} \subset L^p(\mathbb{R}, \lambda)$. Show that i) and ii) following are equivalent. i) For some K and all n, $\|f_n\|_p<K$ and for all x,

$\lim_{n\to\infty} \int_0^x f_n(t)\,dt = \int_0^x f_0(t)\,dt$. ii) The sequence $\{f_n\}_{n=0}^\infty$ converges weakly to f_0.

364. Assume $1 \leqq p \leqq \infty$ and that $g \in L^p(\mathbb{R}, \lambda)$. Show $\lim_{h\to 0} \|g + g_{(h)}\|_p = 2\|g\|_p$.

365. Show that if $f \in L^p(\mathbb{R}, \lambda)$ and $1 \leqq p < \infty$ there is a sequence $\{a_n\}_{n=1}^\infty$ of positive numbers converging to zero and such that if $|b_n| < a_n$ for all n in $\mathbb{N}$ then $f_{(b_n)} \to f$ a.e. as $n \to \infty$.

366. Find the values of p for which the map

$$f: x, y \mapsto \begin{cases} (xy-1)^{-1}, & \text{if } xy - 1 \neq 0 \\ 0, & \text{if } xy - 1 = 0 \end{cases}$$

is in $L^p(I^2, \lambda)$.

367. Show that if $0 \leqq f, g$, $f \in L^p(\mathbb{R}^n, \lambda)$, $g \in L^q(\mathbb{R}^n, \lambda)$, and for all positive t, $E_t = \{x : g(x) > t\}$ then $\int_0^\infty \left(\int_{E_t} f(x)\,dx\right) dt = \int_{\mathbb{R}^n} f(x)g(x)\,dx$.

368. Let S belong to $\text{End}(L^p(\mathbb{T}, \lambda))$ and assume $S(f_{(t)}) = (Sf)_{(t)}$. Show: i) for all f, g in $L^\infty(\mathbb{T}, \lambda)$, $S(f * g) = S(f) * g = f * S(g)$; ii) there is in $\mathbb{C}$ a sequence $\{a_n\}_{n\in\mathbb{Z}}$ such that $(Sf)\hat{}\,(n) = a_n \hat{f}(n)$ for all f in $L^p(\mathbb{T}, \lambda)$.

369. Assume $1 < p < \infty$, $\{a_n\}_{n=0}^\infty$, $\{b_n\}_{n=0}^\infty \subset \mathbb{C}$, and for all N in $\mathbb{N}$, $|\sum_{n=0}^N a_n b_n|^p \leqq \int_I |\sum_{n=0}^N b_n t^n|^p\,dt$. Show there is in $L^q(I, \lambda)$ a unique f such that for all n, $a_n = \int_I t^n f(t)\,dt$.

370. For each p in $[1, \infty)$ find E_p, the set of extreme points of $B(0, 1)$ in $L^p(X, \mu)$. (For the case $p = 1$ assume that μ is nonatomic.)

371. Show that if $1 < p < \infty$ and if A and B are closed subspaces of $L^p(X, \mu)$ then $A = B$ iff $A^\perp = B^\perp$.

372. Show that $(L^\infty(I, \lambda))^* \backslash L^1(I, \lambda) \neq \emptyset$.

373. Assume $f \in L^\infty(I, \lambda)$ and that for all x in I there is a g_x such that $f = g_x$ a.e. and $\lim_{t\to x} g_x(t)$ exists. Show there is in $C(I, \mathbb{C})$ a g such that $f = g$ a.e.

14. Topological Vector Spaces

Conventions

Vector spaces will be considered as vector spaces over $\mathbb{C}$ unless something else is specified. The symbols Hom(X, Y) resp. Sur(X, Y) will be reserved for sets of continuous homomorphisms resp. surjective homomorphisms; End(X) is the set of continuous endomorphisms and Aut(E) is the set of continuous automorphisms (bijective and bicontinuous endomorphisms).

A map T is conjugate linear iff $T(ax+by) = \bar{a}T(x) + \bar{b}T(y)$. If M is a subset of X^*, $M_\perp = \{x : x^*(x) = 0 \text{ for all } x^* \text{ in } M\}$.

A (locally convex) topological vector space E is a vector space with a Hausforff topology (having a neighborhood basis consisting of convex open sets). It is assumed that $\mathbb{C} \times E \ni (a, x) \mapsto ax$ and $E \times E \ni (x, y) \mapsto x + y$ are continuous. The kernel ker(T) of a morphism T is $T^{-1}(0)$. The image im(T) is $T(X)$. A map $T: X \times Y \mapsto Z$ is separately continuous iff for all x resp. y, $T(x, \cdot)$ resp. $T(\cdot, y)$ is continuous on Y resp. X.

The Banach space $C_0(\mathbb{N}, \mathbb{C})$ is denoted $c_0(\mathbb{N})$.

A sequence $\{x_n\}_{n=1}^\infty$ is a Schauder or S-basis for a topological vector space E iff for all x in E there is in $\mathbb{C}$ a unique sequence $\{a_n\}_{n=1}^\infty$ such that $\sum_n a_n x_n = x$ (convergence of the infinite series with respect to the topology of E). If E is a Banach space $\{x_n\}_{n=1}^\infty$ is a norm-S-basis, a weak S-basis, etc., according as the topology under consideration is norm-induced, $\sigma(E, E^*)$, etc. If Γ is a set and Δ is the set of finite subsets δ of Γ then Δ is partially ordered by inclusion. A set $\{x_\gamma\}_{\gamma\in\Gamma}$ is a basis for a topological vector space E iff for all x in E there is in $\mathbb{C}$ a unique set $\{a_\gamma\}_{\gamma\in\Gamma}$ such that the net $\sum_{\gamma\in\delta} a_\gamma x_\gamma \to x$; $\sum_\gamma a_\gamma x_\gamma$ denotes the limit of the net.

If $\{x_n\}_{n=1}^{\infty}$ is an S-basis, S_N resp. P_n are the maps $x \mapsto \sum_{n=1}^{N} a_n x_n$ resp. $x \mapsto a_n x_n$; if $\{x_\gamma\}_{\gamma \in \Gamma}$ is a basis S_δ and P_γ have corresponding meanings. A set $\{x_\gamma, x_\gamma^*\}_{\gamma \in \Gamma}$ is biorthogonal iff $x_{\gamma_1}^*(x_{\gamma_2}) = \delta_{\gamma_1 \gamma_2}$.

A Hamel basis $\{u_\omega\}_{\omega \in \Omega}$ for a vector space E is a subset maximal with respect to the property of linear independence, i.e., any finite subset is linearly independent and any proper superset contains linearly dependent elements. Zorn's lemma implies that every vector space has a Hamel basis.

If X and Y are (topological) vector spaces then $X \oplus Y$ is the direct sum of X and Y, i.e., the vector space $X \times Y$ with vector space structure derived from "coordinatewise" operations (and with the weakest topology with respect to which the (projections) $(x, y) \mapsto x$ and $(x, y) \mapsto y$ are continuous). If Z is a closed subspace of X the quotient topology for X/Z is the strongest with respect to which the canonical quotient map $x \mapsto x/Z$ is continuous. If X is a normed space the quotient topology is derived from the norm $\|\cdot\cdot\cdot\|: x/Z \mapsto \inf\{\|x_1\|: x_1 \text{ in } x+Z\}$. If $T \in \mathrm{Hom}(X, Y)$ there is in $\mathrm{Hom}(Y^*, X^*)$ a unique T^* such that for all x in X and y^* in Y^*, $T^*(y^*)(x) = y(T(x))$; T^* is the adjoint of T. The symbol *id* denotes the identity map in $\mathrm{End}(\cdot)$.

If f is a map between normed vector spaces X and Y the differential (or derivative) df, if it exists, at x_0 is an element of $\mathrm{Hom}(X, Y)$. By definition, $\lim_{h \neq 0, \|h\| \to 0} \|f(x_0+h) - f(x_0) - df(x_0)(h)\|/\|h\| = 0$. (See Functions from $\mathbb{R}^n$ to $\mathbb{R}^m$, Conventions.)

If A is a ring $A[x]$ is the set of all polynomials having coefficients in A: $A[x] = \{\sum_{n=0}^{N} a_n x^n : a_n \text{ in } A, N \text{ in } \mathbb{N}\}$. If $p \in A[x]$ and $p = \sum_{n=0}^{N} a_n x^n$, $a_N \neq 0$, then $\deg(p) = N$.

374. Assume X and Y are Banach spaces and that $\{T_n\}_{n=1}^{\infty} \subset \mathrm{Hom}(X, Y)$. Show that if $g \in Y^*$ implies $\sup_n |g(T_n(x))| < \infty$ for all x in X then $\sup_n \|T_n\| < \infty$.

375. Let E be a Banach space and assume T resp. S is a not necessarily continuous endomorphism of E resp. E^*. Show that if $x^*(T(x)) = (S(x^*))(x)$ for all x in E and all x^* in E^* then T and S are norm-continuous.

376. Assume T is a not necessarily continuous homomorphism of the Banach space X onto the Banach space Y and that $\ker(T) = T^{-1}(0)$ is norm-closed. Prove or disprove that T is continuous; that T is open.

377. Let Y be a closed subspace of a Banach space X. Show that if $x \in X \setminus Y$ and $a > 0$ there is in $\mathrm{span}(Y, x)$ a z such that $\|z\| \leqq 1$ and $d(z, Y) = \inf\{\|z - y\|: y \in Y\} > 1 - a$.

378. Assume E is a Banach space. Show there is no "involution" $\# : E \ni X \mapsto x^\# \in E$ such that $\#$ is idempotent conjugate linear, and for all x^* in E^*, $x^*(x^\#) = \overline{x^*(x)}$.

379. Show that if E is a Banach space and $M \subset E^*$ then $(M_\perp)^\perp$ is the weak* closure of span(M).

380. Show that if M is a convex subset of a Banach space X and M is norm-closed then M is also weakly closed. Show also that if M is convex then its norm-closure and weak closure are the same.

381. Show that if M is a norm-closed subspace of the Banach space X then $(M^\perp)_\perp = M$.

382. Give an example of a Banach space X and a norm-closed subspace M of X^* such that $(M_\perp)^\perp \supsetneqq M$.

383. Let E be a normed vector space and let F be a finite-dimensional subspace. Show that there is in End(E) a projection P (i.e., $P^2 = P$) such that $P(E) = F$.

384. Show that if f is a not necessarily continuous linear functional mapping the Banach space E into $\mathbb{C}$ then f is continuous iff ker(f) is closed. Show also that either ker(f) is closed or ker(f) is a proper dense subset of E.

385. Show that if $\{x_\gamma\}_{\gamma \in \Gamma}$ is a Hamel basis for the infinite-dimensional Banach space E then at least one of the coefficient maps $x \mapsto a_\gamma$ is not continuous.

386. Assume that E and F are Banach spaces, M is a subspace of F, $\dim(F/M) < \infty$, and there is in Hom(E, F) a K such that $K(E) = M$. Show M is closed. Show that the conclusion fails if the hypothesis *re* the existence of K is dropped.

387. Show that if E is a Banach space and E^* is norm-separable then E is norm-separable. Give a counterexample to the converse.

388. Show that if E is a Banach space and E^* is separable then the σ-algebra generated in 2^{E^*} by the norm-open sets is the same as the σ-algebra generated by the weak*-open sets.

389. Let E be a separable infinite-dimensional normed space. Show how to construct in E a subset A that is dense in E and contains no finite linearly dependent subset.

390. Let X be a Banach space and assume $\{x_n\}_{n=1}^\infty \subset X$ and that for all x^* in X^*, $\sum_n |x^*(x_n)| < \infty$. Prove or disprove: if $\{a_n\}_{n=1}^\infty \in c_0(\mathbb{N})$ then $\{\sum_{n=1}^N a_n x_n\}_{N=1}^\infty$ is a norm-convergent sequence.

391. Let X be a Banach space and assume $\{x_n\}_{n=1}^\infty$ is dense in $B(0, 1)$. Show: i) if T is $l^1(\mathbb{N}) \ni \{b_n\}_{n=1}^\infty \mapsto \sum_n b_n x_n$ then $T \in \mathrm{Sur}(l^1(\mathbb{N}), X)$; ii) that $l^1(\mathbb{N})/\ker(T)$ in its quotient-norm topology and X are isometrically isomorphic.

392. Let A be a norm-compact subset of a Banach space X and let K be the norm-closure of the convex hull C of A. Show: i) K is norm-compact; ii) for all f in X^*, $|f(x)||_K$ achieves its maximum value on A; iii) for all x in K there is a complex Borel measure μ_x such that $\|\mu_x\|=1$ and for all f in X^*, $f(x)=\int_A f(y)\,d\mu_x(y)$; iv) μ_x is positive.

393. Show that if E, F, and G are Banach spaces and $B: E\times F \mapsto G$ is bilinear then B is continuous if: i) for all x, y, $\|B(x, y)\| \leqq C\|x\|\cdot\|y\|$ for some constant C; or ii) B is separately continuous.

394. Show that if E and F are Banach spaces, $f\in F^E$, and $df=0$ then f is constant.

395. Show that if E is a Banach space and T is a not necessarily continuous endomorphisn of E then T is norm-continuous iff whenever $x_n \to 0$ weakly as $n\to\infty$ then so also $T(x_n)\to 0$ weakly as $n\to\infty$.

396. Show that if E is a normed vector space, $f\in \mathrm{Hom}(E, \mathbb{C})$, and $H=\ker(f)$ then for all x in E, $d(x, H)\cdot\|f\|=|f(x)|$.

397. Assume E is a Banach space, $T\in \mathrm{Aut}(E)$, and $\|x_n\|\to\infty$ as $n\to\infty$. Show $\|T(x_n)\|\to\infty$ as $n\to\infty$.

398. Show that if $\{x_n\}_{n=1}^{\infty}$ is a norm-basis for the Banach space E then each map (coefficient functional) $x\mapsto a_n$ is in X^*.

399. Show that if $\{x_\gamma\}_{\gamma\in\Gamma}$ is a weak basis for the Banach space E then $\{x_\gamma\}_{\gamma\in\Gamma}$ is a norm-basis for E.

400. Show that if $\{x_n\}_{n=1}^{\infty}$ is a basis for the Banach space E then $S_N^2=S_N$ and $P_n^2=P_n$ and there are constants S, P such that for all N, n, $\|S_N\|\leqq S$, $\|P_n\|\leqq P$.

401. Show that for any Banach space E there is in $E\times E^*$ a maximal biorthogonal set $\{x_\gamma, x_\gamma^*\}_{\gamma\in\Gamma}$.

402. Show that if $\{x_\gamma\}_{\gamma\in\Gamma}$ is a basis for the Banach space E then (see Problem 399) $\{x_\gamma, x_\gamma^*\}_{\gamma\in\Gamma}$ is a maximal biorthogonal set.

403. Show that if $\{x_\gamma, x_\gamma^*\}_{\gamma\in\Gamma}$ is a maximal biorthogonal set for the Banach space E then $\mathrm{span}(\{x_\gamma\}_{\gamma\in\Gamma})$ is dense in E.

404. Show that if $\{x_n, x_n^{\infty}\}_{n=1}^{\infty}$ is a biorthogonal set for a Banach space E, if $\mathrm{span}(\{x_n\}_n)$ is dense, and if the norms of the maps $S_N: E\ni x\mapsto \sum_{n=1}^{N} x_n^*(x)x_n$ are bounded, say by M, then $\{x_n\}_n$ is a basis.

405. Show that $\{x_n\}_{n=1}^{\infty}$ is a basis for the Banach space E iff $\mathrm{span}(\{x_n\}_{n=1}^{\infty})$ is dense in E and there is an M such that if $m\leqq n$ and $\{a_i\}_{i=1}^{n}\subset\mathbb{C}$ then $\|\sum_{i=1}^{m} a_ix_i\|\leqq M\|\sum_{i=1}^{n} a_ix_i\|$.

406. Let $\{f_n\}_{n=1}^{\infty}$ be a complete orthonormal set in the Hilbert space $\mathfrak{H}$. Show that if $0 \leqq c < 2^{-1/2}$ and $\|g_n - f_n\| \leqq c^n$ then $\{g_n\}_{n=1}^{\infty}$ is a norm-basis for $\mathfrak{H}$ and if $\mathfrak{H} \ni f = \sum_{n=1}^{\infty} a_n g_n$ then $\sum_{n=1}^{\infty} |a_n|^2 < \infty$.

407. Show that if $\{x_n\}_{n=1}^{\infty}$ is a norm-basis for the Banach space E there is in $(0, \infty)$ a sequence $\{a_n\}_{n=1}^{\infty}$ such that if $\|x_n - y_n\| \leqq a_n$, n in $\mathbb{N}$, then $\{y_n\}_{n=1}^{\infty}$ is a norm-basis for E.

408. Show that if $\{x_\gamma\}_{\gamma \in \Gamma}$ is a Hamel basis for the Banach space X and $\text{card}(\Gamma) \leqq \text{card}(\mathbb{N})$ then X is finite-dimensional.

409. Give an example of a normed infinite-dimensional vector space X having a countable Hamel basis.

410. Let $\{x_n\}_{n=1}^{\infty}$ be a norm-basis for the Banach space E. If $t \in I$ let $\sum_n e_n 2^{-n}$ be a binary representation of t (for all t off a null set the representation is unique). For each x in E let $C(x)$ be $\{t: t \in I, \sum_{n=1}^{\infty} e_n x_n^*(x) x_n$ converges in norm$\}$. Show: i) for all x, $C(x)$ is measurable; ii) if $C = \bigcap_{x \in E} C(x)$ then C is measurable and is dense in I; iii) either $\lambda(C) = 0$ or $\lambda(C) = 1$; iv) either $\lambda(C) = 0$ or $C = I$.

411. If $a > 0$ let E_a be

$$\{f: f \in \mathbb{C}^{\mathbb{C}}, f(0) = 0, L(f) = \sup_{s \neq t} |f(s) - f(t)|/|s - t|^a < \infty\}.$$

Show that L is a norm for E_a and that E_a thus normed is a Banach space. (The multiplication by constants.)

412. (See Problem 411.) If $a = 1$ let F_1 be $\{f: f \in \mathbb{C}^I, f(0) = 0, L(f) = \sup_{s \neq t} |f(s) - f(t)|/|s - t| < \infty\}$. Let N be $F_1 \ni f \mapsto \|f\|_\infty + L(f)$. Show that (the norm) N is not equivalent to $\|\cdot\cdot\cdot\|_\infty$, i.e., there is no constant B such that for all f in F_1, $N(f) \leqq B\|f\|_\infty$.

413. Show that if f in $\mathbb{R}^{\mathbb{R}}$ is uniformly continuous and bounded and if K is a compact subset of $\mathbb{R}$ then $\{f_{(t)}: t \text{ in } K\}$ is a normal set, i.e., its elements are uniformly bounded and equicontinuous.

414. If $n = 1, 2$ and $0 < a < n$ let k be a measurable function from $\mathbb{R}^n$ to $\mathbb{R}$. Assume $|k(x)| \leqq c|x|^{a-n}$. Let K be $C_{00}(\mathbb{R}^n, \mathbb{R}) \ni f \mapsto k * f$. Show: i) if $1 - a/n < 1/q$, $0 < b$, and y is fixed then $\int_{B(0,b)} |k(x-y)|^q \, dx < \infty$; ii) K has a unique extension to $L^1(B(0, b), \lambda)$ and the extension is in $\text{Hom}(L^1(B(0, b), \lambda), L^q(B(0, b), \lambda))$ if $1 - a/n < 1/q < 1$.

415. Show that if $K = \{f: f \text{ in } \mathbb{C}[x], \deg(f) \leqq 10, \int_I |f(x)| \, dx \leqq 1\}$ then K is a norm-compact subset of $L^1(I, \lambda)$.

416. Show that $(L^1(\mathbb{R}, \lambda))^\wedge$ is dense in $C_0(\mathbb{R}, \mathbb{C})$.

417. Let P_m be $\{p: p \text{ in } \mathbb{R}[x], \deg(p) \leqq m\}$. Show that if $L: P_{2n+1} \mapsto \mathbb{R}$ is a linear functional such that $L(p^2) > 0$ if $p \neq 0$ and $\deg(p) \leqq n$ then there is in P_{n+1} a p_{n+1} such that: i) $L(p_{n+1} \cdot q) = 0$ for all q in P_n; ii) p_{n+1} has $n+1$ distinct real zeros; iii) if i) and ii) obtain for $\tilde{p}_{n+1}$ then $\tilde{p}_{n+1}$ is a constant multiple of p_{n+1}.

418. Assume that $f:\mathbb{R}\ni x\mapsto f(x)\in\mathfrak{H}$ is such that $F:\mathfrak{H}\ni y\mapsto (f(x), y)\in\mathbb{C}$ is, for each y, differentiable with respect to x. Show that for all x_0, $\|f(x)-f(x_0)\|\to 0$ as $x\to x_0$.

419. With abuse of notation, $E=L^{1/2}(I,\lambda)$ is a topological vector space with respect to the weakest topology making the map $f\mapsto\int_I|f(x)|^{1/2}\,dx=\|f\|_{1/2}^{1/2}$ continuous. Show that $E^*=\{0\}$.

420. For the measure situation $(\mathbb{R}, \mathsf{S}_\beta, \mu)$ in which $\mu(\mathbb{R})<\infty$ let X be $(L^2(\mathbb{R},\mu))^n$ with norm $\|\cdot\cdot\cdot\|:(f_1, f_2, \ldots, f_n)\mapsto(\sum_{k=1}^n\|f_k\|_2^4)^{1/4}$. Show that X is a Banach space and describe X^*.

421. Show that if X is a Banach space and $T\in\mathrm{Sur}(X, l^1(\mathbb{N}))$, then: i) there is in $\mathrm{End}(X)$ a projection P $(P^2=P)$ such that $P(X)=\ker(T)$; and ii) $\ker(T)\oplus(id-P)(X)\ni(x,y)\mapsto x+y\in X$ is continuous and open.

422. Show that if E is a normed vector space, $\{y_k\}_{k=1}^K\subset E$, and M is a norm-closed subspace of E then $\mathrm{span}(\{y_k\}_{k=1}^K, M)$ is closed.

15. Miscellaneous Problems

Conventions

The sets $\{f: f \in \mathbb{C}^{\mathbb{R}}, f$ of bounded variation on $\mathbb{R}\}$ resp. $\{f: f \in \mathbb{C}^{\mathbb{R}}, f$ absolutely continuous on $\mathbb{R}\}$ are denoted $BV(\mathbb{R}, \mathbb{C})$ resp. $AC(\mathbb{R}, \mathbb{C})$; similar meanings are attached to $BV(I, \mathbb{C})$ resp. $AC(I, \mathbb{C})$. If $f \in BV(\mathbb{R}, \mathbb{C})$ then $T_f: x \mapsto \mathbb{R}$ is the total variation of f on $(-\infty, x]$, i.e., $T_f(x) = \sup\{\sum_{i=1}^{n} |f(x_{i+1}) - f(x_i)|: n$ in $\mathbb{N}, -\infty < x_1 < \cdots < x_{n+1} = x\}$ and $T_f(\mathbb{R}) = \sup_x T_f(x)$; similar meanings are given to $T_f(x)$ and $T_f(I)$ if $f \in BV(I, \mathbb{C})$. The sets $[x_1, x_2), [x_2, x_3), \ldots, [x_n, x]$ constitute a partition P of $[x_1, x]$ and $|P| = \sup_i |x_{i+1} - x_i|$; $T_{fP} = \sum_{i=1}^{n} |f(x_{i+1}) - f(x_i)|$; μ_f is the Borel measure such that $\mu_f([a, b)) = T_f([a, b))$.

If f is a map between topological spaces, cont(f) is the set of continuity points of f. If $f \in \mathbb{R}^{\mathbb{R}}$, x is a strict maximum (minimum) point of f iff for all y in some neighborhood of x, if $y \neq x$ then $f(y) < f(x)$ $(f(y) > f(x))$.

If (X, S, μ) is a measure situation and $\mu(x) = 1$, a set $\{f_\gamma\}_{\gamma \in \Gamma}$ of measurable $\mathbb{R}$-valued functions is independent iff for every finite subset σ of Γ whenever $\{A_\gamma\}_{\gamma \in \sigma} \subset \mathsf{S}_\beta(\mathbb{R})$, $\mu(\bigcap_{\gamma \in \sigma} f_\gamma^{-1}(A_\gamma)) = \prod_{\gamma \in \sigma} \mu(f_\gamma^{-1}(A_\gamma))$. (The usual context for such notions is in the field of probability. There the genesis of the idea of independence of "events" is the intuitive one, i.e., independent events do not "influence" one another. The mathematical formulation appears to be useful.) If $f \in L^1(X, \mu)$, $E(f) = \int_X f(x)\, d\mu(x)$; if $f \in L^2(X, \mu)$, $\text{var}(f) = E((f - E(f))^2)$.

In earlier conventions $\mathbb{T}$ was taken to be $\{z: z$ in $\mathbb{C}$ and $|z| = 1\}$ and was regarded as a group under ordinary multiplication. Since the map $\exp: [-\pi, \pi) \mapsto e^{ix} \in \mathbb{T}$ is bijective and continuous it may be used to endow $\mathbb{T}$ with an invariant measure derived from λ on $[-\pi, \pi)$. Furthermore the group operation on $\mathbb{T}$ corresponds to addition modulo 2π on $[-\pi, \pi)$. Thus in the discussion of Fourier series the domain of all functions will be $[-\pi, \pi)$ with Lebesgue measure and with addition interpreted as addition modulo 2π. In particular if $f \in L^1([-\pi, \pi), \lambda)$ then $\hat{f}$ will be $\mathbb{Z} \ni n \mapsto (2\pi)^{-1} \int_{-\pi}^{\pi} f(x)\, e^{-inx}\, dx$.

If X is a set and $\Gamma \ni \gamma \mapsto x_\gamma \in X$ is a net, x_γ is eventually in a subset S iff there is some γ_0 such that if $\gamma > \gamma_0$ then $x_\gamma \in S$; x_γ is frequently in S iff for each γ_0 there is in Γ a γ such that $\gamma > \gamma_0$ and $x_\gamma \in S$. Thus x_γ converges to x iff for each neighborhood U of x, x_γ is eventually in U; x is a cluster point of $\{x_\gamma\}_{\gamma \in \Gamma}$ iff for each neighborhood U of x, x_γ is frequently in $U \backslash \{x\}$. A tail of Γ is a subset Δ of the form $\{\gamma: \gamma > \gamma_0\}$; a cofinal subset of Γ is a set Λ such that for each γ in Γ there is in Λ a λ such that $\lambda > \gamma$. Thus x_γ is eventually in S means that for some tail Δ, $\{x_\gamma\}_{\gamma \in \Delta} \subset S$; x_γ is frequently in S means that for some cofinal subset Λ, $\{x_\gamma\}_{\gamma \in \Lambda} \subset S$.

A subset A of a metric space X is scattered iff A contains no (nonempty) perfect subset.

If J is a set the set $\mathcal{M}(J)$ is $\mathbb{K}^{J \times J}$. If $M \in \mathcal{M}(J)$ and if supp(M) is finite (compact) and if $N \in \mathcal{M}(J)$ then $MN(i, j) = \sum_k M(i, k)N(k, j)$ and $NM(i, j) = \sum_k N(i, k)M(k, j)$ (matrix multiplication). The set $\mathcal{U}(J)$ of matrix units consists of those matrices that have at most one nonzero entry and the nonzero entry, if it exists, is one.

A Banach algebra A is a Banach space and an algebra over $\mathbb{C}$. It is assumed that if $x, y \in A$ then $\|xy\| \leqq \|x\| \cdot \|y\|$. A derivation D of a Banach algebra is a linear map $A \mapsto A$ such that $D(xy) = D(x) \cdot y + x \cdot D(y)$. The set of regular maximal ideals of A is denoted $\sigma(A)$ ("spectrum of A"); if A is commutative, $\bigcap_{M \in \sigma(A)} M =$ radical of $A = \{x: \|x^n\|^{1/n} \to 0 \text{ as } n \to \infty\} =$ set of generalized nilpotents of A.

423. Assume f is monotone increasing on $\mathbb{R}$ and that $g = f^2$. Show $\mu_g \ll \mu_f$ and calculate $d\mu_g / d\mu_f$.

424. Assume f is not constant on $[a, b]$. Show that if $f' = 0$ a.e. then $f \notin \text{Lip}(1)$ on $[a, b]$.

425. Construct on $\mathbb{R}$ a monotone increasing function f such that $f' = 0$ a.e. and f is constant on no nonempty interval (a, b).

426. Assume E is Lebesgue measurable and $E \subset I$. Let f_n be $\mathbb{R} \ni x \mapsto n \int_0^{1/n} \chi_E(x + t)\, dt$. Show: i) $f_n \in AC(\mathbb{R}, \mathbb{R})$; ii) $f_n \to \chi_E$ a.e. as $n \to \infty$; iii) $0 \leqq f_n \leqq 1$; iv) $\|f_n - \chi_E\|_1 \to 0$ as $n \to \infty$.

427. Show that if F is $\{f: f \in AC(I, \mathbb{C}), \int_I (|f(x)|^2 + |f'(x)|^2)\, dx \leqq 1\}$ then the closure of F is compact in $C(I, \mathbb{C})$.

428. Assume $g \in BV([-1, 1], \mathbb{C})$ and that for every even continuous function f, $\int_{-1}^{1} f(x)g(x)\, dx = 0$. Show that if g is continuous at a and at $-a$ then $g(a) + g(-a) = 0$.

429. Show that if f is absolutely continuous and strictly monotone increasing on $[a, b]$ and if $f([a, b]) = [c, d]$ then for every Borel set E in $[c, d]$, $\int_{f^{-1}(E)} f'(x)\,dx = \lambda(E)$.

430. Show that if E a null subset of I there is an absolutely continuous monotone increasing function f such that $f' = \infty$ on E.

431. Assume f' exists everywhere on $\mathbb{R}$. Show that f is strictly increasing iff $f' \geqq 0$ and $\{x : f'(x) = 0\}$ is totally disconnected.

432. Exhibit an f such that for all a in $(0, 1)$, $f \in BV(I, \mathbb{C}) \cap AC([0, a), \mathbb{C})$ and such that $f(1) \neq \lim_{x \to 1} f(x)$.

433. Exhibit an f such that for all a in $(0, 1)$ $f \in AC([0, a), \mathbb{C})$ and such that $f \in C(I, \mathbb{C}) \backslash AC(I, \mathbb{C})$.

434. Exhibit an f in $BV(I, \mathbb{C}) \cap C(I, \mathbb{C})$ and such that for some null set A, $f(A)$ is Lebesgue measurable and $\lambda(f(A)) > 0$.

435. Show that $f \in AC(I, \mathbb{C})$ iff all the following obtain: i) $f \in C(I, \mathbb{R})$; ii) $f \in BV(I, \mathbb{R})$; iii) if E is a null set so is $f(E)$.

436. Show that if any one of Problem 435 i), ii), or iii) is omitted then there is an f satisfying the other two and yet not in $AC(I, \mathbb{C})$.

437. Show that if $f \in AC(I, \mathbb{C})$ and A is Lebesgue measurable then $f(A)$ is Lebesgue measurable.

438. Assume that $f \in C(I, \mathbb{C}) \cap BV(I, \mathbb{C})$ and that $T_f(I) = V$. Show that if $W < V$ there is a positive a such that if P is a partition of I and $|P| < a$ then $T_{fP} > W$. Give a counterexample to the conclusion if the hypothesis no longer asserts that f is continuous.

439. Show that if $f \in BV(I, \mathbb{R})$ and for some g, $f = g'$ then f is continuous.

440. (Banach) Show that if $f \in BV(I, \mathbb{R})$ and M is

$$y \mapsto \begin{cases} \operatorname{card}(f^{-1}(y)), & \text{if } f^{-1}(y) \text{ is finite} \\ \infty, & \text{otherwise} \end{cases}$$

then M is Lebesgue integrable and $T_f(I) = \int_{\mathbb{R}} M(y)\,dy$.

441. Show that if f and g are nonnegative Lebesgue measurable functions on $\mathbb{R}$, $A_y = \{x : f(x) \geqq y\}$, and h is $y \mapsto \int_{A_y} g(x)\,dx$ then $\int_{\mathbb{R}} f(x)g(x)\,dx = \int_0^\infty h(y)\,dy$.

442. Show that $f \in \operatorname{Lip}(1)$ on I iff there is in $L^\infty(I, \lambda)$ a g such that $f(x) - f(0) = \int_0^x g(t)\,dt$.

443. Prove or disprove: if $f \in C(I, \mathbb{C})$ there is a measure situation $(I, \mathsf{S}_\beta, \mu)$, μ complex and such that $\mu([a, b]) = f(b) - f(a)$ for all closed subintervals $[a, b]$ of I.

444. Show that if f is Lebesgue measurable on $\mathbb{R}$, there is in $\mathbb{C}$ a sequence $\{a_n\}_{n=1}^{\infty}$ and there is in S_λ a sequence $\{E_n\}_{n=1}^{\infty}$ such that $f=\sum_n a_n\chi_{E_n}$.

445. Show that if $f\in L^1(\mathbb{R},\lambda)$ there is in $C_{00}(\mathbb{R},\mathbb{C})$ a sequence $\{f_n\}_{n=1}^{\infty}$ such that $f_0=\sum_n f_n$ exists and $f=f_0$ a.e.

446. Let $\{f_\gamma\}_{\gamma\in\Gamma}$ be a set of functions integrable in the context of the measure situation (X,S,μ). Assume that for each positive a there is a positive b such that if E is measurable and $\mu(E)<b$ then for all γ, $|\int_E f_\gamma(x)\,d\mu(x)|<a$, i.e., $\{f_\gamma\}_\gamma$ is uniformly integrable. Show $\{|f_\gamma|\}_\gamma$ is also uniformly integrable.

447. Show that if f is Lebesgue measurable on I then $f\in L^2(I,\lambda)$ iff: $f\in L^1(I,\lambda)$ and there is a monotone increasing function g such that for all closed intervals $[a,b]$ in I, $|\int_a^b f(x)\,dx|^2\leqq(g(b)-g(a))|b-a|$.

448. Prove that if $g\in L^2(I,\lambda)$ and G is $x\mapsto\int_0^x g(t)\,dt$ then unless $g=0$, $\|G\|_2<\|g\|_2$.

449. In the construction of *the* Cantor set C on I, open intervals are deleted in stages, one at the first, two at the second, ..., 2^{n-1} at the nth, etc. Assume $f:I\mapsto\mathbb{R}$ is such that

$$f(x)=\begin{cases}0, & \text{if } x\in C\\ n, & x\in \text{an interval deleted at the } n\text{th stage.}\end{cases}$$

Show that $f\in L^1(I,\lambda)$ and evaluate $\int_I f(x)\,dx$.

450. For the measure situation $(\mathbb{R}^2,\mathsf{S}_\beta(\mathbb{R}^2),\mu)$ assume $\mu(\mathbb{R}^2)<\infty$ and that for all A in $\mathsf{S}_\beta(\mathbb{R})$, $\nu(A)=\mu(A\times\mathbb{R})$. Show there is a map $\mathsf{S}_\beta\ni B\mapsto f_B\in L^1(I,\nu)$ such that if $B_1\cap B_2=\varnothing$ then $f_{B_1\cup B_2}=f_{B_1}+f_{B_2}$ and for all B in $\mathsf{S}_\beta(\mathbb{R})$, $\mu(A\times B)=\int_A f_B(x)\,d\nu(x)$.

451. Constrict in I a null set $(\lambda)E$ such that for each function f Riemann integrable on I, $\text{cont}(f)\cap E\neq\varnothing$.

452. Prove or disprove: if $\{p_n\}_{n=1}^{\infty}$ is a sequence of polynomials and $p_n\to f$ uniformly on $\mathbb{R}$ as $n\to\infty$, then f is a polynomial.

453. Assume $\{r_k\}_{k=1}^{\infty}=I\cap\mathbb{Q}$ and that f is $x\mapsto\sum_k k^{-2}|x-r_k|^{-1/2}$. Show $f<\infty$ a.e.

454. Prove or disprove: if $\{c_n\}_{n=1}^{\infty}\subset\mathbb{R}$ and if there is in $\mathsf{S}_\lambda(\mathbb{R})$ on A such that $\lambda(A)>0$ and for all t in A, $\lim_{n\to\infty}e^{ic_nt}$ exists then $\lim_{n\to\infty}c_n$ exists.

455. Assume $f\in L^\infty(\mathbb{R},\lambda)$, $1\leqq p<\infty$, and $-\infty<a<b<\infty$. Prove $(\int_a^b|f(x)|\,dx)^p\leqq(b-a)^{p-1}\int_a^b|f(x)|^p\,dx$.

456. Assume $c\in(0,1)$, $f\in L^\infty(\mathbb{R},\lambda)$, and for all closed intervals $[a,b]$, $|\int_a^b f(x)\,dx|^p\leqq c(b-a)^{p-1}\int_a^b|f(x)|^p\,dx$. Show $f=0$ a.e.

457. Find in $\mathbb{R}$ a Lebesgue measurable set E such that $E+E$ is not Lebesgue measurable. (See Problem 274.)

458. Show that if $A \subset \mathbb{R}$, $\text{card}(A) > \text{card}(\mathbb{N})$, and $D = \{x_0 : x_0 \in A$, for every neighborhood U of x_0, $\text{card}(A \cap U) > \text{card}(\mathbb{N})\}$ then $\text{card}(D) > \text{card}(\mathbb{N})$. Show also that there is in A a y_0 such that for every positive number a, $\text{card}(A \cap (y_0, y_0 + a)) > \text{card}(\mathbb{N})$.

459. Show that if $f \in \mathbb{R}^{\mathbb{R}}$ and if $S = \{x : x$ is a strict maximum of $f\}$ then $\text{card}(S) \leqq \text{card}(\mathbb{N})$.

460. Show that if $a \in \mathbb{R} \backslash \mathbb{Q}$ and if $A = \{m + na : m, n$ in $\mathbb{Z}\}$ then A is dense in $\mathbb{R}$. (See Problem 249.)

461. Let f be in $C(\mathbb{R}, \mathbb{R})$ and let Δ_n be $x \mapsto 2^n (f_{(2^{-n})}(x) - f(x))$. Show that if $\|\Delta_n\|_\infty \leqq M < \infty$ and for all x, $\lim_{n \to \infty} \Delta_n(x) = 0$, then f is constant.

462. What is the distinction between the concepts: i) a function continuous a.e.; and ii) a function equal a.e. to a continuous function?

463. (Wirtinger) Show that if $f \in C^1([0, \pi], \mathbb{C})$ and $f(0) = f(\pi) = 0$ then there is a constant K, independent of f and such that $\|f\|_2^2 \leqq K \|f'\|_2^2$. (See Problem 76.)

464. Assume $f \in C^\infty(\mathbb{R}, \mathbb{R})$ and that for all n and all x, $f^{(n)}(x) \geqq 0$. Show f is real analytic.

465. Show that if $f \in C(I, \mathbb{R})$, $f(0) = 0$, and $f'(0)$ exists then $x \mapsto x^{-3/2} f(x)$ $(x \neq 0)$ is in $L^1(I, \lambda)$.

466. Let (X, S, μ) be a measure situation, E be a Banach space, and T be a linear map $E \mapsto L^1(X, \mu)$. Assume that for all A in S the map $T_A : f \mapsto \int_A (Tf(x))\, d\mu(x)$ is in E^*. Prove that $T \in \text{Hom}(E, L^1(X, \mu))$.

The following problems (467–479) offer a development of some important results in the theory of probability. These are couched in terms of a measure situation (X, S, μ) for which $\mu(X) = 1$ and in terms of the concepts of independence, expected value $(E(f))$ and variance $(\text{var}(f))$.

In connection with the notion of independence the special conventions described below are useful.

If $\{f_\gamma\}_{\gamma \in \Gamma}$ is a set of measurable $\mathbb{R}$-valued functions on X, for each γ if $X_\gamma = X$, $\mathsf{S}_\gamma = \mathsf{S}$, $\mu_\gamma = \mu$, $Y_\gamma = \mathbb{R}$, $\mathsf{S}_{\beta\gamma} = \mathsf{S}_\beta(\mathbb{R})$, and $\Gamma_1 \subset \Gamma$ let $\prod_{\gamma \in \Gamma_1} (X_\gamma, \mathsf{S}_\gamma, \mu_\gamma)$ be $(X_{\Gamma_1}, \mathsf{S}_{\Gamma_1}, \mu_{\Gamma_1})$ and let $\prod_{\gamma \in \Gamma_1} (Y_\gamma, \mathsf{S}_{\beta\gamma})$ be $(Y_{\Gamma_1}, \mathsf{S}_{\beta\Gamma_1})$. Correspondingly there are measurable maps (see Problem 469) $T_{\Gamma_1} : X \ni x \mapsto (\cdots f_\gamma(x_\gamma) \cdots)_{\gamma \in \Gamma_1} \in Y_{\Gamma_1}$ and $S_{\Gamma_1} : X_{\Gamma_1} \ni (\cdots x_\gamma \cdots)_{\gamma \in \Gamma_1} \mapsto (\cdots f_\gamma(x_\gamma) \cdots)_{\gamma \in \Gamma_1}$. The independence of the set $\{f_\gamma\}_{\gamma \in \Gamma}$ is equivalent (see problem 469) to the statement that for every finite subset σ of Γ if, for each γ in σ, A_γ is a Borel set in $\mathbb{R}$ then $\mu(T_\sigma^{-1}(\prod_{\gamma \in \sigma} A_\gamma)) = \mu_\sigma(S_\sigma^{-1}(\prod_{\gamma \in \sigma} A_\gamma)) = \prod_{\gamma \in \sigma} \mu(f_\gamma^{-1}(A_\gamma))$. Halmos [13] suggests the mnemonic:

$\mu T_\sigma^{-1} = \prod_{\gamma\in\sigma} \mu f_\gamma^{-1}$. Note that T_{Γ_1} and S_{Γ_1} depend on the set $\{f_\gamma\}_{\gamma\in\Gamma_1}$. The function f_{γ_0} may well be identified with $\tilde{f}_{\gamma_0}: X_\Gamma \ni (\cdots x_\gamma \cdots) \mapsto f_{\gamma_0}(x_{\gamma_0})$, in which case $\int_X f_{\gamma_0}(x)\, d\mu(x) = \int_{X_\Gamma} \tilde{f}_{\gamma_0}((\cdots x_\gamma \cdots))\, d\mu_\Gamma((\cdots x_\gamma \cdots))$. For simplicity $\tilde{x}_{\Gamma_1}$ denotes the vector $(\cdots x_\gamma \cdots)_{\gamma\in\Gamma_1}$.

If $\{B_\gamma\}_{\gamma\in\Gamma} \subset \mathsf{S}$ the set is independent iff $\{\chi_{B_\gamma}\}_\gamma$ is independent.

467. Show that if $\{B_\gamma\}_\gamma$ is an independent set, then so is $\{X\backslash B_\gamma\}_\gamma \cup \{B_\gamma\}_\gamma$ an independent set.

468. Show that if $\{A_\gamma\}_\gamma$ is a set of Borel sets in $\mathbb{R}$ and if $\{f_\gamma\}_\gamma$ is an independent set of functions, then $\{f_\gamma^{-1}(A_\gamma) = B_\gamma\}_\gamma$ is an independent set (of sets).

469. i) Show that for any set Γ_1 the maps T_{Γ_1} and S_{Γ_1} are measurable.

ii) Validate the mnemonic suggested by Halmos.

iii) Let $\{\gamma_n\}_{n=1}^\infty$ be a countable subset of Γ. Assume the independent set $\{f_{\gamma_n}\}_n \subset L^1(X, \mu)$ and that $\{\sum_{n=1}^N f_{\gamma_n}\}_{N=1}^\infty$ is a Cauchy sequence in $L^1(X, \mu)$. Show that if $\Gamma_N = \{1, 2, \ldots, N\}$ then $\int_{X_{\Gamma_N}} \sum_{m=n+1}^\infty \tilde{f}_{\gamma_m}(\tilde{x}_{\Gamma_N})\, d\mu_{\Gamma_N}(\tilde{x}_{\Gamma_N}) = \sum_{m=n+1}^\infty \tilde{f}_{\gamma_m}$ (convergence in $L^1(X, \mu)$).

470. Show that if $\{\{f_{pq}\}_{q=1}^{Q_p}\}_{p=1}^P$ is an independent set and if $\{g_p\}_{p=1}^P$ are Borel measurable maps on $\mathbb{R}^{Q_p}$ then $\{g_p(f_{p1}, \ldots, f_{pQ_p})\}_{p=1}^P$ is an independent set.

471. Show that if f is constant resp. $A = \varnothing$ and g is measurable resp. $B \in \mathsf{S}$ then $\{f, g\}$ resp. $\{A, B\}$ is an independent set.

472. Show that if f and g are independent and integrable, then fg is integrable and $\int_X f(x) d\mu\,(x) . \int_X g(x)\, d\mu\,(x) = \int_X f(x) g(x)\, d\mu\,(x)$.

473. Show that if the measure situation $\{\mathbb{N}, 2^{\mathbb{N}}, \nu\}$ is such that $\nu(n) = 2^{-n!}$ if $n \geqq 2$ and $\nu(1) = 1 - \sum_{n=2}^\infty 2^{-n!}$ then there are no nonconstant independent functions f, g.

474. For the situation $(I, \mathsf{S}_\lambda, \lambda)$ assume f is measurable, that for some nonempty subinterval (a, b) of I, $f^{-1}(f((a, b))) = (a, b)$, and that f^{-1} is measurable on $f((a, b))$. Show that if f and g are independent then g is constant a.e.

475. Show that if $\{f_\gamma\}_{\gamma\in\Gamma}$ is an orthonormal and independent set in $L^2(X, \mu)$ and if $\text{card}(\Gamma) \geqq 3$ then $\dim(\{f_\gamma\}_\gamma)^\perp \geqq 1$ and if Γ is infinite, then $\dim(\{f_\gamma\}_\gamma)^\perp = \infty$.

476. (Borel) Show that if $\{A_n\}_{n=1}^\infty$ is an independent set of sets, then $\mu(\limsup_{n=\infty} A_n) = 0$ iff $\sum_n \mu(A_n) < \infty$ and $\mu(\limsup_{n=\infty} A_n) = 1$ iff $\sum_n \mu(A_n) = \infty$.

477. (Kronecker) Show that if $\{a_n\}_{n=1}^\infty \subset \mathbb{C}$ and $\sum_{n=1}^\infty a_n/n$ converges, then $\lim_{N\to\infty} N^{-1} \sum_{n=1}^N a_n = 0$.

478. (Strong law of large numbers.) Show that if $\{f_n\}_{n=1}^{\infty}$ is an independent set, $E(f_n)=0$, $\operatorname{var}(f_n)=\sigma_n^2$, and $\sum_n \sigma_n^2/n^2<\infty$, then $N^{-1}\sum_{n=1}^{N} f_n \to 0$ a.e. as $N\to\infty$.

479. If $n\in\mathbb{N}$ and $n>1$ then each t in I may be written $\sum_{k=1}^{\infty} p_k n^{-k}$, p_k in $\mathbb{N}$, $0\leqq p_k\leqq n-1$. If two such representations exist, one, the preferred, must be of the form in which for some k_0, $p_k=n-1$ if $k\geqq k_0$. For $k=0,1,2,\ldots,n-1$ and t in I let $k(t,N)$ be N^{-1} (number of k's among $p_1,p_2,\ldots,p_N$) and let E_k be $\{t\colon \lim_{N\to\infty} k(t,N)/N$ exists and is $n^{-1}\}$. Show E_k is measurable for each k and that $\lambda(E_k)=1$. (For each n almost all numbers in I are normal.)

480. Assume $f\in L^1(\mathbb{T},\lambda)$ and $g\in L^{\infty}(\mathbb{T},\lambda)$. Show that

$$\lim_{n\to\infty}\int_{\mathbb{T}} f(t)g(nt)\,dt=2\pi\hat{f}(0)\,.\,\hat{g}(0).$$

Problems 481–484 constitute the important steps in the proof of the individual ergodic theorem of G. D. Birkhoff. The argument is that of F. Riesz and uses his running water lemma (see Problems 110, 111, and 341).

481. Let (X,S,μ) be a measure situation and let $T\colon X\mapsto X$ be a bijection that preserves measurability and measure, i.e., if $E\in\mathsf{S}$, $T^{-1}(E)$ and $T(E)\in\mathsf{S}$ and $\mu(E)=\mu(T(E))$. If $f\in L^1(X,\mu)$ let s_n be $x\mapsto\sum_{k=0}^{n-1} f(T^k(x))$ and let A be $\{x\colon \sup_n s_n(x)>0\}$. Show that if $E\in\mathsf{S}$ and $T(E)=E$ then $\int_{A\cap E} f(x)\,d\mu(x)\geqq 0$.

482. Show that if $A_a=\{x\colon \sup_n\,(n^{-1}s_n(x))>a\}$, $E\in\mathsf{S}$, and $T(E)=E$, then $\int_{A_a\cap E} f(x)\,d\mu(x)\geqq a\,.\,\mu(A_a\cap E)$.

483. Let $\bar{F}$ resp. $\underline{F}$ be $\limsup_{n=\infty} n^{-1}s_n$ resp. $\limsup_{n=\infty} n^{-1}s_n$. Show: i) $\bar{F}(T(x))=\bar{F}(x)$, $\underline{F}(T(x))=\underline{F}(x)$; ii) $\bar{F}=\underline{F}$ a.e.; iii) $\{n^{-1}s_n\}_n$ is uniformly integrable.

484. Show that if $F=\bar{F}=\underline{F}$, $E\in\mathsf{S}$, and $T(E)=E$ then $\int_E f(x)\,d\mu(x)=\int_E F(x)\,d\mu(x)$.

485. For z_0 in $\mathbb{T}$ and f in $C(\mathbb{T},\mathbb{C})$ show that $\lim_{n\to\infty} n^{-1}\sum_{k=0}^{n-1} f(z_0^k)$ exists and that there is a measure situation $(\mathbb{T},\mathsf{S}_\beta(\mathbb{T}),\mu_{z_0})$ such that the limit is $\int_{\mathbb{T}} f(z)\,d\mu_{z_0}(z)$.

486. Assume $\{a_n\}_{n=1}^{\infty}\in l^1(\mathbb{N})$, $\{f_n\}_n\subset L^1(I,\lambda)$, and $\|f_n\|_1\leqq M<\infty$, n in $\mathbb{N}$. i) Show that $F\colon I\ni x\mapsto\sum_n a_nf_n(x)$ is in $L^1(I,\lambda)$. ii) Characterize pairs $(\{a_n\}_n,\{f_n\}_n)$ in $\mathbb{C}^{\mathbb{N}}\times(L^1(I,\lambda))^{\mathbb{N}}$ and such that $\|F\|_1=\sum_n|a_n|\,.\,\|f_n\|_1$ (*). iii) Characterize the sequences $\{a_n\}_n$ such that (*) holds for all sequences $\{f_n\}_n$. iv) Characterize the sequences $\{f_n\}_n$ such that (*) holds for all sequences $\{a_n\}_n$.

487. Show that if X is a vector space and $\{x_\gamma\}_{\gamma\in\Gamma}$, $\{y_\lambda\}_{\lambda\in\Lambda}$ are Hamel bases for X then they have the same cardinality.

488. Show that if in problem 487, X is Hilbert space $\mathfrak{H}$ and "Hamel bases" is replaced by "maximal orthonormal sets", then the conclusion remains valid.

489. Show that the cardinality of a Hamel basis of a Banach space either is less than card($\mathbb{N}$) or is at least card($\mathbb{R}$).

490. Show that any two separable infinite-dimensional Banach spaces are (not necessarily continuously) isomorphic (see problem 376).

491. Let the nth Rademacher function r_n be $I \ni x \mapsto \operatorname{sgn}(\sin(2^n \pi x))$, $n = 0, 1, \ldots$. Show that in the context of $(I, \mathsf{S}_\lambda, \lambda)$ they constitute an independent set.

492. For m in $\mathbb{N}$ there is in $\mathbb{N} \cup \{0\}$ a unique finite set $\{n_1, n_2, \ldots, n_{K_m}\}$ such that $m = \sum_{k=1}^{K_m} 2^{n_k}$. The mth Walsh function $W_m = \prod_{k=1}^{K_m} r_{n_k}$. Show $\{W_m\}_{m=1}^{\infty}$ is a maximal orthogonal set in $L^2(I, \lambda)$ (see problem 475).

493. Show that if $a < b$, then $\{x \mapsto e^{inx}\}_{n=-\infty}^{\infty}$ is a linearly independent set on $\{a, b\}$.

494. Show that if $n \geqq 2$ and F is a closed proper subset of $\mathbb{R}^n$ and ∂F is scattered (the boundary of F contains no nonempty perfect subset), then F is at most countable.

495. Show that if $\gamma: I \mapsto \mathbb{R}^n$ is a rectifiable curve and $n \geqq 2$ then $\lambda_n(\gamma(I)) = 0$.

496. Show that if $\gamma: I \mapsto \mathbb{R}^n$ is a rectifiable curve and $n \geqq 2$ then $(\gamma(I))^0 = \emptyset$.

497. If $\mathbb{R}$ is given the topology generated by the set of all intervals $[a, b)$ and $(c, d]$, is $\mathbb{R}$ separable?

498. Show that $\mathbb{Q}$ is not the intersection of a countable sequence of open sets ($\mathbb{Q}$ is not a G_δ).

499. Show that if X is a topological space and card$(X) = \mathscr{A}$ then every open cover of X contains a subcover of cardinality not exceeding $\mathscr{A}$.

500. If X is a topological space and $\Gamma \ni \gamma \mapsto x_\gamma \in X$ is a net, let $\mathscr{F}$ be $\{S: S \subset X,\ x_\gamma$ is eventually in $S\}$. A point p is a cluster point of the net iff x_γ is frequently in each deleted neighborhood of p. A point p is a cluster point of $\mathscr{F}$ iff $p \in \bigcap\{\bar{S} : S \in \mathscr{F}\}$. Show that the cluster points of the net and of $\mathscr{F}$ constitute the same set.

501. Let S be a compact semigroup, i.e., S is a compact Hausdorff space, there is a continuous map $S \times S \ni (x, y) \mapsto xy \in S$, and for all x, y, z, $x(yz) = (xy)z$. Assume that whenever $xy = xz$ then $y = z$ and whenever $yx = zx$ then $y = z$ (S has a two-sided cancellation law). Prove that S is a compact group, i.e., there is in S an identity e such that for all x, $ex = xe = x$, for all x there is an x^{-1} such that $xx^{-1} = x^{-1}x = e$, and the map $x \mapsto x^{-1}$ is well-defined and continuous.

502. Let S be a semigroup with a two-sided cancellation law. Assume that (S, S, μ) is a measure situation such that $0<\sup\{\mu(A)\colon A \text{ in } \mathsf{S}\}=M<\infty$, for all x in S and A in S, $x+A\in\mathsf{S}$ and $\mu(x+A)=\mu(A)$, and the map $\theta\colon S\times S\ni(x, y)\mapsto(x, xy)$ preserves measurability with respect to product measure. Show S is a group.

503. Let X be a completely regular topological space, i.e., if F is closed and $p\notin F$ there is in $C(X, I)$ an f such that $f(p)=1$ and $f(F)=0$. Prove there is a topological group $F(X)$ containing X topologically and such that if $f\colon X\mapsto G$ is a continuous map of X into the topological group G then there is a continuous homomorphism $h\colon F(X)\mapsto G$ and h restricted to X is f.

504. Let S be a semigroup such that there is in S an element denoted 0 such that $0\,.\,x=x\,.\,0=0$ for all x. Assume further that if $x\neq 0\neq z$ the relation $xyz\neq 0$ has precisely one solution. Show that there is a set J such that $\mathscr{U}(J)$ (the set of matrix units in $\mathscr{M}(J)$) and S are isomorphic.

505. Let A be a commutative Banach algebra with identity e and let D be a continuous derivation on A. Show $D(A)\subset\bigcap_{M\in\sigma(A)} M$ ($D(A)\subset$ radical of A).

506. Let X be a Hausdorff space and let $\mathscr{U}=\{U\}$ be a basis of open sets for the topology of X. Assume that for all U in $\mathscr{U}$, V_U is an open set containing ∂U. Show that $A=X\backslash\bigcup_{U\in\mathscr{U}} V_U$ is either empty, a single point or totally disconnected.

507. Give an example of a metric space X, a countable subset B such that $\bar{B}$ is nowhere dense ($(\bar{B})^0=\varnothing$) and card $(\bar{B})>\text{card}(\mathbb{N})$.

508. Let (X, d) be a metric space without isolated points. Assume that each f in $C(X, \mathbb{R})$ is uniformly continuous. Show X is compact.

The next set of problems (509–517) is designed for the study of an important set theoretical operation ($\mathscr{A}$) having applications to some of the earlier problems (267, 437).

If $\mathscr{M}$ is a set $\{M\}$ of sets and if f is a map $\mathbb{N}^{\mathbb{N}}\ni\nu=\{n_1, n_2, \ldots\}\mapsto\{f(\nu)\}_{k=1}^{\infty}\in\mathscr{M}^{\mathbb{N}}$ then $M_f=\bigcup_\nu\bigcap_k f(\nu)_k$. If $\mathscr{F}$ is the set $\{f\}$ of all such maps then $\mathscr{A}(\mathscr{M})=\{M_f\colon f \text{ in } \mathscr{F}\}$. It is often convenient to denote $f(\nu)_k$ by $M_{n_1,n_2,\ldots,n_k}$ and thereby indicate not merely the kth element of $f(\nu)$ but also part of the sequence from which the kth element is derived. The map f is regular iff for all ν and all k, $f(\nu)_{k+1}\subset f(\nu)_k$, i.e., $M_{n_1,n_2,\ldots,n_{k+1}}\subset M_{n_1,n_2,\ldots,n_k}$.

509. Show that $\mathscr{A}(\mathscr{A}(\mathscr{M}))=\mathscr{A}(\mathscr{M})$ ("$\mathscr{A}^2=\mathscr{A}$").

510. Show that if $\{M_n\}_{n=1}^{\infty}\subset\mathscr{M}$ then i) $\bigcap_n M_n$ and ii) $\bigcup_n M_n$ are in $\mathscr{A}(\mathscr{M})$.

511. Show that if $\mathcal{M}$ is closed with respect to the formation of finite intersections, i.e., if M, N are in $\mathcal{M}$ so is $M \cap N$ in $\mathcal{M}$, then for every f in $\mathcal{F}$ there is in $\mathcal{F}$ a regular S such that $M_g = M_f$.

512. Show that if f is regular then: i) $\bigcup_{m=1}^{\infty} \bigcup_{\nu} \bigcap_k M_{n_1,n_2,\ldots,n_i,m,n_{i+1},\ldots,n_{i+k}} = \bigcup_{\nu} \bigcap_k M_{n_1,n_2,\ldots,n_i,n_{i+1},\ldots,n_{i+k}}$; ii) the union $\bigcup_{\nu} \bigcup_k M_{n_1,n_2,\ldots,n_k}$ is the countable union of sets in $\mathcal{M}$; iii) if $M_{n_1,n_2,\ldots,n_k}$ is denoted simply by M when $k=0$ then $M \backslash M_f \subset \bigcup_{\nu} \bigcup_{k=0}^{\infty} (M_{n_1,n_2,\ldots,n_k} \backslash \bigcup_{m=1} M_{n_1,n_2,\ldots,n_k,m})$.

513. Show that if $\mathcal{M} = \mathsf{F}(I)$ (the set of closed sets in I) then $\mathsf{S}_\beta(I) \subset \mathcal{A}(\mathcal{M})$ (every Borel set is in $\mathcal{A}(\mathsf{F}(I))$, the set of Suslin or analytic sets in I).

514. Show that if $\mathcal{M} = \mathsf{F}(I)$ and $H \in C(I, \mathbb{R})$ then for all E in $\mathcal{A}(\mathcal{M})$, $H(E) \in \mathcal{A}(\mathsf{F}(\mathbb{R}))$.

515. Show that $\mathcal{A}(\mathsf{S}_\lambda(I)) = \mathsf{S}_\lambda(I)$, i.e., the set of Lebesgue measurable sets is invariant under $\mathcal{A}$.

516. Show that if $S \in \mathcal{A}(\mathsf{F}(I))$ then $\text{card}(S) \leqq \text{card}(\mathbb{N})$ or $\text{card}(S) = \text{card}(\mathbb{R})$ (see Problem 267).

517. Show that if $E \in \mathsf{S}_\beta(I)$ and $H \in C(I, \mathbb{R})$ then $H(E) \in \mathsf{S}_\lambda(\mathbb{R})$ (see Problem 437).

518. Assume $f, g \in C(I, I)$ and that $f \circ g = g \circ f$. Show there is in I an x_0 such that $f(x_0) = g(x_0)$.

Solutions

1. Set Algebra

1. If $\{A_n\}_{n=1}^{\infty} \subset \mathsf{M}$ and if $B_m = \bigcup_{n=1}^{m} A_n$, for m in $\mathbb{N}$, then $B_m \subset B_{m+1}$, $B_m \in \mathsf{M}$ and thus $\bigcup_n A_n = \bigcup_m B_m \in \mathsf{M}$. The proof for countable intersections proceeds *mutatis mutandis*. □

2. Since $\sigma\mathsf{R}(\mathsf{R})$ is monotone it contains $\mathsf{M}(\mathsf{R})$. The conclusion will follow if $\mathsf{M}(\mathsf{R})$ is shown to be a σ-ring. For C in M let $\mathsf{B}(C)$ be $\{D: D\backslash C, C\backslash D, C \cup D \text{ are in } \mathsf{M}\}$. Then: i) $D \in \mathsf{B}(C)$ iff $C \in \mathsf{B}(D)$; ii) for all C in M, $\mathsf{B}(C)$ is monotone; iii) if C and D are in R then $C \in \mathsf{B}(D)$ and so for all D in R, $\mathsf{R} \subset B(D)$. Hence for D in R, $\mathsf{M}(\mathsf{R}) \subset \mathsf{B}(D)$. If $C \in \mathsf{M}(\mathsf{R})$ and $D \in \mathsf{R}$ then $C \in \mathsf{M}(\mathsf{R}) \subset \mathsf{B}(D)$ and so $D \in \mathsf{B}(C)$, i.e., for all C in $\mathsf{M}(\mathsf{R})$, $\mathsf{M}(\mathsf{R}) \subset \mathsf{B}(C)$. The last assertion implies that $\mathsf{M}(\mathsf{R})$ is a ring and by Solution 1 $\mathsf{M}(\mathsf{R})$ is a σ-ring. □

Let the metric for the metric space X be $d: X \times X \ni (x, y) \mapsto d(x, y) \in [0, \infty)$.

3. For F in $\mathsf{F}(X)$ let U_n be $\bigcup_{y \in F} \{x: d(x, y) < 1/n\}$, $n = 1, 2, \ldots$. Each U_n is open, $U_n \supset U_{n+1}$, and $\bigcap_n U_n = F$, whence $\mathsf{F}(X) \subset \mathsf{M}$. □

4. If $a \leqq b$ the interval $[a, b) = \bigcup_{b-1/n \geqq a} [a, b - 1/n]$ and thus all such intervals $[a, b)$ are in M. By the same argument, any finite union of such intervals is in M. Since the set of all finite unions of such intervals is a ring R, $\mathsf{M} \supset \sigma\mathsf{R}(\mathsf{R})$, by virtue of Solution 2. Since every open set is the countable union of disjoint open intervals and since every open interval may be written $\bigcup_{a+1/n < b-1/n} [a + 1/n, b - 1/n]$, and since each of $\mathsf{O}(\mathbb{R})$ and $\mathsf{F}(\mathbb{R})$ is contained in the σ-ring generated by the other, it follows that $\sigma\mathsf{R}(\mathsf{O}(\mathbb{R})) = \sigma\mathsf{R}(\mathsf{F}(\mathbb{R})) = \sigma\mathsf{R}(\mathsf{K}(\mathbb{R})) \subset \mathsf{M}$. □

5. Call a subset A of S disjoint if its elements are pairwise disjoint. The idea of the proof is that *either*: i) $\mathsf{S}=\sigma\mathsf{R}(\mathsf{A})$ for some finite disjoint set A; *or* ii) S contains an infinite disjoint set A. If i) holds, then $\text{card}(\mathsf{S})=2^{\text{card}(\mathsf{A})}<\text{card}(\mathbb{N})$. If ii) holds then $\text{card}(\mathsf{S})\geqq 2^{\text{card}(\mathsf{A})}>\text{card}(\mathbb{N})$. Thus assume there are *no* infinite disjoint sets A.

Partially order by inclusion the set of all disjoint subsets of S. The Hausdorff maximality principle implies there is a maximal linearly ordered set $\{\mathsf{A}_\gamma: \gamma\in\Gamma\}$ of distinct disjoint sets A_γ.

If Γ is infinite there is an infinite disjoint set A since there is either an ascending chain $\{\mathsf{A}_1<\mathsf{A}_2<\cdots\}$ or a descending chain $\{\mathsf{A}_1>\mathsf{A}_2>\cdots\}$ among the A_γ. In either case there emerges the contradiction of the assumption that there are no infinite disjoint sets since each of the symmetric difference sets $\mathsf{A}_n\Delta\mathsf{A}_{n+1}$ is nonempty and if $A_n\in\mathsf{A}_n\Delta\mathsf{A}_{n+1}$ then $\{A_n: n=1,2,\ldots\}$ is a disjoint infinite set.

Thus Γ is finite and there is a largest A_γ, denoted A_m, consisting of sets A_n, $n=1,2,\ldots,N$. If $B\in\mathsf{S}$ then $B\subset\bigcup_{n=1}^N A_n$ since otherwise $\mathsf{A}_m\cup\{B\backslash\bigcup_{n=1}^N A_n\}\gneqq\mathsf{A}_m$. Introduce partitions P as follows: (a) P is a finite disjoint set; (b) each A_n is the union of the elements of P that are in A_n. Partially order the P by refinement, i.e., $\mathsf{P}_2>\mathsf{P}_1$ iff every P_2 of P_2 is a subset of some P_1 of P_1. Apply the Hausdorff maximality principle and get a maximal linearly ordered set $\{\mathsf{P}_\lambda:\lambda\in\Lambda\}$. The preceding argument applied to the partitions shows Λ is finite and if P_m is the largest partition of the set then $\mathsf{S}=\sigma\mathsf{R}(\mathsf{P}_m)$ and i) obtains. □

6. Let D be $\{D: D\in\sigma\mathsf{R}(\mathsf{E}_0), \mathsf{E}_0\subset\mathsf{E}, \text{card}(\mathsf{E}_0)\leqq\text{card}(\mathbb{N})\}$. Then $\sigma\mathsf{R}(\mathsf{E})\supset\mathsf{D}\supset\mathsf{E}$. If $\{A_n\}_{n=1}^\infty\subset\mathsf{D}$ and if $A_n\in\sigma\mathsf{R}(\mathsf{E}_{on})$ then $\bigcup_n A_n$ and $A_1\backslash A_2$ are in $\sigma\mathsf{R}(\bigcup_n \mathsf{E}_{on})$ whence D is a σ-ring, $\mathsf{D}=\sigma\mathsf{R}(\mathsf{E})$ from which the result follows. □

7. For each sequence $\{p, q, r, \ldots\}$ of positive integers let $\{B_p, A_{pq}, B_{pqr}, \ldots\}$ be a sequence contained in 2^X. Assume: i') $B_p=\bigcup_q A_{pq}, A_{pq}=\bigcap_r B_{pqr},\ldots$; ii') ii) with A and B interchanged. Let B be the set of all countable intersection of sets B_p. Then every element of B is an A_p and so $\mathsf{A}=\mathsf{B}$. Thus A is closed with respect to the formation of countable unions and intersections of its members. If $A\in\mathsf{A}$ then $A'=\bigcap_p A_p', A_p'=\bigcup_q B_{pq}',\ldots$. Condition iii) implies that $A'\in\mathsf{B}=\mathsf{A}$, whence A is a σ-algebra and $\mathsf{A}\supset\sigma\mathsf{A}(\mathsf{E})$. If the inclusion is proper let A be in $\mathsf{A}\backslash\sigma\mathsf{A}(\mathsf{E})$. Then $A=\bigcup_p A_p$ and some $A_p\notin\sigma\mathsf{A}(\mathsf{E})$, whence some $B_{pq}\notin\sigma\mathsf{A}(\mathsf{E}),\ldots$. Hence, *a fortiori*, ii) and ii') are contradicted and the result follows. □

(See also Problems 153 and 188.)

2. Topology

8. If $f: [0, 1) \ni x \mapsto (1/(1-x)) \sin(1/(1-x))$, then f is continuous and $f([0, 1)) = \mathbb{R}$. □

9. If $x \sim y$ iff $x - y \in \mathbb{Q}$, then $\sim$ is an equivalence relation that decomposes $[0, 1]$. Each equivalence class is countable and hence there are uncountably many equivalence classes in $[0, 1]$. If E consists of exactly one element from each equivalence class then $E - E \subset (\mathbb{Q}\backslash\{0\})'$ and thus $(E-E)^0 = \varnothing$. □

10. Let $d: [0, 1) \times [0, 1) \ni (x, y) \mapsto |x - y|$ be the standard metric in $[0, 1)$ and let $D: [0, 1) \times [0, 1) \ni (x, y) \mapsto |x/(1-x) - y/(1-y)|$ be a new metric. Their related topologies are the same. A Cauchy sequence $\{x_n\}_{n=1}^{\infty}$ fails to converge with respect to d iff $x_n \to 1$ as $n \to \infty$. Direct calculation shows that for such a sequence $D(x_n, x_m) \not\to 0$ as $n, m \to \infty$. On the other hand if $\{x_n\}_{n=1}^{\infty}$ is "D-Cauchy", the preceding remarks show that $x_n \to x$ in $[0, 1)$ metrized by d and then $D(x_n, x) \to 0$ as $n \to \infty$. Thus $[0, 1)$ is "D-complete". □

11. Since $[0, 1]$ is connected so is $A \times B$ and therefore so are A and B. If a_1 and a_2 are two elements of A and b_1 and b_2 are two elements of B the four different connected sets $f^{-1}(\{a_i\} \times B)$, $f^{-1}(A \times \{b_j\})$, $i, j = 1, 2$, are nondegenerate closed intervals I_m, $m = 1, 2, 3, 4$, and each pair of distinct intervals have exactly one point in common. No such configuration of such intervals can exist in $[0, 1]$ and so either A or B is a single point. □

12. If f maps $\prod_n D_n$ according to the rule $f((\varepsilon_1, \varepsilon_2, \ldots)) \mapsto \sum_{n=1}^{\infty} 2\varepsilon_n 3^{-n}$ then f is a surjective homeomorphism. If $\varepsilon > 0$, $k \in \mathbb{N}$, $a, b \in C$ and $|b - a| < 3^{-(k+2)}$ then $f_k(b) = f_k(a)$ and hence f_k is continuous. □

13. It may be assumed that Γ is the (well-ordered) et of all ordinal numbers not exceeding some ordinal number θ and that $F \subset [0, 1]$. Let γ_1 be the least index γ such that $\sup F \in (a_\gamma, b_\gamma]$. If $\{\gamma_\beta : \beta < \eta\}$ has been defined and if $F \setminus \bigcup_{\beta<\eta} (a_{\gamma_\beta}, b_{\gamma_\beta}] = F_\eta \neq \varnothing$ let γ_η be the least index γ such that $\sup(F \setminus \bigcup_{\beta<\eta} (a_{\gamma_\beta}, b_{\gamma_\beta}]) \in (a_\gamma, b_\gamma]$. (Note that F_η is closed.) Furthermore, let δ be the least ordinal number such that $F \subset \bigcup_{\beta \leqq \delta} (a_{\gamma_\beta}, b_{\gamma_\beta}]$. Then $\delta \leqq \theta$. If $\operatorname{card}(\{\beta : \beta \leqq \delta\}) > \operatorname{card}(\mathbb{N})$, then, since $a_{\gamma_{\beta+1}} < a_{\gamma_\beta}$, $\sum_{\beta<\delta} (a_{\gamma_{\beta+1}} - a_{\gamma_\beta}) = -\infty$. Thus for some β less than δ, $a_{\gamma_\beta} < 0$. If β_0 is the least such β, then $F \subset \bigcup_{\beta<\beta_0+1} (a_{\gamma_\beta}, b_{\gamma_\beta}]$ and so $\beta_0 + 1 > \delta$, a contradiction. Thus $\operatorname{card}(\{\beta : \beta < \delta\}) \leqq \operatorname{card}(\mathbb{N})$. □

14. The sequence $\{x_n : x_n \in \mathbb{R}, x_n = n, n = 1, 2, \ldots\}$ is such that $d(x_n, x_m) \to 0$ as $n, m \to \infty$. Nevertheless there is in $\mathbb{R}$ no x such that $d(x_n, x) \to 0$ as $n \to \infty$. □

15. If $u \notin S$ there is in S a sequence $\{x_n\}_{n=1}^{\infty}$ such that $x_n \to u$ as $n \to \infty$ and $x_n < x_{n+1}$, $n = 1, 2, \ldots$. Hence $x_n / x_{n+1} \in S$ and $x_n / x_{n+1} \leqq u$, and, by induction, $0 \leqq x_n < u^p x_{n+p}$, $n = 1, 2, \ldots, p = 1, 2, \ldots$. Since $u < 1$, all x_n are zero and thus $u = 0$, $u \in S$ and a contradiction emerges. □

16. The map

$$f : \mathbb{R} \setminus \mathbb{Q} \ni t \mapsto \begin{cases} \frac{1}{2} + t/(2t+2), & \text{if } t > 0 \\ \frac{1}{2} + t/(2-2t), & \text{if } t < 0 \end{cases}$$

maps $\mathbb{R} \setminus \mathbb{Q}$ homeomorphically onto $(0, 1) \setminus \mathbb{Q}$. □

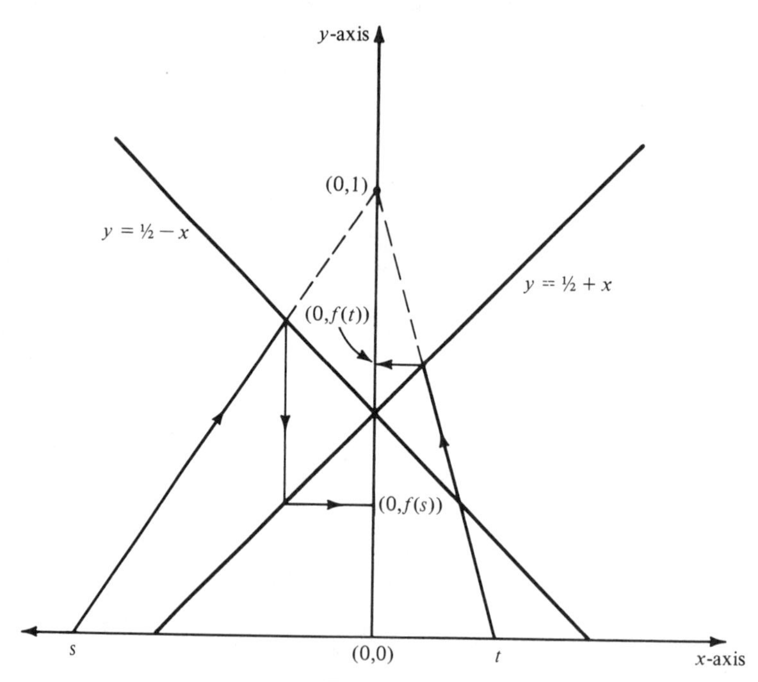

Figure 1

17. If E has no isolated points it is perfect and hence $\text{card}(E) = \text{card}(\mathbb{R})$. and so E is not countable. (In a complete metric space X with metric d, e.g., $\mathbb{R}^2$, the cardinality of every compact nonempty perfect set is $\text{card}(\mathbb{R})$. The following is a sketch of the proof: the compact perfect set E is contained in some finite union of closed balls $B(y_i, 1)$, $i = 1, 2, \ldots, k_1$. The sets $E \cap B(y_i, 1)$ are compact and can be covered by finitely many balls $B(y_{ij}, \frac{1}{2})$, etc. If E is nonempty as well then $\text{diam}(E) > 0$ and since the ball diameters decrease to zero, at some stage two disjoint ones emerge, at a later stage two disjoint ones in each of those, etc. At every stage only balls meeting E are used. Thus this *dyadic* construction permits the choice for each dyadic rational number of two distinct points in E. They may be indexed by finite sequences $\varepsilon_1, \varepsilon_2, \ldots, \varepsilon_n$, $\varepsilon_k = 0$ or 1, so that two points having the first $n-1$ indices the same and the last index different are in disjoint balls and all their successors are in the same two disjoint balls. Because E is compact, X is complete, and the ball diameters converge to 0, all the sequences have limits, all the limits are in E, and distinct sequences have distinct limits, whence $\text{card}(E) \geqq \text{card}(\mathbb{R})$. Since E is a compact metric space $\text{card}(E) \leqq \text{card}(\mathbb{R})$.□) □

18. If $\mathbb{R}^2 \neq \overline{\bigcup_{n=1}^{\infty} F_n^0}$ then $\mathbb{R}^2 \backslash \overline{\bigcup_n F_n^0}$ contains a nonempty compact ball B. Regard B as a compact metric space that is the union of the sets $B \cap (F_n \backslash F_n^0)$, $n = 1, 2, \ldots$. If x is an interior point of one of them, there is an open set $U(x)$ in $\mathbb{R}$, containing x, and such that $U(x) \cap B \subset B \cap (F_n \backslash F_n^0)$. Since $U(x) \cap B \neq \varnothing$, there is in $\mathbb{R}$ an open set V contained in $B \cap (F_n \backslash F_n^0)$, a contradiction since $(F_n \backslash F_n^0)^0 = \varnothing$. Thus the compact space B is the countable union of closed sets having empty interior, a contradiction. (If X is a compact Hausdorff space it is not the union of countably many closed sets having empty interior. A sketch of the proof follows: If the H_n are the closed sets in question, there is in $X \backslash H_1$ a point x_1 and an open neighborhood $U(x_1)$ such that $\overline{U(x_1)} \cap H_1 = \varnothing$. Inductively construct points x_k and neighborhoods $U(x_k)$ for some of the sets H_{n_k}, $k = 2, 3, \ldots$ so that $\overline{U_{k+1}(x_{k+1})} \cap ((X \backslash U_k(x_k)) \cup H_{n_k}) = \varnothing$. The closed sets $\overline{U_k(x_k)}$ have the finite intersection property and any point in their nonempty intersection is in some H_m whereas, if the integers n_k are chosen so that each is the least such that H_{n_k} meets $U(x_k)$, it follows that $U_{k_{m+1}}(x_{k_{m+1}}) \cap H_m = \varnothing$.□) □

19. If, for every x in A, there is a $U(x)$ such that $U(x) \cap A$ is countable, let $\{U_n\}_{n=1}^{\infty}$ be a countable basis of open sets for the topology of $\mathbb{R}^2$. Each $U(x)$ is the union of some of the U_n and so A, which is covered by the union of all $U(x) \cap A$, is covered by the union of some of the $U_n \cap A$. Each of the last is a countable set and therefore A is countable in contradiction of the hypothesis. □

20. Let X be $\mathbb{T}$ in its standard topology and let T be $\mathbb{T} \ni e^{2\pi i\theta} \mapsto e^{2\pi i(2\theta/(1+\theta))}$. For all x in X, $T^n x \to 1$ as $n \to \infty$. If d is any metric compatible with the topology of $\mathbb{T}$ and if $d(Tx, Ty) = d(x, y)$ for all x and y in $\mathbb{T}$, then $d(T^n x, T^n y) = d(x, y)$ and so $d(Tx, Ty) = d(x, y)$ iff $x = y$. □

21. Let $\{T_n\}_{n=1}^{\infty}$ be a sequence in $\{T_\gamma\}_{\gamma \in \Gamma}$, let d be the metric for X, and let D be the standard metric for $C(X, X)$, i.e., for f, g in $C(X, X)$, $D(f, g) = \sup_x d(f(x), g(x))$. If $F_n = f \circ T_n$ for n in $\mathbb{N}$ and if $\{x_m\}_{m=1}^{\infty}$ is dense in X then $\{F_n(x_1)\}_{n=1}^{\infty}$ contains a convergent subsequence $\{F_{n_k}(x_1)\}_{k=1}^{\infty}$; $\{F_{n_k}(x_2)\}_{k=1}^{\infty}$ contains a convergent subsequence $\{F_{n_{k_r}}(x_2)\}_{r=1}^{\infty}$; etc. Relabel $F_{n_1}, F_{n_{k_2}}, \ldots$ as $G_1, G_2, \ldots$. Then $\{G_p(x_m)\}_{p=1}^{\infty}$ is convergent for all m in $\mathbb{N}$. If $\varepsilon_1 > 0$ there are positive ε_2 and ε_3 such that if $d(x, y) < \varepsilon_2$ then $d(f(x), f(y)) < \varepsilon_1/3$ and if $d(x, y) < \varepsilon_3$ then for all n in $\mathbb{N}$, $d(T_n x, T_n y) < \varepsilon_2$. The sets $\{y : |y - x_m| < \varepsilon_3\}$, $m = 1, 2, \ldots$ constitute an open covering of X and thus for some M in $\mathbb{N}$, $X = \bigcup_{m=1}^{M} \{y : |y - x_m| < \varepsilon_3\}$. Let n_0 be such that if $p, q > n_0$ and $1 < m < M$ then $d(G_p(x_m), G_q(x_m)) < \varepsilon_1/3$. Then for any x there is some m not exceeding M and such that $x \in \{y : |y - x_m| < \varepsilon_3\}$. If $p, q > n_0$, $d(G_p(x), G_p(x)) \leqq d(G_p(x), G_p(x_m)) + d(G_p(x_m), G_q(x_m)) + d(G_q(x_m), G_q(x))$. Each of the last three terms is less than $\varepsilon_1/3$. Since n_0 is independent of x it follows that $D(G_p, G_q) \to 0$ as $p, q \to \infty$ and the result follows. □

22. If f is continuous the map $F : X \ni x \mapsto (x, f(x)) \in X \times Y$ is continuous. Since X is compact the graph is compact and is thus closed. Conversely, if the graph is closed and if f is not continuous there is in X an x and a sequence $\{x_n\}_{n=1}^{\infty}$ such that $x_n \to x$ as $n \to \infty$ and $f(x_n) \not\to f(x)$ as $n \to \infty$. Since Y is compact there is a subsequence $\{x_{n_k}\}_{k=1}^{\infty}$ and in Y a y different from $f(x)$ and such that $f(x_{n_k}) \to y$ as $k \to \infty$. Thus the graph of f is not closed. □

23. Let Y be $X \times X$ and metrize Y according to $D : Y \times Y \ni ((a_1, b_1), (a_2, b_2)) \mapsto \sqrt{(d(a_1, a_2))^2 + (d(b_1, b_2))^2}$. Define F by $F : Y \ni (x_1, x_2) \mapsto (f(x_1), f(x_2))$. If $\varepsilon > 0$ and $y \in Y$ there is an n depending on y and such that $D(F^n(y), y) < \varepsilon$. Otherwise, for all n, $D(F^{n+k}(y), F^k(y)) \geqq \varepsilon$, $k = 1, 2, \ldots$. Hence $\{F^n(y)\}_{n=1}^{\infty}$ contains no Cauchy sequence and thereby the compactness of Y is contradicted. Thus for all (a, b) in Y, since $d(f(a), f(b)) \leqq d(f(a), a) + d(a, b) + d(b, f(b))$, if $y = (a, b)$ and $D(F^n(y), y) < \varepsilon$, then $d(f(a), f(b)) < d(a, b) + 2\varepsilon$. Since ε is arbitrary the result follows. □

24. There is in $\mathsf{O}(X)$ a countable set $\{U_n\}_{n=1}^{\infty}$ such that every U in $\mathsf{O}(X)$ is a union of some U_n. Thus $\operatorname{card}(\mathsf{O}(X)) \leqq \operatorname{card}(\mathbb{R})$. Since the correspondence $\mathsf{O}(X) \ni U \mapsto U' \in \mathsf{F}(X)$ is bijective the result follows. □

25. i) The maps $f_n : [0, 1] \ni x \mapsto x/n$, $n = 1, 2, \ldots$ are all in A_i whereas $f_n \to 0$ uniformly as $n \to \infty$ and so A_i is not closed. ii) If $\{f_n\}_{n=1}^{\infty} \subset A_s$ and if $f_n \to f$ uniformly as $n \to \infty$, then for every y in $[0, 1]$ and for every n in $\mathbb{N}$ there is an x_n in $[0, 1]$ such that $f_n(x_n) = y$. If x_0 is a limit point of $\{x_n\}_{n=1}^{\infty}$ then

$f(x_0)=y$ and thus A_s is closed. iii) The maps f_n defined according to

$$f_n(x)=\begin{cases} 3(n-2)x/2n, & 0\leqq x\leqq \frac{1}{3} \\ 6(x-\frac{1}{2})/n+\frac{1}{2}, & \frac{1}{3}<x\leqq \frac{2}{3} \\ f_n(x-\frac{2}{3})+\frac{1}{2}+1/n, & \frac{2}{3}<x\leqq 1 \end{cases} \qquad n=3,4,\ldots$$

and illustrated in Figure 2, are all in A_{is} whereas their uniform limit f as $n\to\infty$ is not and so A_{is} is not closed. iv) Since A is convex it is connected. v) The maps $f_n: x\mapsto x^n$, $n=1,2,\ldots$ admit no uniformly convergent subsequence and thus A is not compact. □

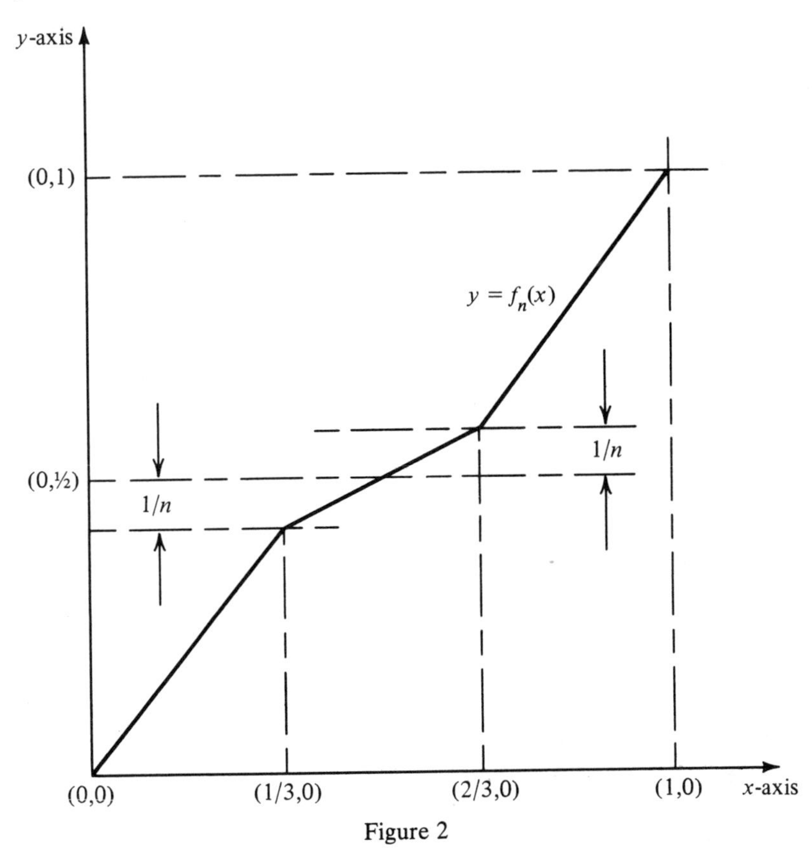

Figure 2

26. (See Problem 17.) If $p_0\in U$ there is some $B(p_0,r_0)^0$ inside U. Because there are no isolated points there are two points p_{00} and p_{01} in $B(p_0,r_0)^0$ and for some positive r_1, $B(p_{00},r_1)\cap B(p_{01},r_1)=\varnothing$. By induction there emerges for each dyadic rational number $\sum_{k=1}^{K}\varepsilon_k 2^{-k}$ a point $P_{\varepsilon_1\varepsilon_2\ldots\varepsilon_K}$ and for each K a positive r_K such that $r_K\downarrow 0$ and such that $B(p_{\varepsilon_1\varepsilon_2\ldots\varepsilon_K 0},r_K)\cap B(p_{\varepsilon_1\varepsilon_2\ldots\varepsilon_K 1},r_K)=\varnothing$. The set of limit points of the set of all $p_{\ldots}$ has cardinality at least that of $\mathbb{R}$. □

27. If $D=f^{-1}(C)$ and D is not connected there are closed sets F_1 and F_2 such that $F_1\cap F_2\cap D=\varnothing$, $(F_1\cap D)\cup(F_2\cap D)=D$, and $F_i\cap D\neq\varnothing$, $i=1,2$. Furthermore $f(F_1\cap D)\cap f(F_2\cap D)=\varnothing$ because if $x_i\in F_i\cap D$, $i=1,2$ and $f(x_1)=f(x_2)=y$ then x_1 and x_2 are in the connected set $f^{-1}(y)$ and yet $f^{-1}(y)=(f^{-1}(y)\cap F_1)\cup(f^{-1}(y)\cap F_2)$ in which both (relatively closed) summands are nonempty. If $H_i=f^{-1}(f(F_i))$, then H_i is compact, $i=1,2$. Furthermore, $H_i\cap D=F_i\cap D$, $i=1,2$, since, e.g., if $x\in H_1\cap D\cap F_2$, then $f(x)\in f(F_i\cap D)$, $i=1,2$, in contradiction of their disjointness. Thus $C=f(H_1\cap D)\cup f(H_2\cap D)=(f(H_1)\cap C)\cup(f(H_2)\cap C)$ and so *via* the compact sets $f(H_i)$ the connected set C is the union of two nonempty relatively closed disjoint sets $f(H_i)\cap C$, $i=1,2$, and a contradiction results.

If $X=[0,1]$ in its standard topology and $Y=[0,1]$ in its weakest topology (only $\varnothing$ and Y are open) then $f\colon X\ni x\mapsto x\in Y$ provides a counterexample since every subset of Y is connected. ☐

28. For positive ε there are in $\mathbb{R}$ open U_n such that $\operatorname{diam}(U_n)<\varepsilon$, $n=1,2,\ldots$, and $\bigcup_{n=1}^{\infty}U_n=\mathbb{R}$. Hence for some N, $X=\bigcup_{n-1}^{N}f^{-1}(U_n)$, and so there are finitely many basic neighborhoods, each contained in some $f^{-1}(U_n)$, $n=1,2,\ldots N$, and whose union is X. Let them be $\{V_p\}_{p=1}^{P}$ and by induction define Z_1 as $V_1,\ldots,Z_k$ as $(V_1\cup\cdots\cup V_k)\setminus(V_1\cup\cdots\cup V_{k-1})$, $k=1,2,\ldots,P$. The Z_k are pairwise disjoint and their union is X. If $Z_k\neq\varnothing$, choose $x^{(k)}$ in Z_k and let $f(x^{(k)})$ be a_k. Since each x is in precisely one Z_k, define g by the formula: $g(x)=a_k$ iff $x\in Z_k$. There are finitely many indices γ_{p_i}, $i=1,2,\ldots,n_p$ that are determining for each V_p and hence if $x,y\in X$ and $x_{\gamma_{p_i}}=y_{\gamma_{p_i}}$, $i=1,2,\ldots,n_p$, $p=1,2,\ldots,P$, then x and y are in the same Z_k and $g(x)=g(y)=a_k$, whereas $|f(x)-a_k|<\varepsilon$ if $x\in Z_k$. In sum, $\sup_x|f(x)-g(x)|<\varepsilon$ and g is finitely determined.

In the argument above if ε is set equal to $1/n$, n in $\mathbb{N}$, countably many indices γ_{p_i}, $i=1,2,\ldots,n_p$, $p=1,2,\ldots$ are determined and for each n a g_n is defined to serve for $1/n$ as g served for ε. The standard triangle inequality argument then shows that if $x_{\gamma_{p_i}}=y_{\gamma_{p_i}}$ for all indices γ_{p_i} then for all n in $\mathbb{N}$, $|f(x)-f(y)|<2/n$, whence f is countably determined. ☐

29. If, for each M in $\mathbb{R}$, there is in X an x_M such that $f(x_M)>M$, then if $M'>M$, $x_{M'}\in f^{-1}([M,\infty))$; whence the closed sets $f^{-1}([M,\infty))$, $M\in\mathbb{R}$, have the finite intersection property and thus there is an $\bar{x}$ in $\bigcap_M f^{-1}([M,\infty))$ and so $f(\bar{x})>M$, for all M in $\mathbb{R}$, a contradiction.

For each n in $\mathbb{N}$ let F_n be $f^{-1}([\sup_x f(x)-1/n,\infty))$. Then each F_n is nonempty, closed, and $F_n\supset F_{n+1}$. If x_0 is in the necessarily nonempty set $\bigcap_n F_n$, then $\sup_x f(x)\geqq f(x_0)>\sup_x f(x)-1/n$ for all n and the result follows. ☐

30. Let L be $K\setminus V$ and let M be $K\setminus U$. Then L and M are compact, $L\subset U$, $M\subset V$, $L\cap M=\varnothing$, and there are disjoint open sets U_1 and V_1 such that $L\subset U_1\subset U$ and $M\subset V_1\subset V$. If $K_U=K\setminus V_1$ and $K_V=K\setminus U_1$ then $K_U\subset U$, $K_V\subset V$, K_U and K_V are compact and $K_U\cup K_V=K$. ☐

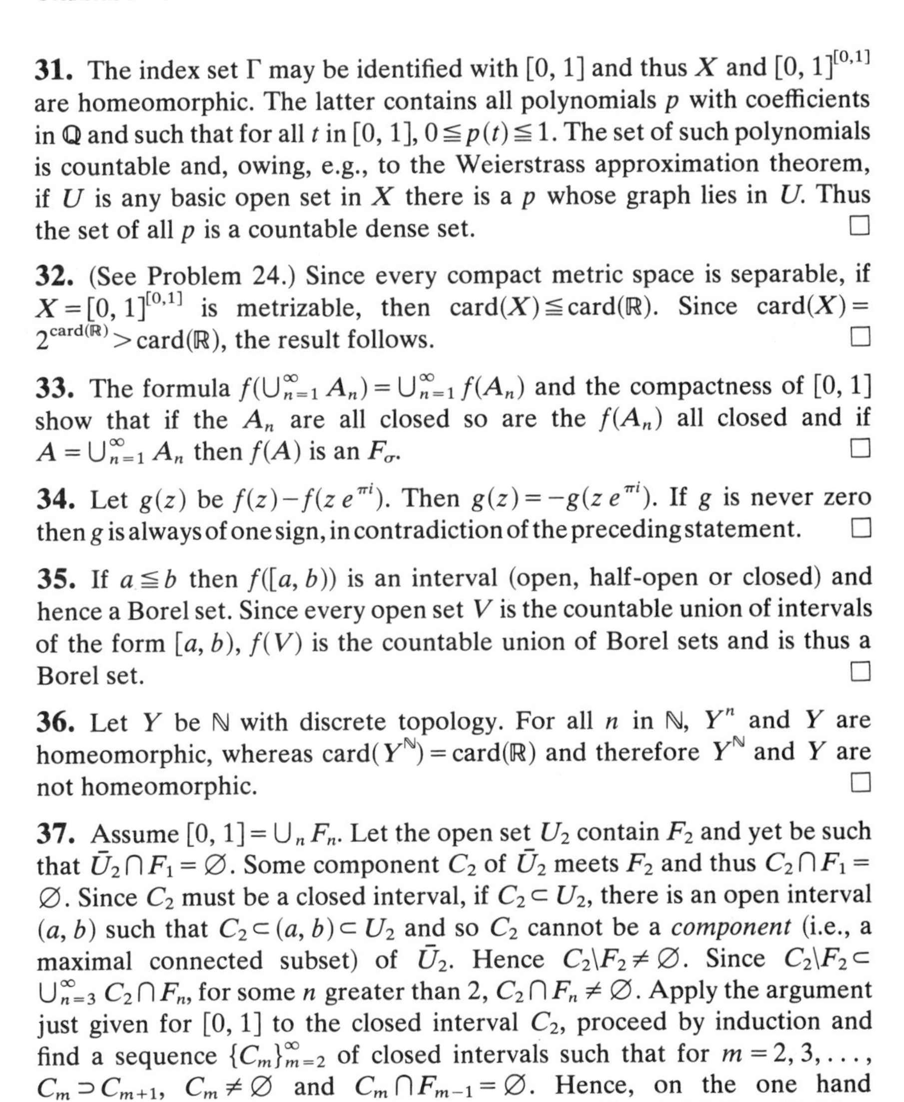

31. The index set Γ may be identified with $[0, 1]$ and thus X and $[0, 1]^{[0,1]}$ are homeomorphic. The latter contains all polynomials p with coefficients in $\mathbb{Q}$ and such that for all t in $[0, 1]$, $0 \leqq p(t) \leqq 1$. The set of such polynomials is countable and, owing, e.g., to the Weierstrass approximation theorem, if U is any basic open set in X there is a p whose graph lies in U. Thus the set of all p is a countable dense set. □

32. (See Problem 24.) Since every compact metric space is separable, if $X = [0, 1]^{[0,1]}$ is metrizable, then $\text{card}(X) \leqq \text{card}(\mathbb{R})$. Since $\text{card}(X) = 2^{\text{card}(\mathbb{R})} > \text{card}(\mathbb{R})$, the result follows. □

33. The formula $f(\bigcup_{n=1}^{\infty} A_n) = \bigcup_{n=1}^{\infty} f(A_n)$ and the compactness of $[0, 1]$ show that if the A_n are all closed so are the $f(A_n)$ all closed and if $A = \bigcup_{n=1}^{\infty} A_n$ then $f(A)$ is an F_σ. □

34. Let $g(z)$ be $f(z) - f(z\, e^{\pi i})$. Then $g(z) = -g(z\, e^{\pi i})$. If g is never zero then g is always of one sign, in contradiction of the preceding statement. □

35. If $a \leqq b$ then $f([a, b))$ is an interval (open, half-open or closed) and hence a Borel set. Since every open set V is the countable union of intervals of the form $[a, b)$, $f(V)$ is the countable union of Borel sets and is thus a Borel set. □

36. Let Y be $\mathbb{N}$ with discrete topology. For all n in $\mathbb{N}$, Y^n and Y are homeomorphic, whereas $\text{card}(Y^{\mathbb{N}}) = \text{card}(\mathbb{R})$ and therefore $Y^{\mathbb{N}}$ and Y are not homeomorphic. □

37. Assume $[0, 1] = \bigcup_n F_n$. Let the open set U_2 contain F_2 and yet be such that $\bar{U}_2 \cap F_1 = \emptyset$. Some component C_2 of $\bar{U}_2$ meets F_2 and thus $C_2 \cap F_1 = \emptyset$. Since C_2 must be a closed interval, if $C_2 \subset U_2$, there is an open interval (a, b) such that $C_2 \subset (a, b) \subset U_2$ and so C_2 cannot be a *component* (i.e., a maximal connected subset) of $\bar{U}_2$. Hence $C_2 \backslash F_2 \neq \emptyset$. Since $C_2 \backslash F_2 \subset \bigcup_{n=3}^{\infty} C_2 \cap F_n$, for some n greater than 2, $C_2 \cap F_n \neq \emptyset$. Apply the argument just given for $[0, 1]$ to the closed interval C_2, proceed by induction and find a sequence $\{C_m\}_{m=2}^{\infty}$ of closed intervals such that for $m = 2, 3, \ldots$, $C_m \supset C_{m+1}$, $C_m \neq \emptyset$ and $C_m \cap F_{m-1} = \emptyset$. Hence, on the one hand $\bigcap_m C_m \neq \emptyset$, and on the other hand, for n in $\mathbb{N}$, $(\bigcap_m C_m) \cap F_n = \emptyset$. Thus $(\bigcap_m C_m) \cap (\bigcup_n F_n) = (\bigcap_m C_m) = \emptyset$. The contradiction implies the result. □

3. Limits

38. Let f_n be $x \mapsto 1+(1-x)+\cdots+(1-x)^{n-1}$. Then $1-(1-x)^n = xf_n(x)$ and thus

$$\sum_{k=1}^{\infty}(1-(1-2^{-k})^n) = \sum_{k=1}^{\infty} 2^{-k}\left(\sum_{p=0}^{n-1}(1-2^{-k})^p\right) = \sum_{k=1}^{\infty} 2^{-k} f_n(2^{-k}).$$

From graphical considerations

$$\frac{1}{2}\sum_{k=2}^{\infty} 2^{-k} f_n(2^{-k}) \leqq \int_0^1 f_n(x)\,dx \leqq \sum_{k=1}^{\infty} 2^{-k} f_n(2^{-k})$$

and by direct integration $\int_0^1 f_n(x)\,dx = \sum_{p=1}^{n} p^{-1}$. Graphical considerations also show $\log n \leqq \sum_{p=1}^{n} p^{-1} \leqq 1+\log n$. Thus

$$\log n \leqq \sum_{p=1}^{n} p^{-1} \leqq \int_0^1 f_n(x)\,dx \leqq \sum_{k=1}^{\infty} 2^{-k} f_n(2^{-k}) = \tfrac{1}{2} f_n(\tfrac{1}{2}) + \sum_{k=2}^{\infty} 2^{-k} f_n(2^{-k})$$

$$< 1 + 2\int_0^1 f_n(x)\,dx \leqq 3 + 2\log n \leqq ((3/\log 2)+2)\log n. \qquad \square$$

39. If $x \in (0, 1)$ define $n_1, n_2, \ldots$ inductively as follows: n_1 is the least n such that $a_n \leqq x$; then $\sum_{k=n_1}^{\infty} a_k \geqq x$ since otherwise $a_{n_1-1} < x$; if $\sum_{k=n_1}^{\infty} a_k = x$, stop; otherwise let n_2 be the greatest n such that $\sum_{k=n_1}^{n} a_k \leqq x$; if equality holds, stop; otherwise let n_3 be the first n greater than n_2 such that $\sum_{k=n_1}^{n_2} a_k + a_n \leqq x$, etc. Note that $n_3 > n_2 + 1$ and that $\sum_{k=n_1}^{n_2} a_k + \sum_{k=n_3}^{\infty} a_k \geqq x$ for reasons like those given earlier. If the process does not stop after finitely many steps then the differences between x and the finite sums are all of the form $\sum_{k=m}^{\infty} a_k$. These sums approach 0 as $m \to \infty$. $\square$

40. For large enough n, $|a_n/b_n|<1$ and so for such large n,

$$a_n/(a_n+b_n)=a_nb_n^{-1}(1-a_n/b_n)/(1-(a_n/b_n)^2)$$
$$=a_nb_n^{-1}(1-a_n/b_n)\left(1+\sum_{k=1}^{\infty}(a_n/b_n)^{2k}\right).$$

For large n, $\sum_{k=1}^{\infty}(a_n/b_n)^{2k}$, denoted δ_n, is small. Hence $\sum_{n=1}^{\infty}a_n/(a_n+b_n)=\sum_{n=1}^{\infty}(a_n/b_n-(a_n/b_n)^2)(1+\delta_n)$; the hypotheses and the fact that the δ_n are all nonnegative and are ultimately small yield the result. □

41. Let a_n be $(1+(-1)^n/\log(n+1))/n$, $n=1,2,\ldots$. Then $\sum_{n=1}^{\infty}(-1)^na_n=\sum_{n=1}^{\infty}(-1)^n/n+\sum_{n=1}^{\infty}1/n\log(n+1)$. The first member is a convergent and the second member is a divergent series. Clearly $na_n\to 1$ as $n\to\infty$. □

42. It may be assumed that $b_1>1$. If $\varepsilon_n=((-1)^n\sum_{k=1}^{n}b_k)^{-1}$, then Abel's theorem shows that $\sum_{n=1}^{\infty}(-1)^n\varepsilon_nb_n$ diverges whereas the alternating series theorem shows that $\sum_{n=1}^{\infty}(-1)^nb_n$ converges. If $a_n=b_n(1+\varepsilon_n)$ the result follows. □

43. Let q_n be $[0,1]\ni x\mapsto x^{n-1}$, $n=1,2,\ldots$ and let $\{f_n\}_{n=1}^{\infty}$ be the set of orthonormal polynomials derived from the q_n *via* the Gram–Schmidt process. Then for n in $\mathbb{N}$, $p_n=\sum_{m=1}^{M}a_{nm}f_m$, and $a_{nm}=\int_0^1p_n(x)f_m(x)\,dx$. Since all p_n are orthogonal to all f_m if $m>M$ and since $p_n\to f$ uniformly as $n\to\infty$, it follows that $f=\sum_{m=1}^{M}a_mf_m$ for $a_m=\int_0^1f(x)f_m(x)\,dx$. Thus f is a polynomial. Since $q_n\to\chi_{\{1\}}$ as $n\to\infty$, pointwise convergence is insufficient to imply that f is a polynomial. □

44. Let $F(x)$ be $x\mapsto\int_x^{\infty}e^{-t^2/2}\,dt\,.\,e^{x^2/2}$. Then F has continuous derivatives of all orders ($F\in C^{\infty}([0,\infty),\mathbb{R})$) and for all x, $F'(x)=-1+xF(x)$. From graphical considerations it follows that for all x in $[0,\infty)$, $F(x)\leqq e^{x^2/2}\sum_{k=0}^{\infty}e^{-(x+k)^2/2}$ whence $F(x)\leqq e^{-x}\sum_{k=0}^{\infty}e^{-k^2/2}$ and so $F(x)\to 0$ and $xF(x)\to 0$ as $x\to\infty$.

To show F is monotone decreasing on $[0,\infty)$ it suffices to show that $F'(x)<0$ for all x in $[0,\infty)$. The equation $F'(x)=-1+xF(x)$ and the previous remarks imply that $F'(0)=-1$ and $F'(x)\to -1$ as $x\to\infty$. Hence F' has a finite maximum M on $[0,\infty)$. If $M=-1$ then $F'<0$ on $[0,\infty)$. If $M\neq -1$ let $F'(a)=M$. Then $a\neq 0$, $F''(a)=0$, and, since $F''(x)=-x+(1+x^2)F(x)$, $F(a)=a/(1+a^2)$. Thus $F'(a)=M=-1/(1+a^2)<0$ and the result follows. □

45. Let ε_n be $\sum_{k=n+1}^{\infty}(k!)^{-1}$, $n=1,2,\ldots$. There is a sequence δ_n such that $\delta_n\downarrow 0$ and $\varepsilon_n=(1+\delta_n)/(n+1)!$. Thus $n\sin(2\pi en!)=n\sin(2\pi\varepsilon_nn!)=(n(1+\delta_n)/(1+n))((\sin(2\pi(1+\delta_n)/(n+1))/((1+\delta_n)/(1+n))$. The last expression approaches 2π as $n\to\infty$. □

46. For n in $\mathbb{N}$ let E_n be $\{x:|\sin(x+r_n\pi/n)|=1\}$ and let δ_n be positive. Each E_n is countable and is contained in an open set U_n such that $\lambda(U_n)<\delta_n$.

Hence off $\bigcup_{k=1}^{\infty} U_k$, $(\sin(x+r_n\pi/n))^n \to 0$ as $n\to\infty$. By the dominated convergence theorem, $\int_{[0,\infty)\setminus(\bigcup_k U_k)} e^{-x}\,(\sin(x+r_n\pi/n))^n\,dx \to 0$ as $n\to\infty$. If e.g., $\delta_n = \delta/2^n$ for a fixed positive δ, then $\int_{[0,\infty)\cap\bigcup_k U_k} e^{-x}\,(\sin(x+r_n\pi/n))^n\,dx < \delta$, from which estimates the result follows. □

47. It may be assumed that $0<\varepsilon<1$. Then $|1-e^{(\varepsilon x)^2}| \leqq 2\varepsilon x\, e^{(\varepsilon x)2}$ and $|1-e^{(\varepsilon x)^2}|\, e^{-x^3} \leqq 2\varepsilon x\, e^{-x^2(x-\varepsilon^2)}$. Thus if $x>2$, the integrand is dominated by $2\varepsilon x\, e^{-x^2}$ and if $0\leqq x\leqq 2$ the integrand is bounded. As $\varepsilon\to 0$ the integrand converges to 0 and the result follows from the dominated convergence theorem. □

48. The dominated convergence theorem implies directly that the limit is 0. □

49. The variable change $y=1-x$ produces: $\int_\varepsilon^{1-\delta} f(x)\,dx = \int_{1-\varepsilon}^{\delta} ((1-y)\log(1-y))/y\,dy$ which in turn is $-y+\sum_{n=2}^{\infty} y^n/n^2(n-1)|_{1-\varepsilon}^{\delta} = 1-\delta-\varepsilon+\sum_{n=2}^{\infty}\delta^n/n^2(n-1)-\sum_{n=2}^{\infty}(1-\varepsilon)^n/n^2(n-1)$. Since

$$\sum_{n=2}^{\infty} 1/n^2(n-1) < \infty$$

still another application of the dominated convergence theorem permits passage to the limit as $\delta, \varepsilon\to 0$ and the stated identity emerges. □

4. Continuous Functions

50. Let $f([0, 1])$ be $[a, b]$. For any x in $[a, b]$ let $f(y)$ be x and define $G(x)$ to be $g(y)$. This G is well-defined on $[a, b]$ since if $f(z)=x$ then $f(y)=f(z)$ and so $g(y)=g(z)$. If $x_n \to x$ as $n \to \infty$ and $G(x_n) \nrightarrow G(x)$ as $n \to \infty$ then by passage to subsequences as needed it may be assumed that for some positive δ and for n in $\mathbb{N}$, $|G(x_n)-G(x)| \geqq \delta$. If $f(y_n)=x_n$, again it may be assumed that $y_n \to y$ as $n \to \infty$. Thus as $n \to \infty$, $f(y_n) \to f(y)$, $g(y_n) \to g(y)$, and $G(x_n) \to G(x)$, a contradiction, whence G is continuous. The Weierstrass approximation theorem implies that there is a sequence $\{p_n\}_{n=1}^{\infty}$ of polynomials such that $p_n \to G$ uniformly on $[a, b]$ as $n \to \infty$. Translated into a statement about functions on $[0, 1]$ the last sentence provides the desired result. □

51. From Solution 50, it is clear that there is a sequence $\{p_n\}_{n=1}^{\infty}$ of polynomials such that $p_n(f) \to g$ uniformly as $n \to \infty$. The hypothesis implies that $L(p_n(f))=0$ for n in $\mathbb{N}$ and so $L(g)=0$. □

52. If M_N is the $\mathbb{R}$-linear span of S_N, then M_N is a real separating algebra of continuous functions on $[0, 1]$, whence, by the Stone–Weierstrass theorem, $\bar{M}_N = C([0, 1], \mathbb{R})$, and the result follows by "complexification". □

53. If $\int_0^1 x^n f(x)\,dx = 0$, $n=0, 1, 2, \ldots$ the Weierstrass approximation theorem implies that $\int_0^1 g(x)f(x)\,dx = 0$ for any continuous g. In particular $\int_0^1 (f(x))^2\,dx = 0$ and since f is $\mathbb{R}$-valued and continuous, $f=0$.

If $\int_0^1 e^{\pm 2\pi inx} f(x)\,dx = 0$, $n=0, 1, \ldots$, let F be $x \mapsto \int_0^x f(t)\,dt$. Then $F(0)=F(1)=0$, $F \in C([0, 1], \mathbb{R})$ and integration by parts shows that if $n \neq 0$, $\int_0^1 e^{\pm 2\pi inx} F(x)\,dx = 0$. Fejér's theorem implies F is constant, whence $F=0$ and $f=F'=0$. □

54. According to the Stone–Weierstrass theorem, for any k in $\mathbb{N}$, the set of all polynomials in x^k is dense in $C([0, 1], \mathbb{C}) \cap \{f: f(0)=0\}$, denoted in this discussion by A. Let $\{g_n\}_{n=1}^{\infty}$ be dense in A and let polynomials p_n be chosen inductively as follows: $\|g_1-p_1\|_\infty<\frac{1}{2}$; $p_1, p_2, \ldots, p_n$ having been chosen, let $d_{n+1}-1$ be the highest of the degrees of $p_1, p_2, \ldots, p_n$ and choose p_{n+1} so that for all x $|g_{n+1}(x)-p_1(x)-p_2(x^{d_2})-\cdots-p_{n+1}(x^{d_{n+1}})|<(\frac{1}{2})^{n+1}$. Then $\sum_{n=1}^{\infty} p_n(x^{d_n})$ is a power series $\sum_{n=1}^{\infty} a_n x^n$. If $f\in A$ and $g_{n_k}\to f$ as $k\to\infty$ then $\sum_{n=1}^{n_k} a_n x^n \to f(x)$ uniformly on $[0, 1]$ as $k\to\infty$. □

55. For m in $\mathbb{N}$ let A_m be $\{x: mx\in G\}$. Then $D=\bigcap_{n=1}^{\infty}\bigcup_{m=n}^{\infty} A_m$. Since each A_m is open so is each $\bigcup_{m=n}^{\infty} A_m$ open. The following argument shows each of the latter is dense in $[0, \infty)$.

Otherwise there is in $\mathbb{N}$ an n_0 and there are in $(0, \infty)$ an a and a y greater than a and such that $(y-a, y+a)\cap(\bigcup_{m=n_0}^{\infty} A_m)=\emptyset$. Choose m_0 in $\mathbb{N}$ so that $a>1/m_0$. If $m_1>\max(m_0, n_0, (y-a)/2a)$ then $m_1(y+a)>(m_1+1)(y-a)$ and then $\bigcup_{m=m_1}^{\infty} m(y-a, y+a)=(m_1(y-a), \infty)$ which must meet G since G is unbounded. Thus there is in $[m_1, \infty)\cap\mathbb{N}$ an m_2 and in $(y-a, y+a)$ a z such that $m_2 z\in G$, i.e., $z\in A_{m_2}\subset\bigcup_{m=n_0}^{\infty} A_m$. In short, $(y-a, y+a)\cap(\bigcup_{m=n_0}^{\infty} A_m)\neq\emptyset$ and a contradiction emerges.

The Baire category theorem implies D is dense. □

56. If there is a positive δ and a sequence $\{x_n\}_{n=1}^{\infty}$ such that $x_n\to\infty$ as $n\to\infty$ and $|f(x_n)|\geqq\delta$ for all n, then since f is continuous there is for each n an open set U_n containing x_n and on which $|f(x)|>\delta/2$. Thus if $G=\bigcup_{n=1}^{\infty} U_n$ then G is an open unbounded set in $[0, \infty)$ and thus by Solution 55, the corresponding set D is dense.

Hence $(a, b)\cap D\neq\emptyset$ and if $h\in(a, b)\cap D$ then there are in $\mathbb{N}$ sequences $\{m_k\}_{k=1}^{\infty}$ and $\{n_k\}_{k=1}^{\infty}$ such that $m_k h\in U_{n_k}$, $k=1, 2, \ldots$. Hence $|f(m_k h)|>\delta/2$, $k=1, 2, \ldots$, in contradiction of the hypothesis that implies $f(m_k h)\to 0$ as $k\to\infty$. □

57. i) For n in $\mathbb{N}$ let E_n be $\{x: f_n(x)=0\}$. All E_n are closed and their union is $[0, 1]$, whence the Baire category theorem implies that some E_{n_0} contains a closed interval $[a, b]$ of positive length. For x in $[a, b]$, $f_{n_0}(x)=\int_0^x f_{n_0-1}(t)\,dt=0$, whence $f_{n_0-1}(x)=0$ for x in $[a, b]$, whence ..., whence $f_0(x)=0$ for x in $[a, b]$.

ii) If $f_0(0)=\delta>0$, then, for some positive b, $f_0(x)\geqq\delta/2$ if $x\in[0, b]$. Consequently, for n in $\mathbb{N}$, $f_n(b)\geqq\delta b^n/2n$, a contradiction. If $f_0(0)<0$ a similar contradiction is derivable, and so $f_0(0)=0$. If f_0 is not identically zero and yet is of one sign in $(0, b]$ then f_n is never zero in $(0, b]$ for all n in $\mathbb{N}$, again a contradiction. Similar arguments show that for all n either f_n is identically zero in some nondegenerate interval $(0, b]$ or that f_n undergoes infinitely many changes of sign (and hence has infinitely many zeros) in $(0, b]$. □

58. The hypothesis immediately implies that g is one-one and thus is strictly monotone increasing. If, for some x, $g(x)>x$ then for all n in $\mathbb{N}$, $g^n(x)>g^{n-1}(x)>\cdots>g(x)>x$, a contradiction if $n=m$. A similar argument negates the possibility $g(x)<x$ for some x. □

59. i) Let $f(x)$ be

$$\begin{cases} 1/x, 0<x\leqq 1 \\ 0, x=0 \end{cases}.$$

Then f is left-continuous and yet unbounded.

ii) If, for each n in $\mathbb{N}$ there is in $[0, 1]$ an x_n such that $f(x_n)>n$, by passage to a subsequence it may be assumed that $x_n \to x_0$ as $n\to\infty$ whence $\limsup_{y=x_0} f(y)=\infty$ and this contradiction shows that f is bounded above. □

60. Let $f^{\cdot n}$ represent $\underbrace{f.f.\ldots.f}_{n}$. If $\|f\|_\infty \geqq \delta \geqq 1$, then since $f(0)=0$ there is in $(0, 1]$ some x_0 such that $|f(x_0)|=1$ and $|f(x)|<1$ if $x\in[0, x_0)$. Hence on $[0, x_0)$, $|f^{\cdot n}|\to 0$ as $n\to\infty$ whereas $|f^{\cdot n}(x_0)|=1$ for all n; $\{f^{\cdot n}\}_{n=1}^{\infty}$ is not equicontinuous.

Conversely, if $\|f\|_\infty=\varepsilon<1$ and if $\delta>0$ then if $2\varepsilon^{n_0}<\delta$ for x, y in $[0, 1]$ $|f^{\cdot n}(x)-f^{\cdot n}(y)|<\delta$ if $n\geqq n_0$. Thus $\{f^{\cdot n}\}_{n=n_0}^{\infty}$ is equicontinuous and the result follows. □

61. Let $L_n(f)$ be $(\int_0^1 x^n f(x)\,dx)/(\int_0^1 x^n\,dx)$, n in $\mathbb{N}$. Then, for all n, $L_n\in C([0, 1], \mathbb{R})^*$ and $|L_n(f)|\leqq\|f\|_\infty$. Furthermore if p is a polynomial then $L_n(p)\to p(1)$ as $n\to\infty$. Owing to the Weierstrass approximation theorem, $L_n(f)\to f(1)$ as $n\to\infty$. □

62. Since f is continuous, if $\varepsilon, |\delta|>0$, there is in $\mathbb{Q}$ an h_δ such that $h_\delta\neq 0$, $|h_\delta-\delta|\leqq\delta^2$, $|f(c+\delta)-f(c+h_\delta)|<\varepsilon|\delta|$. Thus

$$\left|\frac{f(c+\delta)-f(c)}{\delta}-L\right|\leqq\left|\frac{f(c+\delta)-f(c+h_\delta)}{\delta}\right|+\left|\frac{f(c+h_\delta)-f(c)}{h_\delta}\,.\,h_\delta/\delta-L\right|.$$

For small $|\delta|$ the second term is small since h_δ/δ is near 1 and the first is less than ε. Since ε is arbitrary and positive the result follows. □

63. If $\text{supp}(h)\subset[a, b]$ and if $F(x)=\int_a^x f(t)\,dt$ then integration by parts shows $\int_{-\infty}^{\infty} f(t)h(t)\,dt=-\int_a^b F(t)h'(t)\,dt=-\int_a^b g(t)h'(t)\,dt$, whence $F=g$, and since $F'=f$, g' exists and $g'=f$. □

64. If there is no such K then assume that for each n in $\mathbb{N}$ there is in A an f_n such that $\|f_n''\|_\infty>n$ and hence, for some x_n, $|f_n''(x_n)|>n$. Since $\|f_n'''\|_\infty\leqq 1$, if for any x_0, $|f_n''(x_0)|\leqq n-1$, integration shows that, for all x, $|f_n''(x)|\leqq n$, a contradiction. Thus $|f_n''(x)|>n-1$ for all x. Since f_n'' is continuous it is of one sign, say $f_n''>n-1$. Hence f_n' is strictly increasing. If $f_n'(0)\geqq-(n-1)/2$ integration shows $f_n'(1)>(n-1)/2$; in short $\|f_n'\|_\infty>(n-1)/2$. Let $f_n'(0)$ be a_n, $f_n'(1)$ be b_n. Integration in each of the situations represented by i) $b_n\leqq 0$, ii) $a_n\geqq 0$ and iii) $a_n<0<b_n$ shows that for i) and ii) $n\leqq 5$ and for iii) $n\leqq 16$. All these are contradictions if $n>16$.

Hence, for some K_1 and all f in A, $\|f''\|_\infty\leqq K_1$ and thus for all f in A, $|f'(x)|\leqq|f'(0)|+K_1$, $yf'(0)=f(y)-f(0)-\int_0^y(\int_0^x f''(t)\,dt)\,dx$ whence $|f'(0)|\leqq 2+K_1$ and finally $\|f'\|_\infty\leqq 2(1+K_1)$, which serves for K. □

65. If $\|f_n\|_\infty \not\to 0$ as $n \to \infty$ then for some positive δ, some x_0 and some sequence $\{x_n\}_{n=1}^\infty$, it may be assumed that $|f_n(x_n)| \geqq \delta$ and $x_n \to x_0$ as $n \to \infty$. It may be assumed also that all $f_n(x_n)$ are positive. The formula $f_n(x_n) - f_n(x_0) = \int_{x_0}^{x_n} f'_n(x)\,dx$ and the inequality $\|f'_n\|_\infty \leqq 1$, both valid for all n in $\mathbb{N}$, show that for all n, $f_n(x_0) \geqq f_n(x_n) - |x_n - x_0|$. Thus if, for n greater than some n_0, $|x_n - x_0| < \delta/2$, then for the same n, $f_n(x_0) \geqq \delta/2$. The integral formula now shows that, for x in $[x_0 - \delta/4,\ x_0 + \delta/4]$ and n in (n_0, ∞), $f_n(x) \geqq \delta/4$.

There is in $(0, \delta/8)$ an η such that $\int_{x_0-\delta/4}^{x_0-\delta/4+\eta/2} f_n(x)\,dx + \int_{x_0+\delta/4-\eta/2}^{x_0+\delta/4} f_n(x)\,dx < \delta(\delta - \eta)/8$. If g is the piecewise linear and continuous function defined by:

$$g(x) = \begin{cases} 1 & \text{on } [x_0 - \delta/4 + \eta/4,\ x_0 + \delta/4 - \eta/4], \\ 0 & \text{off } [x_0 - \delta/4,\ x_0 + \delta/4] \end{cases}$$

then $\int_0^1 f_n(x) g(x)\,dx \geqq \delta(\delta - \eta)/8$ for n in (n_0, ∞) and the contradiction proves the result. □

66. Let f be $x \mapsto e^{-x^2}$. If $\{p_n\}_{n=1}^\infty$ is a sequence of polynomials and if $\|f - p_n\|_\infty \to 0$ as $n \to \infty$ then, *via* subsequences as needed, it may be assumed that no p_n is constant. Since $|p_n(x)| \to \infty$ as $x \to \infty$ and $f(x) \to 0$ as $x \to \infty$ it follows that $\|f - p_n\|_\infty \not\to \infty$ as $n \to \infty$. □

67. If $\varepsilon_n \downarrow 0$ then for each n in $\mathbb{N}$ there is in $C^\infty(\mathbb{R}, \mathbb{R})$ a nonnegative g_n such that $\|g_n\|_\infty = 1$ and

$$g_n(x) = \begin{cases} 1 & \text{if } |x| \leqq \varepsilon_n/2 \\ 0 & \text{if } |x| \geqq \varepsilon_n \end{cases}.$$

For each n in $\mathbb{N}$ let there be defined $h_{n,0}, h_{n,1}, \ldots, h_{n,n-1}$ according to the following:

$$h_{n,0}(x) = g_n(x)$$

$$h_{n,1}(x) = \int_0^x h_{n,0}(t)\,dt$$

$$\cdots$$

$$h_{n,n-1}(x) = \int_0^x h_{n,n-2}(t)\,dt$$

and let f_n be $a_n h_{n,n-1}$.

Since $\|h_{n,n-1}\|_\infty \leqq \varepsilon_n^{n-1}/(n-1)!$, if $\varepsilon_n |a_n| \to 0$ as $n \to \infty$ then the function f equal to $\sum_{n=1}^\infty f_n$ is in $C^\infty(\mathbb{R}, \mathbb{R})$ and for all k, $f^{(k)} = \sum_{n=1}^\infty f_n^{(k)}$. Since for all n, $f^{(n-1)}(0) = f_n^{(n-1)}(0) = a_n h_{n,0}(0) = a_n$, the result follows. □

68. Clearly the Riemann integrals of f, f^+ and f^- exist over all finite intervals. Since $|f|$ (which is $f^+ + f^-$) is improperly Riemann integrable, so are f^+ and f^- and thus so also is f (which is $f^+ - f^-$). Hence for all finite intervals (a, b), $\int_a^b f(x)\,dx = \int_a^b f(x)\,d\lambda(x)$ and $\int_a^b |f(x)|\,dx = \int_a^b |f(x)|\,d\lambda(x)$ from which the result follows. □

69. For each a in $\mathbb{R}$ the series $\sum_{n=0}^{\infty} b_n(a)(x-a)^{n+1}/n+1$ converges if $|x-a| \leqq r(a)/2$. Let the sum be $g_a(x)$. Then if $|x-a| \leqq r(a)/2$, $\int_a^x f'(t)\,dt = g_a(x)$ and since f' is continuous, $g'_a = f'$, whence $f = g_a + \text{constant}$ and so f is real analytic. □

70. For n in $\mathbb{N}$ let g_n be $g \cdot \chi_{[-n,n]}$. Then $g_n \in L^2(\mathbb{R}, \lambda)$ and for all h in $L^2(\mathbb{R}, \lambda)$ $\lim_{n\to\infty} \int_{-n}^{n} g_n(x)h(x)\,dx$ exists. The uniform boundedness principle applied to the g_n regarded as elements of $L^2(\mathbb{R}, \lambda)^*$ implies that for some finite M, $\|g_n\|_2 \leqq M$. Since $|g_n|^2 \uparrow |g|^2$ the monotone convergence theorem implies the result. □

71. The image $f(\mathbb{R}^n)$ is closed since if $f(x_m) \to y$ as $m \to \infty$ then $\{x_m\}_{m=1}^{\infty}$ is a convergent sequence, say with limit x and thus $f(x) = y$.

Clearly f is one-one and on any closed (hence compact) ball in $\mathbb{R}^n$, f is a homeomorphism, whence by Brouwer's theorem on the invariance of domain, f is an open map and so $f(\mathbb{R}^n)$ is also open. Since $\mathbb{R}^n$ is connected and $f(\mathbb{R}^n)$ is nonempty, $f(\mathbb{R}^n) = \mathbb{R}^n$. □

72. The Stone–Weierstrass theorem implies that the compactness of F is sufficient. If F is not compact, any nonconstant polynomial in x^2 is unbounded and cannot approximate uniformly any continuous bounded function on F. □

73. If $f(0) = \|f\|_\infty = M$ and if $|f(x_0| < M$ for some x_0 in U, then $M = \oint_{\|y\|=\|x_0\|} f(y)\,dy/2\pi\|x_0\|$. Since f is continuous, $|f(y)| < M$ on an open arc containing x_0; hence the integral is less than M and a contradiction results. □

74. There is a positive δ such that $|f(x)-f(y)| < 1$ if $|x-y| < \delta$. For any x in (a, b), $|x - \frac{1}{2}(a+b)| < \frac{1}{2}(b-a) = [(b-a)/\delta]\frac{1}{2}\delta$. If n_0 in $\mathbb{N}$ is greater than $(b-a)/\delta$, then for any x in (a, b) and for some m not greater n_0 there is a sequence $x_1(=\frac{1}{2}(a+b))$, $x_2, \ldots, x_m(=x)$ such that $|x_k - x_{k+1}| < \delta$, $k = 1, 2, \ldots, m-1$. Hence

$$|f(x) - f(\tfrac{1}{2}(a+b))| \leqq \sum_{k=1}^{m} |f(x_k) - f(x_{k-1})| \leqq n_0$$

and the result follows. □

75. If $f \in A$ let g be the sum of the uniformly convergent series $\sum_{n=-\infty}^{\infty} na_n e^{inx}$. Then g is continuous and if $G(x) = \int_0^x g(t)\,dt$ for x in $[0, 2\pi]$, then $iG(x) = \sum_{n=-\infty}^{\infty} a_n e^{inx}$. Since $\sum_{n=-\infty}^{\infty} |a_n| < \infty$, Solution 53 implies that $f(x) = \sum_{n=-\infty}^{\infty} a_n e^{inx}$ and so $f \in C^1([0, 2\pi], \mathbb{C})$ and indeed $f(0) = f(2\pi)$. Furthermore $\|f\|_\infty + \|f'\|_\infty \leqq 1$ and so if $|x-y| < \varepsilon$, $|f(x) - f(y)| < \varepsilon$. Thus A is an equicontinuous set of functions contained in the closed unit ball of $C([0, 2\pi], \mathbb{C})$. The Arzelà–Ascoli theorem implies $\bar{A}$ is compact. □

76. The hypothesis implies that the Fourier series for f converges to f uniformly in $[0, 2\pi]$. If $\{c_n\}_{n=-\infty}^{\infty}$ is the set of Fourier coefficients for f, then $\|f\|_2^2 = \sum_{n=-\infty}^{\infty} 2\pi|c_n|^2$. The set of Fourier coefficients of f' is $\{inc_n\}_{n=-\infty}^{\infty}$ and so $\|f'\|_2^2 = \sum_{n=-\infty}^{\infty} 2\pi n^2|c_n|^2$ whence $\|f\|_2 \leqq \|f'\|_2$. Equality obtains iff $c_n = 0$ whenever $|n| > 1$. Since $\int_0^{2\pi} f(x)\, dx = 0$, the result follows. □

77. Since $\int_0^1 = \sum_{k=0}^{n-1} \int_{k/n}^{(k+1)/n}$, $n = 1, 2, \ldots$, it follows that

$$\int_0^1 f(x)g(nx)\, dx = (1/n) \sum_{k=0}^{n-1} \int_0^1 f((x+k)/n)g(x)\, dx$$

$$= \int_0^1 g(x) \sum_{k=0}^{n-1} f((x+k)/n) \,.\, 1/n\, dx.$$

However the sum in the last integral converges uniformly on $[0, 1]$ to $\int_0^1 f(x)\, dx$ as $n \to \infty$, whence the result follows. □

78. The trivial case ($f = 0$) aside, it may be assumed that $\|f\|_\infty = 1$. Then A is an equicontinuous subset of the unit ball of $C(\mathbb{T}, \mathbb{R})$ and so the convex hull $\text{conv}(A)$ of A is also an equicontinuous subset of the same unit ball. Let $F(x) = (1/2\pi) \int_0^{2\pi} f(xy)\, dy$. Clearly $F \in \overline{\text{conv}(A)}$ and for all z in $\mathbb{T}$, $F(zx) = F(x)$, whence F is constant.

On the other hand if G is constant and in $\overline{\text{conv}(A)}$ then G is, for any positive δ, δ-approximable by a convex linear combination of translates of f. Integration of the corresponding inequality then shows $|G - F| < \delta$ and the result follows: $G = F$. □

79. For n in $\mathbb{N}$ let U_n be $\{x : \sup_{f \in \mathcal{F}} |f(x)| \leqq n\}$. Then at least one U_n, say U_{n_0}, is not nowhere dense, i.e., $\overline{U_{n_0}}$ contains a nonempty open set V. If $x \in V$, there is in U_{n_0} a sequence $\{x_n\}_{n=1}$ such that $x_n \to x$ as $n \to \infty$. Thus $f(x_n) \to f(x)$ as $n \to \infty$ whence $|f(x)| \leqq n_0$; n_0 and V are respectively the required M and Ω. □

80. Since any L may be viewed as a complex measure μ on X, i.e., for some μ and all f, $L(f) = \int_X f(x)\, d\mu(x)$, the bounded convergence theorem implies the result. □

81. The subset $A_{\mathbb{R}}$ (of A) consisting of sums $\sum_{i=1}^{n} f_i g_i$ in which the functions f_i, g_i are $\mathbb{R}$-valued is a separating algebra and hence is dense in $C(X \times Y, \mathbb{R})$. The result follows by complexification. □

82. For x in C and n in $\mathbb{N}$ let $N_n(x)$ be $\{y : |f(y) - f(x)| < 1/n, |x - y| < 1/n\}$. Then $N_n(x)$ contains an open set $U_n(x)$ containing x. If $G_n = \bigcup_{x \in C} U_n(x)$ then G_n is open and it will be shown that $C = \bigcap_{n=1}^{\infty} G_n$.

Since $G_n \supset C$ for all n it suffices to show that $\bigcap_{n=1}^{\infty} G_n \subset C$. If $y \in (\bigcap_{n=1}^{\infty} G_n) \backslash C$ there is a sequence $\{y_m\}_{m=1}^{\infty}$ and a positive δ such that $y_m \to y$ as $m \to \infty$ and for all m, $|f(y_m) - f(y)| \geqq \delta$. But for each n in $\mathbb{N}$ there is in C an x_n such that $y \in U_n(x_n)$. Hence there is an m_n such that if $m \geqq m_n$ then $y_m \in U_n(x_n)$ whence $|f(y_m)-f(y)|<2/n$. If $2/n<\delta$ a contradiction results. □

83. Let f be an extreme point of B_1. If $g \in B_1$ and $g \neq 0$ then $-g \in B_1$ and $0=\frac{1}{2}g+\frac{1}{2}(-g)$ and so 0 is not an extreme point, in particular $f \neq 0$. Hence at least one of $\max(f)$ and $\min(f)$ is not zero. Let Δ be $\max(f)-\min(f))/3$. If $E_M=\{x: f(x) \geqq \max(f)-\Delta\}$ and $E_m=\{x: f(x) \leqq \min(f)+\Delta\}$ then E_m and E_M are nonempty disjoint compact sets and thus are contained in disjoint compact neighborhoods U_m resp. U_M. Choose in $C_0(X, \mathbb{R})$ an ε such that: i) $\varepsilon \geqq 0$; ii) $\varepsilon=0$ on $U_m \cup U_M$; iii) $\varepsilon \neq 0$; iv) $\|\varepsilon\|_\infty<\Delta/6$. Then if $g_\pm=f \pm \varepsilon$ it follows that $g_\pm \in B_1$ and $f=\frac{1}{2}g_+ + \frac{1}{2}g_-$, whence f is not an extreme point and the contradiction shows that the set of extreme points of B_1 is empty. □

84. i) If $f_C: A \mapsto C$ is a continuous surjection of A onto *the* Cantor set C, then, since C is totally disconnected and $\mathrm{card}(C)=\mathrm{card}(\mathbb{R})$, for each x in C, $f_C^{-1}(x)$ contains a component of A and so $\mathrm{card}(\mathscr{C})=\mathrm{card}(\mathbb{R})$.

ii) Each component of A is either a point or a nondegenerate closed interval. There are at most countably many $\{[a_n, b_n]: 0 \leqq a_n<b_n \leqq 1, 1 \leqq n<N \leqq \infty\}$ of the latter. (If $N=1$ there are no nondegenerate closed intervals among the components.) Thus $B=A \backslash A^0=A \backslash \bigcup_{n<N}(a_n, b_n)$, B is closed and nowhere dense and, by i), $\mathrm{card}(B)=\mathrm{card}(\mathbb{R})$. The Cantor–Bendixson theorem implies B is the disjoint union of a set P, homeomorphic to C, and a countable set D containing no nonempty subset dense in itself. If $U=\bigcup_{n<N}(a_n, b_n)$ then P, D, U are pairwise disjoint and $A=P \cup D \cup U$. Furthermore, for all n in question $f_C([a_n, b_n])$ is a single point whence $f_C(B)=f_C(A)$.

iii) Since any compact metric space K is the continuous image of C [15], there is a continuous surjection $h_K: C \mapsto K$. If $g_K=h_K \circ f_C|_B$ then $g_K: B \rightarrow K$ is a continuous surjection. □

85. If $f_0=0$ and f_n is

$$x \mapsto \begin{cases} nx, & 0 \leqq x \leqq 1/n \\ n(-x+2/n), & 1/n \leqq x \leqq 2/n, \\ 0, & 2/n \leqq x \leqq 1 \end{cases}$$

n in $\mathbb{N}$, then $\|f_n\|_\infty \leqq 1$ and for all x in $[0,1]$, $f_n(x) \rightarrow 0$ as $n \rightarrow \infty$. If $L \in C([0,1], \mathbb{C})^*$ there is complex Borel measure μ such that for all f in $C(0,1], \mathbb{C})$, $L(f)=\int_0^1 f(x)\, d\mu(x)$. The bounded convergence theorem implies $L(f_n) \rightarrow 0$ as $n \rightarrow \infty$ whereas $f_n(1/n)=1$ for all n. □

86. Let L be $C([0,1], \mathbb{C}) \ni f \mapsto \sum_{n=2}^\infty(-1)^n f(1-1/n) 2^{-(n-1)}$. Then if $\|f\|_\infty \leqq 1$, $|L(f)| \leqq \sum_{n=2}^\infty 2^{-(n-1)}=1$ as required. If $k=2,3,\ldots$, there is in

$C([0, 1], \mathbb{C})$ an f_k such that $f_k(1-1/n)=(-1)^n$, $n=2, 3, \ldots k$. Then $L(f_k)=1-2^{-(k-1)}+\sum_{n=k+1}^{\infty}(-1)^n f_k(1-1/n)2^{-n}$. Since the f_k may be chosen so that $\|f_k\|_\infty=1$ and $f_k(1-1/n)-0$ if $n>k$, it follows that $\sup\{|L(f)|=\|f\|_\infty \leqq 1\}=1$.

On the other hand, if $\|f\|_\infty \leqq 1$ and $|L(f)|=1$ then for some θ in $[0, 2\pi)$, $L(f)=e^{i\theta}$ and $L(e^{-i\theta}f)=1$. It follows that $f(1-1/n)=(-1)^n e^{i\theta}$ and so $\lim_{x\to 1} f(x)$ does not exist, i.e., $f\notin C([0, 1], \mathbb{C})$. The contradiction yields the result. □

87. i) If there is in $C([0, 1], \mathbb{C})$ a k such that $A=k\,.\,C([0, 1], \mathbb{C})$ then $k(\frac{1}{2})=0$. If $|k|=r$ then $r^{1/2}\in A$ whence for some g in $C([0, 1], \mathbb{C})$, $r^{1/2}=kg=r|g|$. But then for x different from $\frac{1}{2}$, $|g(x)|=(r(x))^{-1/2}$, in contradiction of the boundedness of g.

ii) If h is $x\mapsto x-\frac{1}{2}$, then the Stone–Weierstrass theorem implies $A=\overline{hC([0, 1], \mathbb{C})}$. □

88. If B is a bounded set in $C^1([0, 1], \mathbb{C})$ and if $\{f_n\}_{n=1}^{\infty}\subset B$ then the f_n are both uniformly bounded and equicontinuous (see Problem 75) and the compactness of $\overline{T(B)}$ is implied by the Arzelà–Ascoli theorem. □

89. Clearly $E_1=L^2([0, 1], \lambda)$ and $D(C^1([0, 1], \mathbb{C})\subset L^2([0, 1], \lambda)$. Furthermore $\|D(f)\|'\leqq\|f\|''$ and so D may be extended to a linear continuous map $\tilde{D}: E_2\to L^2([0, 1], \lambda)$.

If $\tilde{D}(f)=0$ there is in $C^1([0, 1], \mathbb{C})$ a sequence $\{f_n\}_{n=1}^{\infty}$ such that $\|f_n-f\|''\to 0$ and $\|D(f_n)\|'\to 0$ as $n\to\infty$. The Schwarz inequality implies that $\int_0^1 |f_n'(x)|\,dx\to 0$ as $n\to\infty$ and the assumption *re* the f_n implies $\int_0^1 [|f_n(x)-f_m(x)|^2+|f_n'(x)-f_m'(x)|^2]\,dx\to 0$ as $n, m\to\infty$. *Via* subsequences as needed it may be assumed that $\lim_{n\to\infty} f_n=f$ a.e. and that $f_n'\to 0$ a.e. as $n\to\infty$. If c_n is such that $f_n(x)=\int_0^x f_n'(t)\,dt+c_n$ for all x in $[0, 1]$ and all n in $\mathbb{N}$, then $|c_n-c_m|\leqq|f_n(x)-f_m(x)|+\int_0^1 |f_n'(t)-f_m'(t)|\,dt$. Hence $c=\lim_{n\to\infty} c_n=f$ a.e. since x may be chosen off the null set on which the sequence $\{f_n\}_{n=1}^{\infty}$ fails to converge. □

90. i) If $\{f_n\}_{n=1}^{\infty}\subset X$ and $\|f_n-f\|^{(1)}\to 0$ as $n\to\infty$ then $\|f_n-f\|_\infty\to 0$ as $n\to\infty$ and so $f\in X$.

ii) For all f, $\|f\|_\infty\leqq\|f\|^{(1)}$. The injection T (see Problem 88), when restricted to X is, by virtue of i) a surjection of Banach spaces, whence T is open. Thus T^{-1} is continuous and so for some positive constant M and all f in X, $\|f\|^{(1)}\leqq M\|f\|_\infty$. Let k be $1/M$ and K be 1.

iii) In Solution 88 T is shown to be compact whence the set $\{f: \|f\|^{(1)}<1\}$ in X is mapped by T into an open set (see ii)) having compact closure. Thus, since a Banach space is finite-dimensional iff some nonempty open set in the space has compact closure [1], X is finite-dimensional. □

91. Clearly $B_K\subset A_K$ and since A_K is closed it follows that $\overline{B_K}\subset A_K$. Let $\{V_n\}_{n=1}^{\infty}$ be a sequence of open sets such that for all n, $V_n\supset V_{n+1}$ and such that $\bigcap_{n=1}^{\infty} V_n=K$. Furthermore for each n let f_n be in A_K and be such that

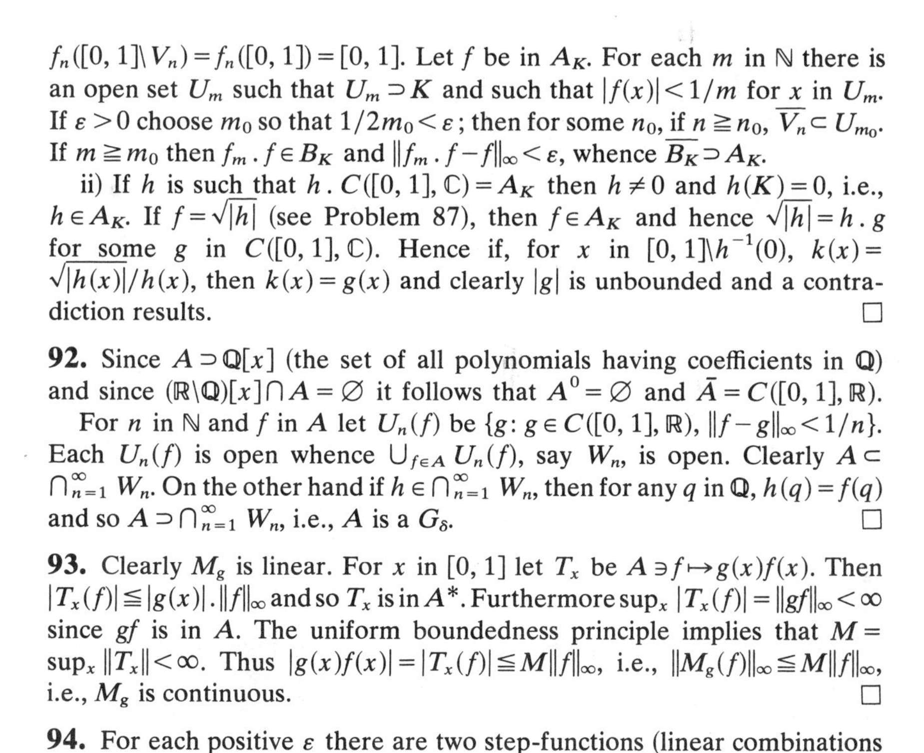

$f_n([0, 1]\setminus V_n) = f_n([0, 1]) = [0, 1]$. Let f be in A_K. For each m in $\mathbb{N}$ there is an open set U_m such that $U_m \supset K$ and such that $|f(x)| < 1/m$ for x in U_m. If $\varepsilon > 0$ choose m_0 so that $1/2m_0 < \varepsilon$; then for some n_0, if $n \geqq n_0$, $\overline{V_n} \subset U_{m_0}$. If $m \geqq m_0$ then $f_m . f \in B_K$ and $\|f_m . f - f\|_\infty < \varepsilon$, whence $\overline{B_K} \supset A_K$.

ii) If h is such that $h . C([0, 1], \mathbb{C}) = A_K$ then $h \neq 0$ and $h(K) = 0$, i.e., $h \in A_K$. If $f = \sqrt{|h|}$ (see Problem 87), then $f \in A_K$ and hence $\sqrt{|h|} = h . g$ for some g in $C([0, 1], \mathbb{C})$. Hence if, for x in $[0, 1]\setminus h^{-1}(0)$, $k(x) = \sqrt{|h(x)|}/h(x)$, then $k(x) = g(x)$ and clearly $|g|$ is unbounded and a contradiction results. □

92. Since $A \supset \mathbb{Q}[x]$ (the set of all polynomials having coefficients in $\mathbb{Q}$) and since $(\mathbb{R}\setminus\mathbb{Q})[x] \cap A = \varnothing$ it follows that $A^0 = \varnothing$ and $\bar{A} = C([0, 1], \mathbb{R})$.

For n in $\mathbb{N}$ and f in A let $U_n(f)$ be $\{g : g \in C([0, 1], \mathbb{R}), \|f - g\|_\infty < 1/n\}$. Each $U_n(f)$ is open whence $\bigcup_{f \in A} U_n(f)$, say W_n, is open. Clearly $A \subset \bigcap_{n=1}^\infty W_n$. On the other hand if $h \in \bigcap_{n=1}^\infty W_n$, then for any q in $\mathbb{Q}$, $h(q) = f(q)$ and so $A \supset \bigcap_{n=1}^\infty W_n$, i.e., A is a G_δ. □

93. Clearly M_g is linear. For x in $[0, 1]$ let T_x be $A \ni f \mapsto g(x)f(x)$. Then $|T_x(f)| \leqq |g(x)| . \|f\|_\infty$ and so T_x is in A^*. Furthermore $\sup_x |T_x(f)| = \|gf\|_\infty < \infty$ since gf is in A. The uniform boundedness principle implies that $M = \sup_x \|T_x\| < \infty$. Thus $|g(x)f(x)| = |T_x(f)| \leqq M\|f\|_\infty$, i.e., $\|M_g(f)\|_\infty \leqq M\|f\|_\infty$, i.e., M_g is continuous. □

94. For each positive ε there are two step-functions (linear combinations of characteristic functions of intervals) l_ε and u_ε such that $l_\varepsilon \leqq f \leqq u_\varepsilon$ and such that $\|l_\varepsilon - u_\varepsilon\|_\infty < \varepsilon$. Hence the Jordan content of the graph of f is less than ε, and since ε is arbitrary (and positive) the result follows. □

5. Functions from $\mathbb{R}^n$ to $\mathbb{R}^m$

95. Assume that $p<q$. The Heine–Borel theorem implies there is a finite set $\{x_k\}_{k=1}^K$ and positive numbers $\{\delta_k\}_{k=1}^K$ such that: i) $x_1-\delta_1<p<x_1<\cdots<x_K<q<x_K+\delta_K$; ii) $\bigcup_{k=1}^K (x_k-\delta_k, x_k+\delta_k)\supset[p,q]$; iii) if $x_{k-1}-\delta_{k-1}<a_k<x_{k-1}<b_k<x_{k-1}+\delta_{k-1}$, then $f(a_k)\leqq f(b_k)$, $k=2,3,\ldots,K+1$. Hence $f(p)\leqq f(q)$ as required. □

96. The functions $\psi: x\mapsto x^2$ and $\varphi: x\mapsto e^{-x}$ are convex whereas $\varphi\circ\psi: x\mapsto e^{-x^2}$ is not. On $[0,\infty)$ ψ is also monotone increasing. □

97. The function $\varphi: x\mapsto 1+|x|$ is convex and positive whereas $\log\varphi$ is not convex for the following reason. If $x>0$ then $(\log\varphi)'(x)=1/(1+x)$, $(\log\varphi)''(x)=-1/(1+x)^2$ although a convex function must have a nonnegative second derivative on any open set where the second derivative exists (see Problem 105). □

98. If $x, y\in\mathbb{R}$, $0\leqq\alpha,\beta$, $\alpha+\beta=1$, then $\log(\varphi)(\alpha x+\beta y)\leqq \alpha\log(\varphi)(x)+\beta\log(\varphi)(y)$. Exponentiating both members of the last inequality and applying the basic "arithmetic vs. geometric mean" inequality (for u, v positive, $u^\alpha v^\beta\leqq\alpha u+\beta v$) yield the desired result. □

99. If $a<c<b$ and if $\alpha=(b-c)/(b-a)$, $\beta=(c-a)/(b-a)$, then $\alpha,\beta>0$, $\alpha+\beta=1$ and $c=\alpha a+\beta b$. Thus

$$(\varphi(b)-\varphi(c))/(b-c)\geqq(\varphi(b)-\alpha\varphi(a)-\beta\varphi(b))/(b-\alpha a-\beta b)$$

$$=(\varphi(b)-\varphi(a))/(b-a).$$

Similarly $(\varphi(c)-\varphi(a))/(c-a)\leqq(\varphi(b)-\varphi(a))/(b-a)$. These inequalities applied first with $x=a$, $x'=c$ and $y=b$ and then with $x'=a$, $y=c$ and $y'=b$ respectively yield the result. The sketch in Figure 3 should prove helpful. □

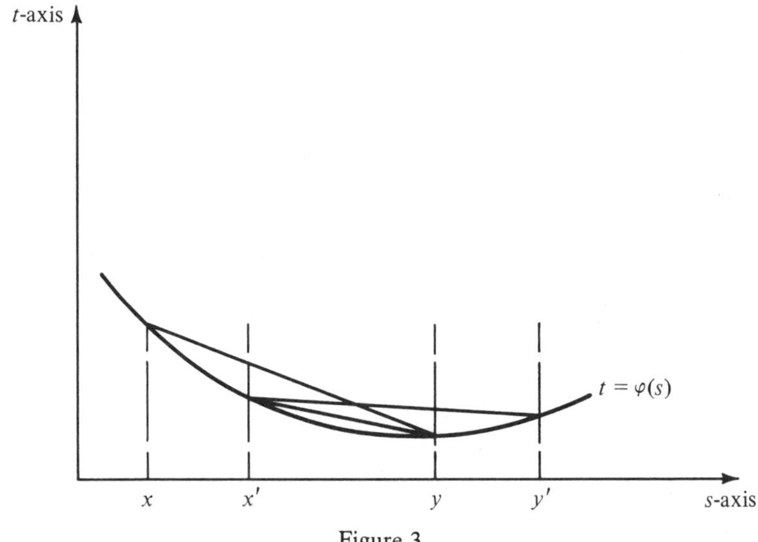

Figure 3

100. For any interval (a, b) if $a<c\leqq x<y\leqq d<b$ then by Solution 99,

$$(\varphi(c)-\varphi(a))/(c-a)\leqq(\varphi(y)-\varphi(x))/(y-x)\leqq(\varphi(b)-\varphi(d))/(b-d)$$

and so $\varphi\in\text{Lip}(1)$ on $[c, d]$, from which the result follows.

The inequalities in Solution 99 also show that the difference quotients $(\varphi(x+h)-\varphi(x))/h$ are monotone (increasing functions of h and so the right- and left-hand derivatives $D_{\pm}\varphi$, of φ exist everywhere. □

101. If φ is neither monotone increasing nor monotone decreasing there must be three points a, b, c, such that $a<b<c$ and either $\varphi(a)<\varphi(b)>\varphi(c)$ or $\varphi(a)>\varphi(b)<\varphi(c)$. Convexity rules out the first possibility and shows that in the remaining possibility if $a''<a'<a$, then $\varphi(a')\geqq\varphi(a)$ (whence $\varphi(a'')\geqq\varphi(a')$) and if $b''>b'>b$ then $\varphi(b')\geqq\varphi(b)$ (whence $\varphi(b'')\geqq\varphi(b')$). Hence on $(-\infty, a)$ φ is monotone decreasing and on (b, ∞) φ is monotone increasing. Let p be $\sup\{a: \varphi \text{ is monotone decreasing on } (-\infty, a)\}$ and let q be $\inf\{b: \varphi \text{ is monotone increasing on } (b, \infty)\}$. Then $p\leqq q$ and $[p, q]$ is the required interval. Examples: $x\mapsto e^x$ is convex and monotone increasing; $x\mapsto e^{-x}$ is convex and monotone decreasing; $x\mapsto x^2$ is convex and neither monotone increasing nor monotone decreasing and its associated p and q are both zero. □

102. Since $D_+\varphi\geqq D_-\varphi$, for each a in $\mathbb{R}$ let γ_a be in $[D_-\varphi(a), D_+\varphi(a)]$. Then $\gamma_a(x-a)+\varphi(a)\leqq\varphi(x)$ ("the curve lies above the supporting line"). Since φ is continuous $\varphi(f)$ is measurable. If $a=\int_0^1 f(t)\,dt$ then $\varphi(f(t))\geqq\gamma_a(f(t)-a)+\varphi(a)$ and so

$$\int_0^1 \varphi(f(t))\,dt\geqq\gamma_a\int_0^1 (f(t)-a)\,dt+\varphi(a)=0+\varphi\left(\int_0^1 f(t)\,dt\right). \qquad \square$$

103. There is a sequence $\{x_n\}_{n=1}^{\infty}$ such that $x_n \downarrow 0$ and $g(x) \geqq n$ if $x \leqq x_n$. Let $\varphi(x_n)$ be $n-1$, $n=1, 2, \ldots$, and let φ be continuous, piecewise linear and nonnegative. Then φ is convex, monotone decreasing, $\varphi \leqq g$, and $\varphi(x) \to \infty$ as $x \to 0$. □

104. Let g be $x \mapsto \log(1+x)$ and assume that the convex function φ is such that $\varphi \leqq g$. Then (see Figure 4), since $-g$ is convex, g is concave and so

$$\varphi(e^n/n) \leqq (\log(1+e^n))/e^n \times e^n/n = 1 + (\log(1+e^{-n}))/n.$$

As $n \to \infty$, $e^n/n \to \infty$ and $1 + (\log(1+e^{-n}))/n \to 1$ and so $\varphi(x) \not\to \infty$ as $x \to \infty$.

□

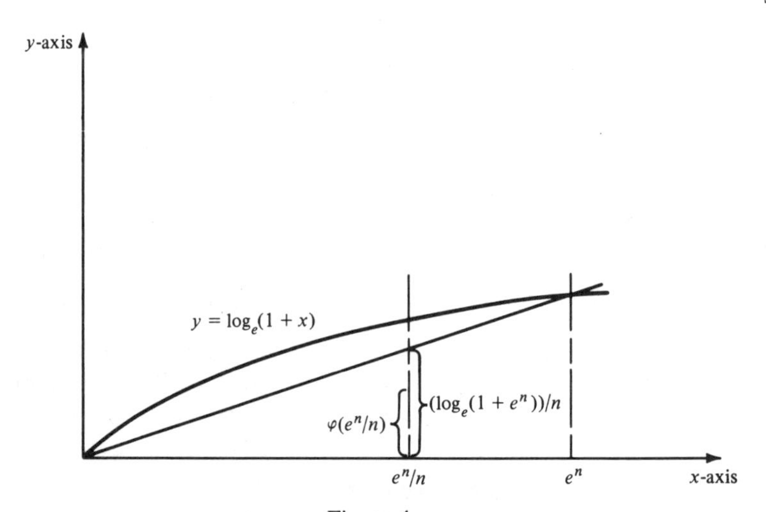

Figure 4

105. i) If $a<p<q<r<b$ and $\varphi(q) > ((r-q)/(r-p))\varphi(p) + ((q-p)/(r-p))\varphi(r)$ then the map $f: x \mapsto \varphi(x) - ((r-x)/(r-p))\varphi(p) - ((x-p)/(r-p))\varphi(r)$ has a positive maximum on $[p, r]$. Since $f(p)=f(r)=0$, this maximum occurs at some s in (p, r) and thus $f''(s) \leqq 0$. Since $f''(s) = \varphi''(s) > 0$, a contradiction results.

ii) If $p<r$ and $\varphi'(p) > \varphi'(r)$, then since convexity of φ implies that $\varphi(r) > \varphi(p) + \varphi'(p)(r-p)$, there follow the inequalities

$$\varphi(r) + (p-r)\varphi'(r) > \varphi(r) + (p-r)\varphi'(p)$$
$$> \varphi(p) + \varphi'(p)(r-p) + \varphi'(p)(p-r) = \varphi(p).$$

Hence $(p, \varphi(p))$ does not lie above the supporting line through $(r, \varphi(r))$ and there is a contradiction. Hence always $\varphi'(p) \leqq \varphi'(r)$ if $p \leqq r$, i.e., φ' is monotone increasing and so $\varphi'' \geqq 0$. □

106. For n in $\mathbb{N}$ let g_n be $f \cdot \chi_{E_n}$. Then $g_n \in L^1([0, 1], \lambda)$ and $\|g_n\|_1 \to 0$ as $n \to \infty$. Hence, by passage to a subsequence as needed, it may be assumed that $g_n \to 0$ a.e. as $n \to \infty$. If $\lambda(E_n) \not\to 0$ as $n \to \infty$, again by passage to a subsequence as needed, it may be assumed that for some positive ε and all n in $\mathbb{N}$, $\lambda(E_n) \geqq \varepsilon$. But then $\limsup_{n=\infty} f \cdot \chi_{E_n} = f \cdot \chi_{\limsup_{n=\infty} E_n} \neq 0$ on a set of positive measure, namely on $(\limsup_{n=\infty} E_n) \cap f^{-1}((0, \infty))$, a set having measure at least ε. Thus g_n fails to converge a.e. to zero, a contradiction. □

107. Define the following maps: $k:[0,\infty) \ni s \mapsto \int_0^s b(r)\, dr$; $g: s \mapsto k(s) + c(s)$; $T: C([0,\infty), \mathbb{R}) \ni f \mapsto (F: s \mapsto \int_0^s a(r) f(r)\, dr)$. Then

$$d(T(y)(s))/ds - a(s)T(y)(s) \leqq a(s)g(s),$$

$$d(e^{-\int_t^s a(r)\, dr} T(y)(s))/ds \leqq e^{-\int_t^s a(r)\, dr} a(s)g(s),$$

$$T(y)(t) \leqq \int_0^t a(s)g(s)\, e^{\int_s^t a(r)\, dr}\, ds,$$

$$\int_0^t a(s)\left[y(s) - g(s)\, e^{\int_s^t a(r)\, dr}\right] ds \leqq 0,$$

from which the result follows. □

Owing to the importance of fixed-point theory, there is given below an alternative solution using this theory.

If M_0 is an endomorphism of a Banach space X and if for all x in X, $\sum_{n=1}^{\infty} M_0^n(x)$ converges, then for each y in X the map $M_y: x \mapsto y + M_0(x)$ has a unique fixed point P, namely $y + \sum_{n=1}^{\infty} M_0^n(y)$. If, to boot, X is a function space and M_0 preserves positivity, then whenever $x \leqq y + M_0(x)$ it follows that $x \leqq P$. The verification of these statements is straightforward. Furthermore direct calculations for the case in which X is $C([0,a], \mathbb{R})$, $0 \leqq t \leqq a$, and $M_0 = T$ show that the results just quoted are applicable. For y read g, for x read y and for P read $t \mapsto g(t) + \int_0^t a(s)(\exp \int_s^t a(r)\, dr) g(s)\, ds$. □

108. The solution involves two steps: i) showing that for a, b in $\mathbb{R}$ the differential equation $y'' + (1+q)y = 0$ has a unique solution y such that $y(0) = a$, $y'(0) = b$; ii) the solution is bounded.

Ad i). If w'' is continuous and if v satisfies: $v(t) = w(t) - \int_0^t \sin(t-s) q(s) v(s)\, ds$, then $v'' + (1+q)v = 0$. Since $q \in L^1([0,\infty), \lambda)$, the map $T: f \mapsto (t \mapsto \int_0^t \sin(t-s) q(s) f(s)\, ds)$ is one for which $\sum_{n=1}^{\infty} \|T^n(f)\|_\infty$ converges if $0 \leqq t \leqq R$ and $f \in C([0,R], \mathbb{R})$. Hence (Problem 107) the map $M_w: f \mapsto w - T(f)$ has a unique fixed point v and if $w(0) = a$, $w'(0) = b$ then also $v(0) = a$ and $v'(0) = b$.

Since the differential equation is homogeneous, the uniqueness of solution question is resolved by examining the case in which $a = b = 0$. The original equation is equivalent to the system

$$y_1' = 0 \,.\, y_1 + 1 \,.\, y_2$$

$$y_2' = -(1+q)y_1 + 0 \,.\, y_2.$$

Any solutions of the system must satisfy the equations

$$\text{(a)} \quad y_1(t) = \int_0^t y_2(s)\,ds,$$

$$\text{(b)} \quad y_2(t) = \int_0^t -(1+q(s))y_1(s)\,ds.$$

The results in Problem 107, applied to (a), show y_1 (hence also y_2) is zero since $y_1(t) \leqq \int_0^t y_2(s)\,ds$ and $-y_1(t) \leqq -\int_0^t (-y_2(s))\,ds$. Thus the uniqueness question is settled.

Ad ii). If $w''+w=0$, $w(0)=a$, $w'(0)=b$, then $|v(t)| \leqq \|w\|_\infty + \int_0^t |q(s)|\,|v(s)|\,ds$ and Problem 107 applied again yields

$$|v(t)| \leqq \|w\|_\infty\, e^{\int_0^t |q(s)|\,ds} \leqq \|w\|_\infty\, e^{\|q\|_1}$$

whence v is bounded. □

109. i) Let $|f|''$ be zero on $f^{-1}(0)$. If $f(x) \neq 0$, then $|f|''(x) = \operatorname{sgn}(f(x))f''(x)$. Since derivatives are limits of measurable functions $|f|''$ as defined is measurable. Since f'' is bounded so is $|f|''$.

ii) Let $\{(a_n, b_n)\}_{n=1}^\infty$ be the sequence of disjoint open intervals such that $(0,1)\backslash f^{-1}(0) = \bigcup_{n=1}^\infty (a_n, b_n)$. Then $\int_0^1 g(x)|f|''(x)\,dx = \sum_{n=1}^\infty \operatorname{sgn}(f(\frac{1}{2}(a_n+b_n))) \int_{a_n}^{b_n} g(x) f''(x)\,dx$, since on each interval (a_n, b_n), $\operatorname{sgn}(f)$ is constant. Integration by parts then leads to the equation (in which $c_n = \frac{1}{2}(a_n+b_n)$)

$$\int_0^1 g(x)|f|''(x)\,dx = \sum_{n=1}^\infty \operatorname{sgn}(f(c_n)) g(x) f'(x)\Big|_{a_n}^{b_n} - \sum_{n=1}^\infty \operatorname{sgn}(f(c_n)) g'(x) f(x)\Big|_{a_n}^{b_n}$$

$$+ \sum_{n=1}^\infty \operatorname{sgn}(f(c_n)) \int_{a_n}^{b_n} f(x) g''(x)\,dx.$$

If $\operatorname{sgn}(f(c_n)) = 1$ then, since $f(a_n) = f(b_n) = 0$, $g(b_n)f'(b_n) \leqq 0$, and $g(a_n)f'(a_n) \geqq 0$; if $\operatorname{sgn}(f(c_n)) = -1$ the inequalities are reversed. Hence in the displayed equation the second sum in the right member is zero and the first sum is nonpositive, and the result follows. □

110. For each n in D let n' be the first index greater than n and such that $s_{n'} > s_{n-1}$. If $n < m < n'$ then $s_{n'} - s_{n-1} = s_{n'} - s_{m-1} + s_{m-1} - s_{n-1} > 0$ and $s_{m-1} - s_{n-1} \leqq 0$ whence $s_{n'} > s_{m-1}$. In other words, $n, n+1, \ldots, n'$ constitute a block or a part of a block.

Thus start with n_1, denoted $\bar{n}_1$, continue to $\bar{n}_1'$; let $\bar{n}_2$ be the first distinguished index after $\bar{n}_1'$, continue to $\bar{n}_2'$, etc. In this way enumerate the elements of D. The argument of the preceding paragraph shows that if $n^\#$ is the last member of a block to which n belongs and if $n < n^\#$ then $s_n^\# > s_{n-1}$.

□ [21]

111. If $S=\varnothing$ then S is open. If $S\neq\varnothing$, $x\in S$, $\limsup_{y=x}f(y)=L_x$, $x'>x$, and $f(x')>L_x$, there is an open set U containing x and not x' and such that $\sup\{f(y)\colon y\in U\}<f(x')$ whence $U\subset S$ and so S is open.

If $S\neq\varnothing$ and $x\in(a_n, b_n)$ assume $f(x)>L_{b_n}$. Since $b_n\notin S$, if $x''>b_n$ then $L_{b_n}\geqq f(x'')$. Thus $L_x\geqq f(x)>f(x'')$ and so all x' such that $x'>x$ and $f(x')>L_x$ are in $(x, b_n]$. Let c be the supremum of all such x'. Then $x<c\leqq b_n$. If $c=b_n$ then $f(c)\geqq L_x$ as claimed. If $c<b_n$ there is a c' such that $c'>c$ and $f(c')>L_c$. Then $f(c')>L_c\geqq f(c)\geqq L_x$, and hence $c'\in(x, b_n]$, $c'>c$, a contradiction. Thus $c=b_n$ and the result follows. □

REMARKS. The results in Problem 110 resp. Problem 111 were found and used by F. Riesz to give particularly simple and perspicuous proofs of the individual ergodic theorem of G. D. Birkhoff and the "almost everywhere differentiability of monotone functions" theorem [21], [11]. In particular, Problem 111 is sometimes called the "running water" ("eau courante") lemma. The accompanying figure for the case of f continuous suggests the origin of the term.

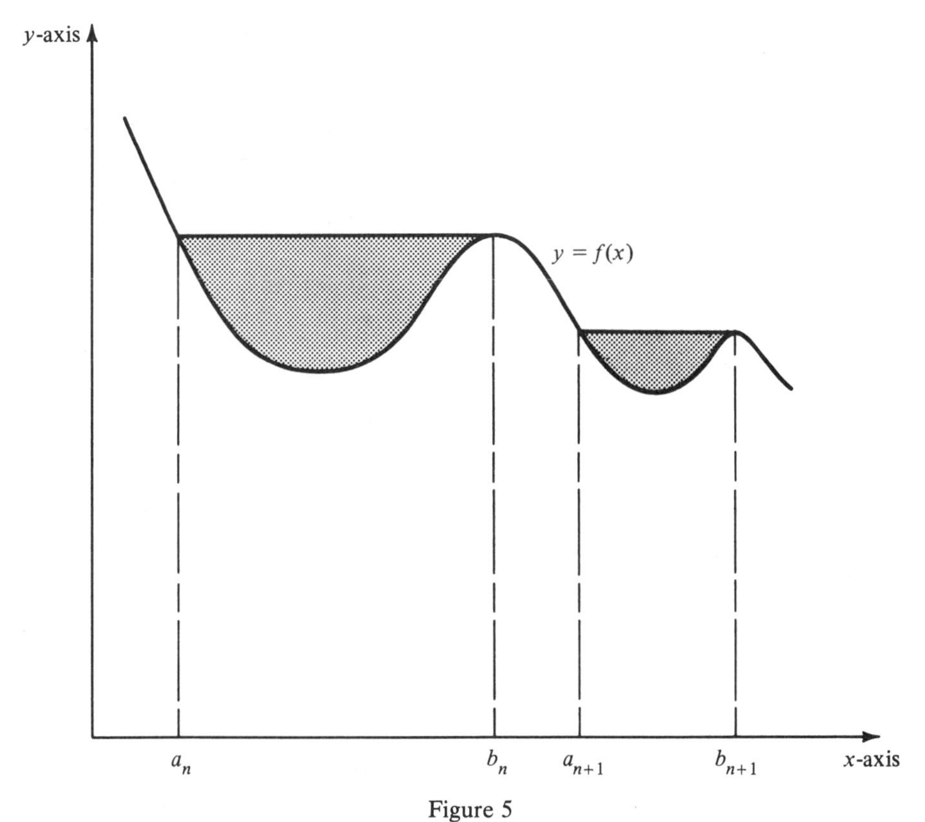

Figure 5

112. Let E be the set of all polynomials p over $\mathbb{R}$ and such that $\deg(p)\leqq R+S-1$. Then with respect to the usual vector operations E is an

$(R+S)$-dimensional vector space over $\mathbb{R}$. The map $T:E \ni p \mapsto (p(1), \ldots, p^{(R-1)}(1), p(2), \ldots, p^{(S-1)}(2))$ is in $\mathrm{End}_{\mathbb{R}}(\mathbb{R}^{R+S})$. It suffices to show T is surjective and hence to show $T^{-1}(0)=0$.

If $T(p)=0$ let N be $R+S-1$ and let $p(x)$ be represented by the finite Taylor series $\sum_{k=0}^{N} A_k(x-1)^k$. Then $A_0=A_1=\cdots=A_{R-1}=0$ and so if $1 \leqq s \leqq S$ then $p^{(s-1)}(x)=\sum_{k=R}^{N}(k!/(k-s+1)!)A_k(x-1)^{k-s+1}$ and so $\sum_{k=R}^{N}(k!/(k-s+1)!)A_k=0$, $1 \leqq s \leqq S$. Hence it suffices to show that all A_k are zero, which will be implied by showing that the following matrix is nonsingular:

$$M=\begin{pmatrix} 1/N! & 1/(N-1)! & \cdots & 1/(N-S+1)! \\ 1/(N-1)! & 1/(N-2)! & \cdots & 1/(N-S)! \\ \cdots & \cdots & \cdots & \cdots \\ 1/(N-S+1)! & 1/(N-S)! & \cdots & 1/(N-2S+2)! \end{pmatrix}.$$

To emphasize the dependence of $\det(M)$ on N and S let $\det(M)$ be $D(N, S)$. If the following operations are performed there emerges the recursion formula: $D(N, S)=(-1)^{S-1}(S-1)!D(N-1, S-1)/N!$ (whence $D(N, S)= [(-1)^{(1/2)S(S-1)}(S-1)!\cdots 1!D(N-S+1, 1)]/[N!\cdots(N-S+2)!]$).

i) Multiply the elements of each row by the reciprocal of the first entry in that row (the resulting determinant is $N!\cdots(N-S+1)!D(N, S)$).
ii) Subtract the elements in row $S-1$ from their counterparts in row $S, \ldots,$ the elements in row 1 from their counterparts in row 2.

Since $D(N-S+1, 1)=1/(N-S+1)! \neq 0$ the desired result follows. □

113. i) If f is the map $x \mapsto (1/2)\sin(1/x)$ then $\{2/n\pi\}_{n=1}^{\infty}$ is a Cauchy sequence whereas $\{f(2/n\pi)\}_{n=1}^{\infty}$ is not.

ii) If f is not continuous there is in $(0, 1)$ an x_0 and a sequence $\{x_n\}_{n=1}^{\infty}$ such that $x_n \to x_0$ and $f(x_n) \not\to f(x_0)$. If $y_{2n}=x_n$, $y_{2n-1}=x_0$, $n=1, 2, \ldots$ then $\{y_n\}_{n=1}^{\infty}$ is a Cauchy sequence and $\{f(y_n)\}_{n=1}^{\infty}$ is not. □

114. It suffices to prove that if $x_n \downarrow 0$ then $\{f(x_n)\}_{n=1}^{\infty}$ is a Cauchy sequence. Since $|f(x_n)-f(x_m)| \leqq |g(x_n)-g(x_m)|$ and since $\lim_{n\to\infty} g(x_n)$ exists the result follows. □

115. Let δ be the map

$$x \mapsto \begin{cases} 1, & \text{if } x \geqq 0 \\ 0, & \text{if } x < 0 \end{cases}.$$

Then $\sum_{n=1}^{\infty} a_n\delta(x-b_n)$ is a monotone increasing function.

If $x \in \mathbb{R}\backslash\{b_n\}_{n=1}^{\infty}$ and $\varepsilon>0$ let n_0 be such that $\sum_{n=n_0+1}^{\infty} a_n < \varepsilon/2$. If $\alpha>0$ and $|x-b_n| \geqq \alpha$, $n=1, 2, \ldots, n_0$, then whenever $|y-x|<\alpha$, $|f(y)-f(x)| \leqq \sum_{n=1}^{n_0} a_n|\delta(x-b_n)-\delta(y-b_n)|+2\sum_{n=n_0+1}^{\infty} a_n < 0+2\,.\,(\varepsilon/2)$ and so f is continuous on $\mathbb{R}\backslash\{b_n\}_{n=1}^{\infty}$.

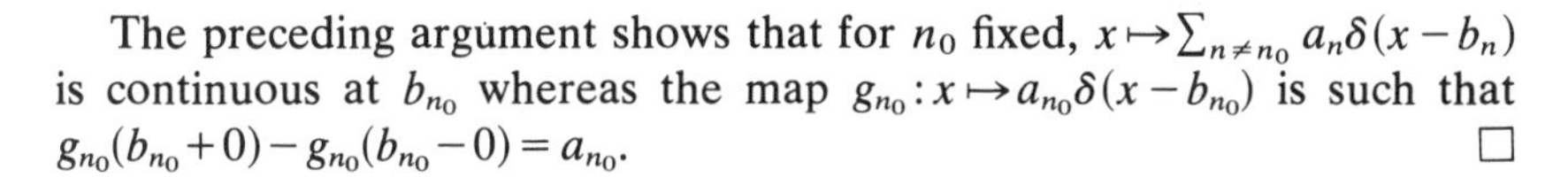
The preceding argument shows that for n_0 fixed, $x \mapsto \sum_{n \neq n_0} a_n \delta(x - b_n)$ is continuous at b_{n_0} whereas the map $g_{n_0}: x \mapsto a_{n_0}\delta(x - b_{n_0})$ is such that $g_{n_0}(b_{n_0}+0) - g_{n_0}(b_{n_0}-0) = a_{n_0}$. □

116. For any trigonometric polynomial $p: t \mapsto \sum_{n=-N}^{N} c_n e^{-int}$ let $L_0(p)$ be $\sum_{n=-N}^{N} a_n c_n$. The hypotheses show that L_0 is a bounded linear functional defined on a dense subset of $C(\mathbb{T}, \mathbb{C})$ and thus L_0 is extendible to a bounded linear functional L on $C(\mathbb{T}, \mathbb{C})$. The Riesz representation theorem shows there is a complex Borel measure μ such that $L(p) = \int_0^{2\pi} p(t)\, d\mu(t)$ and the result follows. □

117. If a μ as described exists, if $f \in C(\mathbb{T}, \mathbb{C})$, and if $c_n = (1/2\pi)\int_0^{2\pi} e^{-int} f(t)\, dt$, $n = 0, \pm1, \pm2, \ldots$, then Fejér's theorem shows $\sum_{n=-N}^{N} c_n(1 - |n|/(N+1))\, e^{int} \to f(t)$ uniformly on $\mathbb{T}$ as $N \to \infty$. Hence $\int_0^{2\pi} f(t)\, d\mu(t) = \lim_{N\to\infty} (1/2\pi)\int_0^{2\pi} \sum_{n=-N}^{N} c_n(1 - |n|/(N+1))\, e^{int} f(t)\, dt = \lim_{N\to\infty} L_N(f)$. The uniform boundedness theorem applied to the L_N shows they are uniformly bounded and hence the measures μ_N associated *via* the Riesz representation theorem to the L_N are also uniformly bounded. For a Borel set E, $\mu_N(E) = (1/2\pi)\int_E \sum_{n=-N}^{N} c_n(1 - |n|/(N+1))\, e^{int}\, dt$, $N = 0, 1, \ldots$, and the result follows.

On the other hand, if the measures μ_N are uniformly bounded, then so are the associated linear functionals L_N. For the maps $p: t \mapsto \sum_{n=-N_0}^{N_0} a_n e^{-int}$, $\lim_{N\to\infty} L_N(p) = \sum_{n=-N_0}^{N_0} c_n a_n$. Since these p constitute a dense subset of $C(\mathbb{T}, \mathbb{C})$, for all f in $C(\mathbb{T}, \mathbb{C})$, $\lim_{N\to\infty} L_N(f)$ exists, call it $L(f)$. Then L is a bounded linear functional and, again *via* the Riesz representation theorem, there is a complex Borel measure μ such that, for all f in $C(\mathbb{T}, \mathbb{C})$, $L(f) = \int_0^{2\pi} f(t)\, d\mu(t)$. In particular for the map $p_{N_0}: t \mapsto e^{-iN_0 t}$, $L(p_{N_0}) = \lim_{N\to\infty} L_N(p_{N_0}) = \lim_{N\to\infty} (1/2\pi)\int_0^{2\pi} c_{N_0}(1 - |N_0|/(N+1))\, e^{iN_0 t} e^{-iN_0 t}\, dt = c_{N_0} = \int_0^{2\pi} e^{-iN_0 t}\, d\mu(t)$, as required. □

118. For each f in $\mathcal{F}$ let N_f be $\{x : f(x) = 0\}$ and let S_f be $X \backslash N_f$. Then for all f in $\mathcal{F}$, $f(\bigcap_{f' \in \mathcal{F}} N_{f'}) = \{0\}$. Thus it suffices to prove $X \backslash \bigcap_{f' \in \mathcal{F}} N_{f'}$, denoted X_0, is finite.

If X_0 is not finite let $\{x_n\}_{n=1}^{\infty}$ be an infinite subset of X_0. There is in $\mathcal{F}$ an f_1 such that $f_1(x_1) \neq 0$. Let n_1 be 1 and if $f_1, f_2, \ldots, f_m$ have been found so that $f_k(x_{n_k}) \neq 0$, $k = 1, 2, \ldots, m$, then $\{x_n\}_{n=1}^{\infty} \backslash \bigcup_{k=1}^{m} S_{f_k}$ is infinite. If n_{m+1} is least index of elements in that set, there is in $\mathcal{F}$ an $f_{n_{m+1}}$ such that $f_{n_{m+1}}(x_{n_{m+1}}) \neq 0$. The construction insures that $S_{f_{m+1}} \backslash \bigcup_{k=1}^{m} S_{f_k} \neq \emptyset$.

Define a map g as follows: if $x \notin \bigcup_{m=1}^{\infty} S_{f_m}$, $g(x) = 0$; on S_{f_1}, g is defined so that $\sum_x f_1(x) g(x) \geqq 1$; if g has been defined on $\bigcup_{k=1}^{m} S_{f_k}$ so that $\sum_x f_k(x) g(x) \geqq k$, $k = 1, 2, \ldots, m$, define g on $S_{f_{m+1}} \backslash \bigcup_{k=1}^{m} S_{f_k}$ so that $\sum_x f_{m+1}(x) g(x) \geqq m+1$. Hence $\sup_{f \in \mathcal{F}} |\sum_x f(x) g(x)| = \infty$ and the contradiction shows X_0 is finite. □

119. Let f be the map $r \mapsto \lambda\{x : \sin x > r,\ x \in [0, 4\pi]\}$. Then f is monotone decreasing, $f([-1, 1]) = f(\mathbb{R}) = [0, 4\pi]$, and on $[-1, 1]$ f is one-one. Thus if

$g = f^{-1}$ on $[0, 4\pi]$ then g is monotone decreasing and $g(x) > r$ iff $x < f(r)$ from which the result follows. □

120. Since $f'_n(x) - f'_n(y) = \int_x^y f''_n(t)\,dt$, it follows that the sequence $\{f'_n\}_{n=1}^\infty$ is equicontinuous. If M' as described does not exist then, *via* passage to subsequences as needed, it may be assumed that $a_n = \|f_n\|_\infty \downarrow 0$ and that $\|f'_n\|_\infty = M'_n \uparrow \infty$. The preceding sentences imply that there is a positive b and for each n an x_n such that $|f'_n(x)| > \frac{1}{2}M'_n$ if $|x - x_n| < b$. Since for some c_n, $f_n(x) = \int_{x_n}^x f'_n(t)\,dt + c_n$ and since $f_n(x_n) = c_n$ it follows that $|c_n| \leqq a_n$ and $|f_n(x)| \geqq \frac{1}{2}M'_n b - a_n$ if $|x - x_n| = \frac{1}{2}b$. Hence $\|f_n\|_\infty \geqq \frac{1}{4}M'_n b - a_n \to \infty$, a contradiction, and the result follows. □

121. Since each f_k is monotone, f'_k exists a.e. and so the set E, $\{x\colon$ for all k in $\mathbb{N}$, $f'_k(x)$ exists and is finite$\}$, is measurable and $\lambda(E) = 1$. Hence for all k, $f_k(1) - f_k(0) \geqq \int_E f'_k(x)\,dx$ and $\limsup_{k=\infty}(f_k(1) - f_k(0)) = 0 \geqq \liminf_{k=\infty} \int_E f'_k(x)\,dx \geqq \int_E \liminf_{k=\infty} f'_k(x)\,dx$. Since $f'_k \geqq 0$ on E the first result follows.

If f_k is the function described by the graph in Figure 6, then $f'_k(\frac{1}{2}) = k$. □

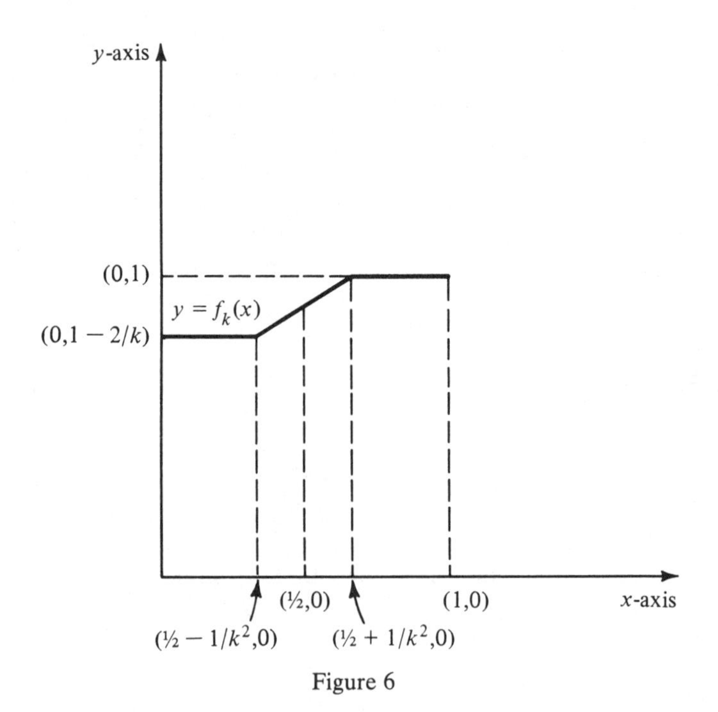

Figure 6

122. Since the set of polynomials over $\mathbb{Z}$ is countable, $\mathscr{A}$ is countable. If $f \in \mathbb{R}^\mathbb{R}$ let E_f be $\{x\colon f$ is continuous at $x\}$. Then if $x \in E_f$ and $n \in \mathbb{N}$ there is a positive $\delta(n, x)$ such that if $|y - x| < \delta(n, x)$, $|f(y) - f(x)| < 1/n$. Thus $\bigcup_{x \in E_f}\{y\colon |y - x| < \delta(n, x)\}$ is an open set U_n and $\bigcap_{n=1}^\infty U_n$ is a G_δ, G, containing E_f. On the other hand if $x \in G$ and if $\varepsilon > 0$, let $1/n$ be less than ε. Then $x \in U_n$ and so if $|y-x| < \delta(n, x)$, $|f(x)-f(y)| < \varepsilon$, i.e., $E_f = G$ and so E_f is a G_δ.

If, furthermore, $E_f = \mathscr{A}$, then E_f is a dense G_δ and therefore a set of the second category (Baire's theorem), whereas $\mathscr{A}$ is countable and hence of the first category. Since $\mathscr{A}$ is an F_σ there is [9] an f such that $\mathbb{R}\backslash E_f = \mathscr{A}$, i.e., $\mathbb{R}\backslash \mathscr{A} = E_f$. □

123. If U is open in $\mathbb{R}$, $f^{-1}(U)$ is the union of two disjoint sets, C_U (the set in $f^{-1}(U)$ of points of continuity of f) and D_U (the set in $f^{-1}(U)$ of points of discontinuity of f). By hypothesis, $\lambda(D_U) = 0$. If $x \in C_U$, then for some positive $\varepsilon(x)$, $\{y : |f(x)-y| < \varepsilon(x)\} \subset U$ and for some positive $\delta(x)$, $f(\{x' : |x'-x| < \delta(x)\}) \subset \{y : |f(x)-y| < \varepsilon(x)\} \subset U$. Thus $f^{-1}(U) = (\bigcup_{x \in C_U} \{x' : |x'-x| < \delta(x)\} \cup D_U$, which is measurable (the union of an open set and a null set) and so f is measurable. □

124. If the equation $f'(x) = \chi_{(-\infty,0]} \circ f(x)$ is valid for all x in some open set U containing 0, then

$$f'(x) = \begin{cases} 0, & \text{if } f(x) > 0 \\ 1, & \text{if } f(x) \leqq 0 \end{cases}$$

and by hypothesis, $f(0) = 0$. The proof proceeds by showing that if $\varepsilon > 0$, f cannot be identically zero on $[0, \varepsilon)$ nor can f be only positive or only negative on $(0, \varepsilon)$. This granted, on each $[0, \varepsilon)$ f' assumes both values, 0 and 1. Since these are the only values f' is permitted, Darboux's theorem *re* the intermediate value property of derivatives provides a contradiction.

If $f = 0$ on $[0, \varepsilon)$ then $f' = 0$ in $(0, \varepsilon)$ whereas by hypothesis $f' = 1$ on $(0, \varepsilon)$. If $f > 0$ in $(0, \varepsilon)$ then $f' = 0$ in $(0, \varepsilon)$. Hence for some positive p, $f = p$ in $(0, \varepsilon)$ and thus the continuity of f at 0 is contradicted.

If $f < 0$ in $(0, \varepsilon)$ then $f' = 1$ in $(0, \varepsilon)$ whence for x in $(0, \varepsilon)$ and some k, $f(x) = x + k$. Since $f(0) = 0$, $k = 0$ and then, by hypothesis, $f' = 0$ on $(0, \varepsilon)$, a contradiction. The result follows. □

125. If $x < y$ and $f(x) > f(y)$, then for some positive ε_0 and all ε in $(0, \varepsilon_0)$, $f(y) < f(x) - \varepsilon(y-x)$. As indicated in Figure 7, the hypothesis implies there is in (x, y) an x_1 such that $f(x_1) \geqq f(x) - \varepsilon(x_1 - x)$. Let x_∞ be $\sup\{x_1 : x < x_1 < y,\ f(x_1) \geqq f(x) - \varepsilon(x_1 - x)\}$. If $x_\infty < y$ then $f(x_\infty) \geqq f(x) - \varepsilon(x_\infty - x)$ and by hypothesis there is in (x_∞, y) a z such that $f(z) \geqq f(x_\infty) - \varepsilon(z - x_\infty) \geqq f(x) - \varepsilon(x_\infty - x) - \varepsilon(z - x_\infty) = f(x) - \varepsilon(z - x)$, a contradiction. Thus $x_\infty = y$ and $f(y) \geqq f(x) - \varepsilon(y - x)$. Since ε is arbitrary in $(0, \varepsilon_0)$, $f(y) \geqq f(x)$, a second contradiction. □

126. If $x_0 < y$ and $f(x_0) > f(y)$ then for any positive ε there is in (x_0, y) an x such that $|x - x_0| < \varepsilon$ and such that $\limsup_{h \downarrow 0} (f(x+h) - f(x))/h \geqq 0$. If ε is sufficiently small, $f(x) > f(y)$. The argument of Solution 125, again implies the existence in (x, y) of an x_1 such that $f(x_1) > f(x) - \varepsilon(x_1 - x)$. The

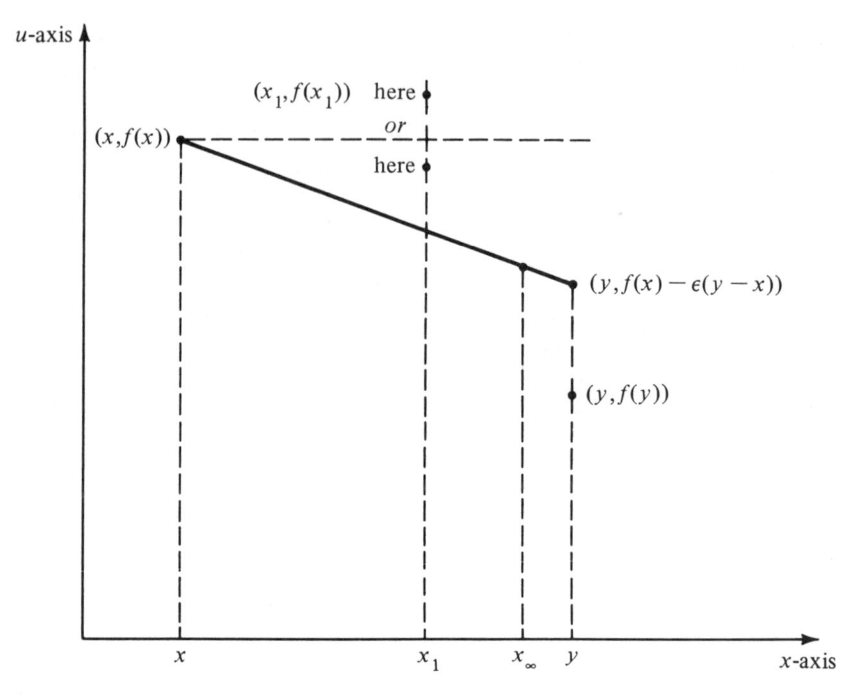

Figure 7. The situation: $x < y, f(x) > f(y)$, and $x_\infty < y$ is shown.

density of the complement of any null set and the continuity of f imply that x_1 may be chosen so that $\limsup_{h\downarrow 0} (f(x_1+h)-f(x_1))/h \geqq 0$. If x_∞ is as defined in Solution 125 and $x_\infty < y$, the density/continuity argument shows a contradiction as deduced in Solution 125. Thus $x_\infty = y$ and the final contradiction is reached. □

127. For x, y in $(0, 1)$ let $d(x, y)$ be $\max(|f(x)-f(y)|, |x-y|)$. A direct check shows that d is a true metric and furthermore that $d(x, y) \geqq |x-y|$. If $|x_k - x| \to 0$ as $k \to \infty$ then $|f(x_k)-f(x)| \to 0$ as $k \to \infty$ and so $d(x_k, x) \to 0$ as $k \to \infty$, whence d and $|\cdots|$ provide the same topology for $(0, 1)$. Finally if $\varepsilon > 0$ then $|f(x)-f(y)| < \varepsilon$ if $d(x, y) < \varepsilon$ and so f is uniformly continuous with respect to d. □

128. If $n = 1$, f may be regarded as the restriction to $(-r, r)$ of a function F analytic in $\{z: z \in \mathbb{C}, |z| < r\}$. In that event, if $f \neq 0$ then $F^{-1}(0)$ is at most countable and the result follows.

Assume the result is true if $n = 1, 2, \ldots, N-1$. Let A be the set $f^{-1}(0) \cap B(0, r)^0$. According to the Fubini theorem, if $A_{x_N} = \{(x_1, x_2, \ldots, x_{N-1}): (x_1, x_2, \ldots, x_{N-1}, x_N) \in A\}$ then $\lambda_N(A) = \int_{-r}^{r} \lambda_{N-1}(A_{x_N})\, dx_N$. However, if $x_N \in (-r, r)$, A_{x_N} in $\mathbb{R}^{N-1}$ is the set on which f, regarded as a function of the variables $x_1, x_2, \ldots, x_{N-1}$ (x_N is fixed), is zero. However, even if $A_{x_N} = \emptyset$, the inductive assumption shows $\lambda_{N-1}(A_{x_N}) = 0$ and the result follows. □

NOTE: The result in Problem 128 shows, e.g., that almost all $n \times n$ matrices have n distinct eigenvalues (hence are diagonable), that almost all $n \times n$ matrices are nonsingular, that for every analytic or algebraic variety V in $\mathbb{R}^n$, $\lambda_n(V) = 0$, etc. The phrase "almost all" is to be interpreted with respect to λ_{n^2} and the identification of the set of all $n \times n$ matrices with $\mathbb{R}^{n^2}$.

129. If $\{f_n\}_{n=1}^\infty \subset C_0(\mathbb{R}^3, \mathbb{R})$ and for all x, $f_n(x) \downarrow 0$, then the dominated convergence theorem implies $\int_{\mathbb{R}^3} f_n(x)\, d\mu(x) \downarrow 0$. Hence $F(f_n) \downarrow 0$ and so *via* the Daniell integral construction there can be produced a Borel measure ν so that for all f in $C_0(R^3, R)$, $F(f) = \int_{\mathbb{R}^3} f(x)\, d\nu(x)$. Hence $\mu = \nu$. If $f \in S$ and $f \geqq 0$ there is in $C_0(\mathbb{R}^3, \mathbb{R})$ a sequence $\{f_n\}_{n=1}^\infty$ such that $f_n \uparrow f$, i.e., $f - f_n \downarrow 0$, $F(f - f_n) \downarrow 0$. The monotone convergence theorem implies $f \in L^1(\mathbb{R}^3, \mu)$ and that $F(f) = \int_{\mathbb{R}^3} f(x)\, d\mu(x)$. Finally, for any f in s, $f^\pm$ are also in S, $f = f^+ - f^-$, $f^\pm \geqq 0$, $F(f) = F(f^+) - F(f^-) = \int_{\mathbb{R}^3} f^+(x)\, d\mu(x) - \int_{\mathbb{R}^3} f^-(x)\, d\mu(x) = \int_{\mathbb{R}^3} f(x)\, d\mu(x)$. □

130. For t_0 fixed in $\mathbb{R}$, the map $g: \Sigma \ni x \mapsto (f(x, t_0))^2 + (\partial f(x, t_0)/\partial t)^2$ is in $C(\Sigma, \mathbb{R})$ and $g > 0$ whence $\min_{x \in \Sigma} g(x, t_0) > 0$ (Σ is compact). The compactness of Σ also implies there is an x_{t_0} such that $g(x_{t_0}, t_0) = \min_{x \in \Sigma} g(x, t_0)$. If, for each n in $\mathbb{N}$ there is a t_n and an x_{t_n} such that $|t_n - t_0| < 1/n$ and $\min_{x \in \Sigma} g(x, t_n) = g(x_{t_n}, t_n) = 0$, by passage to subsequences as needed, it may be assumed that $x_{t_n} \to x_\infty$ in Σ as $n \to \infty$ and thus $g(x_\infty, t_0) = 0$, a contradiction. Thus, for some open set $U(t_0)$ containing t_0, if $t \in U(t_0)$, $\min_{x \in \Sigma} g(x, t) > 0$.

If, for each n in $\mathbb{N}$ there are in $U(t_0)$, t_{1n} and t_{2n} such that $t_{1n} \neq t_{2n}$, $|t_{1n} - t_0| + |t_{2n} - t_0| < 1/n$ and for some x_n in Σ, $f(x_n, t_{1n}) = f(x_n, t_{2n}) = 0$, then by Rolle's theorem there is between t_{1n} and t_{2n} a t_{3n} such that $\partial f(x_n, t_{3n})/\partial t = 0$. Again, by passage to subsequences as needed, it may be assumed that $x_n \to \bar{x}$ in Σ as $n \to \infty$. Thus $\partial f(\bar{x}, t_0)/\partial t = 0$, $f(\bar{x}, t_0) \neq 0$ and for large n, $f(x_n, t_{1n}) \,.\, f(x_n, t_{2n}) \neq 0$, a contradiction. Hence, for some n_0, if $|t_1 - t_0| + |t_2 - t_0| < 1/n_0$ and $t_1 \neq t_2$ then $(f(x, t_1))^2 + (f(x, t_2))^2 > 0$, as required. □

131. Let Σ be as in Problem 130 and let N be $\Sigma \cap f^{-1}(0)$. Then N is compact and $g(x) > 0$ if $x \in N$. Thus for some positive ε, $g(x) \geqq 2\varepsilon$ if $x \in N$. Hence for some open set U containing N, $g(x) \geqq \varepsilon$ if $x \in U$. For some positive δ, $f(x) \geqq \delta$ if x is in the compact set $\Sigma \backslash U$. Since U may be chosen so that $0 \notin U$, and since for all x other than 0, $f(x) = \|x\|^m f(x/\|x\|)$, $g(x) = \|x\|^m g(x/\|x\|)$, it follows that $f(x)/\|x\|^m \geqq \delta$ if $x/\|x\| \in \Sigma \backslash U$ and $g(x)/\|x\|^m \geqq \varepsilon$ if $x/\|x\| \in U$. If $C = 1/\delta$ and $D = 1/\varepsilon$, then $Cf(x) + Dg(x) \geqq \|x\|^m$ if $x \neq 0$. If $x = 0$ the inequality is *a fortiori* true and the result follows. □

132. If $a=(a_1,\ldots,a_n)\in B(0,1)$ let $f(a)$ be $(f_1(a),\ldots,f_n(a))=(f_1(a_1,\ldots,a_n),\ldots,f_n(a_1,\ldots,a_n))$. Then

$$\begin{aligned} f_i(a)-f_i(b) &= \sum_{j=1}^{n} (f_i(a_1,\ldots,a_j,b_{j+1},\ldots,b_n) \\ &\quad - f_i(a_1,\ldots,a_{j-1},b_j,b_{j+1},\ldots,b_n)) \\ &= \sum_{j=1}^{n} \partial f_i/\partial x_j(a_1,\ldots,a_{j-1},c_{ij},b_{j+1},\ldots,b_n)\,.\,(a_j-b_j), \end{aligned}$$

and c_{ij} is between a_j and b_j, There is a p such that if $\|\partial f_i/\partial x_j\|_\infty<p$, $i,j=1,2,\ldots,n$, then $\|f(a)-f(b)\|<\frac{1}{2}\|a-b\|$.

Since $d(id)=id$ it follows that if $g=f-id$ then $dg=df-id$. There is a positive δ such that if $\sup_x\|df(x)-id\|<\delta$ then $\|\partial f_i/\partial x_j-\delta_{ij}\|_\infty<p$ and so it follows that $\|a-b\|-\|f(a)-f(b)\|\leqq\|g(a)-g(b)\|<\frac{1}{2}\|a-b\|$, i.e., $\|f(a)-f(b)\|\geqq\frac{1}{2}\|a-b\|$ whence f is injective. □

NOTES. If $n=1$ and $\delta=1$ then $f'>0$ and f is injective. For any n, p may be chosen to be $1/2n^2$.

133. If $df=T$, T is a map from $\mathbb{R}^n$ to $\operatorname{Hom}(\mathbb{R}^n,\mathbb{R})$, i.e., for each x in $\mathbb{R}^n$, $T(x)$ is a linear functional on $\mathbb{R}^n$. Thus $T(x)$ may be regarded as an element of $\mathbb{R}^n$ since $\operatorname{Hom}(\mathbb{R}^n,\mathbb{R})$ and $\mathbb{R}^n$ are isomorphic vector spaces. In short, since $f\in C^2(\mathbb{R}^n,\mathbb{R})$, $T\in C^1(\mathbb{R}^n,\mathbb{R}^n)$. Thus d^2f $(=dT)$ is a map from $\mathbb{R}^n$ to $\operatorname{End}(\mathbb{R}^n)$. (For each x, $d^2f(x)$ may be represented with respect to the standard basis for $\mathbb{R}^n$ by the matrix $(\partial^2 f(x)/\partial x_i\,\partial x_j)_{i,j=1}^n$, the Hessian.)

By definition, there is a map $\varepsilon:[-1,1]\mapsto(0,\infty)$ such that $\varepsilon(t)\to 0$ as $t\to 0$, and there is a positive δ such that if $\|u\|<\delta$ then $\|T(x_0+tu)-T(x_0)-dT(x_0)(tu)\|=\varepsilon(t)|t|\,.\,\|u\|$. Since $T(x_0)=df(x_0)=0$, $\|T(x_0+tu)-dT(x_0)(tu)\|=\varepsilon(t)|t|\,.\,\|u\|$.

If $T(x_0+tu)=0$ for some nonzero t in $[-1,1]$ and some nonzero u, then since $dT(x_0)^{-1}$ exists, let y be $dT(x_0)(tu)$. If $K=\|dT(x_0)^{-1}\|$ then $K>0$ and then $\|tu\|\leqq K\|y\|$ and so $y\neq 0$ and $\|y\|\leqq\varepsilon(t)K\|y\|$, i.e., $1\leqq\varepsilon(t)K$.

If there is a sequence $\{x_n\}_{n=1}^\infty$ such that $\|x_n-x_0\|<1/n$ and $T(x_n)=0$ for all n, let $1/n_0$ be less than $\delta/2$. If $n>n_0$, there is a u_n such that $\|u_n\|=\delta/2$ and a t_n such that $x_n=x_0+t_nu_n$ and $|t_n|<1/n$. Since $t_n\to 0$ as $n\to\infty$, the last inequality of the preceding paragraph provides a contradiction. Hence there is an open set U containing x_0 and such that $T(y)=df(y)\neq 0$ for all y in $U\backslash\{x_0\}$. □

6. Measure and Topology

134. i) If $0<\varepsilon_1<\varepsilon_2$, then whenever $\operatorname{diam}(U)<\varepsilon_1$, $\operatorname{diam}(U)<\varepsilon_2$ as well, whence $\rho_\varepsilon^p(A)$, for fixed A, is a monotone increasing function of ε and so $\rho^p(A)=\lim_{\varepsilon\to 0}\rho_\varepsilon^p(A)$.

ii) If $\delta>0$, $\{U_{nm}\}_{n,m=1}^\infty$ is a (double) sequence of open sets, $\operatorname{diam}(U_{nm})<\varepsilon$ for all n, m, $\bigcup_{m=1}^\infty U_{nm}\supset A_n$ for all n, and

$$\sum_{m=1}^\infty (\operatorname{diam}(U_{nm}))^p<\rho_\varepsilon^p(A_n)+\frac{\delta}{2^n},$$

then $\rho_\varepsilon^p(\bigcup_{n=1}^\infty A_n)\leqq\sum_{n,m=1}^\infty (\operatorname{diam}(U_{nm}))^p=\sum_{n=1}^\infty (\sum_{m=1}^\infty (\operatorname{diam}(U_{nm}))^p)<\sum_{n=1}^\infty \rho_\varepsilon^p(A_n)+\delta\leqq\sum_{n=1}^\infty \rho^p(A_n)+\delta$. Since δ is an arbitrary positive number the subadditivity of ρ_ε^p follows; since ε is also an arbitrary positive number the subadditivity of ρ^p is also proved, i.e., ρ^p is an outer measure (Hausdorff p-measure).

iii) Let $\gamma([0,1])$ be A. If $\operatorname{length}(\gamma)=\infty$, then $\operatorname{length}(\gamma)\geqq\rho^1(A)$. If $\operatorname{length}(\gamma)=L<\infty$, a change of scale permits the assumption that $L=1$. (The case in which $L=0$ is excluded since γ is simple.) Furthermore it may be assumed that for all t in $[0,1]$, $\operatorname{length}(\gamma|_{[0,t]})=t$, i.e., that the parameter t is itself arc-length.

In these circumstances, if $m=1,2,\ldots,$ and $k=1,2,\ldots,m-1$, then $|\gamma(k/m)-\gamma((k+1)/m)|<1/m$. Hence, if $\{B_r\}_{r=1}^m$ is a sequence of open balls such that B_r is centered at $\gamma((2r-1)/2m)$ and has radius $1/(2m-1)$, then $\bigcup_{r=1}^m B_r\supset A$ and so $\rho_{1/(2m-1)}^1(A)\leqq m\,.\,2/(2m-1)$. As $m\to\infty$ there emerges the inequality: $\rho^1(A)\leqq 1$. In the next paragraph it will be shown that $\rho^1(A)\geqq 1$ from which the result will follow.

First it will be useful to prove that if $\delta>0$ there is a positive ε such that if $0=t_0<t_1<\cdots<t_K=1$ and $\max_k|\gamma(t_{k+1})-\gamma(t_k)|<\varepsilon$, then $\sum_{k=0}^{K-1}|\gamma(t_{k+1})-\gamma(t_k)|>1-\delta$. Indeed, there is a sequence $\{s_p\}_{p=0}^P$ such that

$0=s_0<s_1<\cdots<s_p=1$ and such that $\sum_{p=0}^{P-1}|\gamma(s_{p+1})-\gamma(s_p)|>1-\delta$. Assume $0<\varepsilon_1<\max(\delta/4P, \min_p|\gamma(s_{p+1})-\gamma(s_p)|)$. There is a positive η such that if $|t'-t''|<\eta$ then $|\gamma(t')-\gamma(t'')|<\varepsilon_1$. Hence if $0=t_0<t_1<\cdots<t_K=1$ and $\max_k|t_{k+1}-t_k|<\min(\eta, \min_p|s_{p+1}-s_p|)$, then for appropriate indices $k_1, k_2, \ldots, 0=t_0=s_0<t_1<\cdots<t_{k_1}\leqq s_1<t_{k_1+1}<\cdots<t_{k_2}\leqq s_2<\cdots<t_K=s_P=1$. Consequently

$$1-\delta<\sum_p\left(\sum_{j=k_p+1}^{k_{p+1}-1}|\gamma(t_{j+1})-\gamma(t_j)|\right)+\sum_p(|\gamma(s_p)-\gamma(t_{k_p})|+|\gamma(t_{k_p+1})-\gamma(s_p)|).$$

The second sum is less than $P.2\varepsilon_1$ which is less than $\delta/2$ and so the first (double) sum exceeds $1-3\delta/2$. Since γ is a homeomorphism there is a positive ε such that if $|\gamma(t')-\gamma(t'')|<\varepsilon$ then $|t'-t''|<\min(\eta, \min_p|s_{p+1}-s_p|)$ and the desired conclusion follows.

For δ and ε as in the preceding paragraph and ζ positive let $\{U_m\}_{m=1}^\infty$ be a sequence of open sets such that $\bigcup_{m=1}^\infty U_m\supset A$, $\operatorname{diam}(U_m)<\varepsilon$ for all m, and $\sum_{m=1}^\infty\operatorname{diam}(U_m)<\rho_\varepsilon^1(A)+\zeta$. Since A is compact, there is a finite set $\{U_{m_p}\}_{p=1}^P$ such that $\bigcup_{p=1}^P U_{m_p}\supset A$. Let t_0 be 0 and let V_0 be one of the U_{m_p} containing $\gamma(t_0)$. If $t_1=\sup\{t: \gamma(t)\in V_0\}$, then $t_0<t_1\leqq 1$. If $t_1<1$, let V_1 be one of the U_m containing $\gamma(t_1)$. Since V_0 is open and $t_0<t_1<1$, $\gamma(t_1)\notin V_0$ and so $V_1\neq V_0$. $\cdots$ If $\{t_i^p\}_{i=0}^k$ and $\{V_i\}_{i=0}^k$ have been found so that $\gamma(t_i)\in V_i$, no two V_i are the same, and $t_0<t_1<\cdots<t_k\leqq 1$, the process stops if $t_k=1$; if $t_k<1$, let t_{k+1} be $\sup\{t: \gamma(t)\in V_k\}$ and let V_{k+1} be one of the U_{m_p} containing $\gamma(t_{k+1})$. Then V_{k+1} is different from all V_i chosen before. Since there are only finitely many U_{m_p}, at some stage, say when $k=K$, $t_K=1$. Then $\max|\gamma(t_{k+1})-\gamma(t_k)|<\varepsilon$ and so $\sum_{k=0}^{K-1}|\gamma(t_{k+1})-\gamma(t_k)|>1-\delta$. On the other hand $\sum_{k=0}^{K-1}|\gamma(t_{k+1})-\gamma(t_k)|\leqq\sum_{k=0}^{K-1}\operatorname{diam}(V_k)<\rho_\varepsilon^1(A)+\zeta$. Thus, finally, $1-\delta\leqq\rho^1(A)+\zeta$, from which the result follows. □

135. It suffices to show $\rho_\varepsilon^q(A)\leqq\varepsilon^{q-p}\rho_\varepsilon^p(A)$. However, if $\{U_m\}_{m=1}^\infty$ is a sequence of open sets such that $\bigcup_{m=1}^\infty U_m\supset A$ and $\operatorname{diam}(U_m)<\varepsilon$ for all m, then $\rho_\varepsilon^q(A)\leqq\sum_{m=1}^\infty(\operatorname{diam}(U_m))^q\leqq\sum_{m=1}^\infty(\operatorname{diam}(U_m))^p\,.\,\varepsilon^{q-p}$ and the result follows. □

136. If $E\subset\mathbb{R}^p$ then $\lambda_p^*(E)\leqq(\operatorname{diam}(E))^p$. Hence, if $\{U_m\}_{m=1}^\infty$ is a sequence of open sets such that $\bigcup_{m=1}^\infty U_m\supset A$ and $\operatorname{diam}(U_m)<\varepsilon$ for all m, then $\lambda_p^*(A)\leqq\sum_{m=1}^\infty\lambda_p(U_m)\leqq\sum_{m=1}^\infty(\operatorname{diam}(U_m))^p$ from which follows the inequality: $\lambda_p^*(A)\leqq\rho^p(A)$.

On the other hand, if $\lambda_p^*(A)=\infty$, then for any constant c_p, $\lambda_p^*(A)\geqq c_p\rho^p(A)$. If $\lambda_p^*(A)<\infty$ and $\delta, \varepsilon>0$, there is a sequence $\{U_m\}_{m=1}^\infty$ of open sets (indeed open balls) such that $\bigcup_{m=1}^\infty U_m\supset A$, $\operatorname{diam}(U_m)<\varepsilon$ for all m, and $\lambda_p^*(A)>\sum_{m=1}^\infty\lambda_p(U_m)-\delta=\sum_{m=1}^\infty c_p(\operatorname{diam}(U_m))^p-\delta\geqq c_p\rho_\varepsilon^p(A)-\delta$, whence $\lambda_p^*(A)\geqq c_p\rho^p(A)$, as required. □

NOTE. If $B_p(0, r)$ is the closed ball centered at 0 in $\mathbb{R}^p$ and of radius r, then $\lambda_p(B_p(0,1))=\int_{-1}^1\lambda_{p-1}(B_{p-1}(0,\sqrt{1-x^2}))\,dx$, i.e., $\lambda_p(B_p(0,1))$ may be calculated inductively. Then $c_p=\lambda_p(B_p(0,1))/2^p$.

137. Since ρ^p is countably subadditive, it suffices to prove that $\rho(A \cup B) \geqq \rho^p(A) + \rho^p(B)$. If $0 < \varepsilon < \delta/3$ and if $\{U_m\}_{m=1}^{\infty}$ is a sequence of open sets such that $\bigcup_{m=1}^{\infty} U_m \supset A \cup B$ and $\operatorname{diam}(U_m) < \varepsilon$ for all m, then no U_m meets both A and B. In other words, there are disjoint sequences $\{m_k\}_{k=1}^{K}$, $\{m'_l\}_{l=1}^{L}$, $K, L \leqq \infty$, such that $\bigcup_{k=1}^{K} U_{m_k} \supset A$, $\bigcup_{l=1}^{L} U_{m'_l} \supset B$. If $\eta > 0$, the sequence $\{U_m\}_{m=1}^{\infty}$ can be chosen so that $\rho^p(A \cup B) > \sum_{m=1}^{\infty} (\operatorname{diam}(U_m))^p - \eta \geqq \sum_{k=1}^{K} (\operatorname{diam}(U_{m_k}))^p + \sum_{l=1}^{L} (\operatorname{diam}(U_{m'_l}))^p - \eta \geqq \rho_\varepsilon^p(A) + \rho_\varepsilon^p(B) - \eta$. The result follows. □

138. If A is closed and $\rho^p(S) = \infty$, owing to the subadditivity of ρ^p, $\rho^p(S) = \rho^p(S \cap A) + \rho^p(S \backslash A)$. If $\rho^p(S) < \infty$, it suffices to show $\rho^p(S) \geqq \rho^p(S \cap A) + \rho^p(S \backslash A)$. If, for m in $\mathbb{N}$, $U_m = \{x : d(x, A) > 1/m\}$, then U_m is open, $U_m \subset U_{m+1}$, and $\bigcup_{m=1}^{\infty} U_m = X \backslash A$. Since $((S \backslash A) \backslash U_m) = \bigcup_{k=m+1}^{\infty} (S \cap (U_k \backslash U_{k-1}))$, it follows that $\rho^p((S \backslash A) \backslash U_m)) \leqq \sum_{k=m+1}^{\infty} \rho^p(S \cap (U_k \backslash U_{k-1}))$.

Assume $\sum_{k=1}^{\infty} \rho^p(S \cap (U_k \backslash U_{k-1})) < \infty$. Then $\rho^p((S \backslash A) \backslash U_m) \to 0$ as $m \to \infty$ and the inequalities $\rho^p(S \cap U_m) \leqq \rho^p(S \backslash A) \leqq \rho^p(S \cap U_m) + \rho^p((S \backslash A) \backslash U_m)$ imply that $\rho^p(S \cap U_m) \to \rho^p(S \backslash A)$ as $m \to \infty$. Since

$$\inf\{d(x, y) : x \in S \cap A, y \in S \cap U_m\} > 0,$$

Problem 137 shows $\rho^p(S) \geqq \rho^p(S \cap A) + \rho^p(S \cap U_m)$ and the result follows.

The assumption that $\sum_{k=1}^{\infty} \rho^p(S \cap (U_k \backslash U_{k-1})) < \infty$ is justified as follows: $\sum_{k \leqq n, k \text{ odd}} \rho^p(S \cap (U_k \backslash U_{k-1}) = \rho^p(\bigcup_{k \leqq n, k \text{ odd}} (S \cap (U_k \backslash U_{k-1})) \leqq \rho^p(S)$, and similarly $\sum_{k \leqq n, k \text{ even}} \rho^p(S \cap (U_k \backslash U_{k-1})) \leqq \rho^p(S)$, since the disjoint sets are pairwise a positive distance apart; hence $\sum_{k=1}^{\infty} \rho^p(S \cap (U_k \backslash U_{k-1})) \leqq 2\rho^p(S)$, and the result follows. □

139. According to Problem 138, all Borel sets are ρ^p-measurable, and Problem 136 shows that λ_p and ρ^p are mutually absolutely continuous. For any set S and any x, $\rho^p(x + S) = \rho^p(S)$ and so the Radon/Nikodým derivative of ρ^p with respect to λ_p is a constant K_p, whence for all Borel sets A, $\lambda_p(A) = K_p \rho^p(A)$. Since $\lambda_p^*(S) = \inf\{\lambda_p(U) : U \text{ open}, U \supset S\} = K_p \inf\{\rho^p(U) : U \text{ open}, U \supset S\} \geqq K_p \rho^p(S)$, it suffices to show $\rho^p(S) \geqq \inf\{\rho^p(U) : U \text{ open}, U \supset S\}$.

If U is a ball of diameter ε then for all positive δ, $\rho_\varepsilon^p(U) \leqq (\varepsilon + \delta)^p$, whence $\rho_\varepsilon^p(U) \leqq \varepsilon^p$. Thus, if $\operatorname{diam}(U) = \eta < \varepsilon$, then $\varepsilon^p > \eta^p \geqq \rho_\eta^p(U) \geqq \rho_\varepsilon^p(U)$.

If $\delta > 0$ there is a positive ε and a sequence $\{U_m\}_{m=1}^{\infty}$ of open sets such that $\bigcup_{m=1}^{\infty} U_m \supset S$, $\operatorname{diam}(U_m) < \varepsilon$ for all m, and $\rho^p(S) \geqq \rho_\varepsilon^p(S) > \sum_{m=1}^{\infty} (\operatorname{diam}(U_m))^p - \delta$. According to the preceding argument, $(\operatorname{diam}(U_m))^p \geqq \rho_\varepsilon^p(U_m)$ and so $\rho^p(S) \geqq \sum_{m=1}^{\infty} \rho_\varepsilon^p(U_m) - \delta \geqq \rho_\varepsilon^p(\bigcup_{m=1}^{\infty} U_m) - \delta \geqq \inf\{\rho_\varepsilon^p(V) : V \text{ open}, V \supset S\} - \delta$.

Hence for each n in $\mathbb{N}$ there is an open set V_n containing S and such that $\rho^p(S) \geqq \rho_\varepsilon^p(V_n) - 1/n - \delta$. Allowing ε to approach zero, then

δ to approach zero and finally n to approach ∞ leads to the desired conclusion. □

NOTE. The results above show that the startling Besicovitch example [2], to the effect that there is in $\mathbb{R}^3$ a homeomorphic image S of $B_3(0, 1)\backslash B_3(0, 1)^0$ such that $\lambda_3(S)$ is large while the (two-dimensional) area of S is small, cannot be constructed so that $\rho^2(S)$ is small while $\rho^3(S)$ is large.

140. Let $\mathcal{R}$ be the set of regular Borel sets. It will be shown that $\mathcal{R}$ is a σ-ring and contains $\mathsf{K}(X)$.

If $A, B \in \mathcal{R}$, and $\varepsilon > 0$ let U, V be open sets containing A resp. B and such that $\mu(U) < \mu(A) + \frac{1}{2}\varepsilon$, $\mu(V) < \mu(B) + \frac{1}{2}\varepsilon$; let K, L be compact sets contained in A resp. B and such that $\mu(A) < \mu(K) + \frac{1}{2}\varepsilon$, $\mu(B) < \mu(L) + \frac{1}{2}\varepsilon$. Then $U\backslash L \supset A\backslash B \supset K\backslash V$ and $(U\backslash L)\backslash(K\backslash V) = (V\backslash L) \cap U) \cup ((U\backslash K) \cap L)$, whence $\mu((A\backslash B)\backslash(K\backslash V)) < \varepsilon$ and $\mu((U\backslash L)\backslash(A\backslash B)) < \varepsilon$ and so $A\backslash B \in \mathcal{R}$.

Since $A \cup B = (A\backslash B) \cup B$, this time let U, V be open, K, L be compact and such that $U \supset A\backslash B \supset K$, $V \supset B \supset L$ and such that $\mu(U\backslash(A\backslash B)) + \mu((A\backslash B)\backslash K) < \varepsilon/3$, $\mu(V\backslash B) + \mu(B\backslash L) < \varepsilon/3$. Then $U \cup V \supset (A\backslash B) \cup B \supset K \cup L$ and thus $\mu((A\backslash B) \cup B)\backslash(K \cup L) = \mu((A\backslash B)\backslash K) + \mu(B\backslash L) < 2\varepsilon/3$ and so $A \cup B \in \mathcal{R}$.

Finally if $\{A_n\}_{n=1}^{\infty} \subset \mathcal{R}$, then for all n, $A_n\backslash \bigcup_{k<n} A_k \in \mathcal{R}$. Thus let $\{U_n\}_{n=1}^{\infty}$ resp. $\{K_n\}_{n=1}^{\infty}$ be sequences of open resp. compact sets such that for all n, $U_n \supset A_n\backslash \bigcup_{k<n} A_k \supset K_n$ and such that $\mu(U_n) - \varepsilon/2^n < \mu(A_n\backslash \bigcup_{k<n} A_k) < \mu(K_n) + \varepsilon/2^n$. Then $\mu(\bigcup_{n=1}^{\infty} U_n) \leqq \sum_{n=1}^{\infty} \mu(U_n) < \sum_{n=1}^{\infty} \mu(A_n\backslash \bigcup_{k<n} A_k) + \varepsilon/2^n = \mu(\bigcup_{n=1}^{\infty} A_n) + \varepsilon$. Since μ is finite, if $0 < \eta < \varepsilon$, for some N, $\sum_{n>N} \mu(A_n\backslash \bigcup_{k<n} A_k) < \eta$ and then $\mu(\bigcup_{n=1}^{N} K_n) = \sum_{n=1}^{N} \mu(K_n) > \sum_{n=1}^{N} \mu(A_n\backslash \bigcup_{k<n} A_k) - \varepsilon > \sum_{n=1}^{\infty} \mu(A_n\backslash \bigcup_{k<n} A_k) - \eta - \varepsilon = \mu(A) - \eta - \varepsilon$. Hence $\mathcal{R}$ is a σ-ring.

According to Problem 3, every closed, and in particular every compact, set in X is the countable intersection of open sets (a G_δ). Thus every compact set is in $\mathcal{R}$ and so $\mathcal{R}$ contains $\sigma\mathsf{R}(\mathsf{K}(X))$. (Note that every closed set is outer regular.) □

141. If $\varepsilon \in (0, \mu(X))$, let $\{x_n\}_{n=1}^{\infty}$ be dense in X. For each m in $\mathbb{N}$, $\bigcup_{n=1}^{\infty} B(x_n, 1/m) = X$. If $F_{mN} = \bigcup_{n=1}^{N} B(x_n, 1/m)$, then $F_{mN} \subset F_{m,N+1}$, $\lim_{N\to\infty} F_{mN} = X$ and $\lim_{N\to\infty} \mu(F_{mN}) = \mu(X)$. For each m there is an N_m such that $\mu(X\backslash F_{nN_m}) < \varepsilon/2^m$, whence $\mu(\bigcap_{m=1}^{\infty} F_{mN_m}) \geqq \mu(X) - \sum_{m=1}^{\infty} \mu(X\backslash F_{mN_m}) > \mu(X) - \varepsilon$. It will be shown next that $\bigcap_{m=1}^{\infty} F_{mN_m}$, a closed set denoted K_ε, is compact.

Since X is a metric space, it suffices to show that every sequence $\{y_p\}_{p=1}^{\infty}$ in K_ε contains a convergent subsequence. Since each F_{mN_m} is a finite union of balls of radius $1/m$, there is an infinite subsequence $\{y_{p_k}\}_{k=1}^{\infty}$ contained in some $B(x_{n_1}, 1/1)$, an infinite subsubsequence $\{y_{p_{k_l}}\}_{l=1}^{\infty}$ contained in some $B(x_{n_{1_2}}, 1/2), \ldots$, and the sequence $\{y_{p_1}, y_{p_{k_2}}, \ldots\}$ is a Cauchy sequence. Since X is complete and K_ε is closed, this Cauchy sequence has a limit in K_ε and thus K_ε is compact.

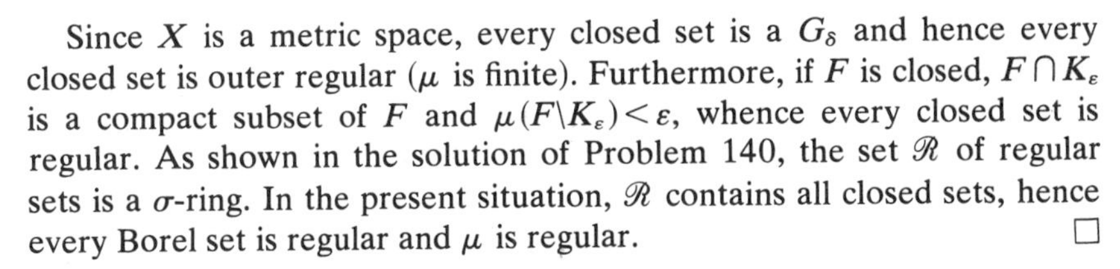

Since X is a metric space, every closed set is a G_δ and hence every closed set is outer regular (μ is finite). Furthermore, if F is closed, $F \cap K_\varepsilon$ is a compact subset of F and $\mu(F \backslash K_\varepsilon) < \varepsilon$, whence every closed set is regular. As shown in the solution of Problem 140, the set $\mathscr{R}$ of regular sets is a σ-ring. In the present situation, $\mathscr{R}$ contains all closed sets, hence every Borel set is regular and μ is regular. □

142. According to Problem 140, μ is regular. Hence for each x there is a positive $\delta(x)$ such that $\mu(B(x, \delta(x))^0) < \varepsilon/2$. There is a finite subset $\{x_i\}_{i=1}^n$ such that $X = \bigcup_{i=1}^n B(x_i, \frac{1}{2}\delta(x_i))^0$. If $\delta = \frac{1}{2}\min_i \{\delta(x_i)\}$, if $\operatorname{diam}(E) < \delta$, and if $E \cap B(x_i, \frac{1}{2}\delta(x_i))^0 \neq \varnothing$ then $E \subset B(x_i, \delta(x_i))^0$ and so $\mu(E) < \varepsilon$. □

143. The result in Problem 141 is applicable to μ restricted to any $B(0, r)$, i.e., μ so restricted is regular. If E is a Borel set, $\varepsilon > 0$ and if, for each r, $E_r = E \cap B\ (0, r)$, then $E = \bigcup_{m=1}^\infty E_m = \bigcup_{m=1}^\infty (E_m \backslash \bigcup_{k<m} E_k)$; for each m there is an open set U_m resp. a compact set K_m such that $U_m \supset E_m \backslash \bigcup_{k<m} E_k \supset K_m$ and such that $\mu(U_m) - \varepsilon/2^m < \mu(E_m \backslash \bigcup_{k<m} E_k) < \mu(K_m) + \varepsilon/2^m$.

If $\mu(E) = \infty$, E is outer regular and $\mu(E) = \lim_{m\to\infty} \mu(E_m) \leqq \lim_{m\to\infty} \mu(\bigcup_{k=1}^m K_k) + \varepsilon$, whence E is also inner regular.

If $\mu(E) < \infty$, then, by an argument like that in the solution of Problem 140, it follows that E is regular. □

144. Let ε be positive and let $\mathscr{U}$ be $\{U : U \text{ open}, \mu(U) = 0\}$, partially ordered by inclusion ($U < U'$ iff $U \subset U'$). Let $\{U_\gamma\}_{\gamma\in\Gamma}$ be a maximal linearly ordered subset of $\mathscr{U}$. It will be shown that if $U_\infty = \bigcup_{\gamma\in\Gamma} U_\gamma$ then $X \backslash U_\infty$, a closed set denoted F, has the required properties.

Since μ is regular and finite there is an open set V containing F and such that $\mu(V) < \mu(F) + \varepsilon$. Then $\{V, U_\gamma\}_{\gamma\in\Gamma}$ is an open cover of X and, since $\{U_\gamma\}_{\gamma\in\Gamma}$ is linearly ordered and X is compact, for some γ_0 in Γ, $V \cup U_{\gamma_0} = X$. Hence $\mu(X) \leqq \mu(V) + \mu(U_{\gamma_0}) = \mu(V) < \mu(F) + \varepsilon$. Thus $\mu(X) = \mu(F)$ and $\mu(U_\infty) = 0$. If F_1 is closed and $\mu(X \backslash F_1) = 0$, then $X \backslash F_1 \in \mathscr{U}$. If $F \not\subset F_1$ then for all γ in Γ, U_γ is a proper subset of $(X \backslash F_1) \cup U_\infty$, and the contradiction of the maximality of $\{U_\gamma\}_{\gamma\in\Gamma}$ results.

Since $F^{-1}(0)$ is closed, $f = 0$ a.e. iff $f^{-1}(0) \supset F$. □

145. If E is a Borel set and if $\varepsilon > 0$ there is a compact set K_ε contained in E and such that $\nu(E) < \nu(K_\varepsilon) + \varepsilon$. If, furthermore, $\{E_n\}_{n=1}^\infty$ is a sequence of pairwise disjoint Borel sets and $E = \bigcup_{n=1}^\infty E_n$, then for each n there is an open set U_n containing E_n and such that $\nu(E_n) > \nu(U_n) - \varepsilon/2^n$. Since for all N, $\nu(E) \geqq \sum_{n=1}^N \nu(E_n)$, it follows that $\nu(E) \geqq \sum_{n=1}^\infty \nu(E_n)$. There is an N_1 such that $\bigcup_{n=1}^{N_1} U_n \supset K_\varepsilon$ and so $\nu(E) < \nu(K_\varepsilon) + \varepsilon \leqq \sum_{n=1}^{N_1} \nu(U_n) + \varepsilon \leqq \sum_{n=1}^\infty \nu(U_n) + \varepsilon \leqq \sum_{n=1}^\infty \nu(E_n) + 2\varepsilon$ and the result follows. □

146. If $\{E_n\}_{n=1}^N$ is a finite sequence of pairwise disjoint Borel sets such that $\bigcup_{n=1}^N E_n = X$ ($\{E_n\}_{n=1}^N$ is a (finite) Borel partition π of X) and if $\{x_n\}_{n=1}^N$ is a finite set σ of N different points, let (σ, π) denote the pair

$(\{x_n\}_{n=1}^N, \{E_n\}_{n=1}^N)$ and write $\sigma \sim \pi$ iff no two x_i belong to the same E_j. Partially order the set $\{(\sigma, \pi): \sigma \sim \pi\}$ according to the rule: $(\sigma_1, \pi_1) < (\sigma_2, \pi_2)$ iff $\sigma_1 \subset \sigma_2$ and every element of π_2 is a subset of some element of π_1 (π_2 refines π_1, i.e., in the natural partial order of partitions, $\pi_1 < \pi_2$); let the set so ordered be $\Gamma = \{\gamma\}$. For each $\gamma = (\{x_n\}_{n=1}^N, \{E_n\}_{n=1}^N)$ and each Borel set E let $\mu_\gamma(E)$ be $\sum_{x_n \in E} \mu(E_n)$.

If $f \in C(X, \mathbb{C})$ and $\varepsilon > 0$, let $\{P_n\}_{n=1}^N$ be a Borel partition of $f(X)$ and such that $\max_n\{\operatorname{diam}(P_n)\} < \varepsilon/2\mu(X)$. If $\pi = \{f^{-1}(P_n)\}_{n=1}^N$, $x_n \in f^{-1}(P_n)$, $n = 1, 2, \ldots, N$, $\sigma = \{x_n\}_{n=1}^N$, then $\sigma \sim \pi$; if $\gamma = (\sigma, \pi)$, $\gamma_1 = (\sigma_1, \pi_1)$, and $\gamma_1 > \gamma$, let π_1 be $\{F_m\}_{m=1}^M$. Then

$$\left| \int_X f(x)\, d\mu_{\gamma_1}(x) - \int_X f(x)\, d\mu(x) \right| \leq \sum_{m=1}^{M} \left| \int_{F_m} f(x)\, d\mu_{\gamma_1}(x) - \int_{F_m} f(x)\, d\mu(x) \right|.$$

Since $\gamma_1 > \gamma$, the mth summand does not exceed $\varepsilon\mu(F_m)/2\mu(X)$ and so the whole sum does not exceed ε. □

147. If $\mu_3 = \mu_1 - \mu_2$ and $\varepsilon > 0$, for some (infinite but countable) Borel partition $\{E_n\}_{n=1}^\infty$ of E, $|\mu_3|(E) < \sum_{n=1}^\infty |\mu_1(E_n) - \mu_2(E_n)| + \varepsilon$. For each n there is a compact set K_n contained in E_n and such that $|\mu_1|(E_n \backslash K_n) + |\mu_2|(E_n \backslash K_n) < \varepsilon/2^n$. Let N be such that $\sum_{N+1}^\infty |\mu_1(E_n) - \mu_2(E_n)| < \varepsilon$. Then $|\mu_3|(E) < \sum_{n=1}^N |\mu_1(K_n) - \mu_2(K_n)| + 3\varepsilon \leqq |\mu_3|(\bigcup_{n=1}^N K_n) + 3\varepsilon$ and so $|\mu_3|$ is inner regular.

Similarly, for each n there is an open set U_n containing E_n and such that $|\mu_1|(U_n \backslash E_n) + |\mu_2|(U_n \backslash E_n) < \varepsilon/2^n$. The argument of the preceding paragraph, *mutatis mutandis*, leads to the conclusion: $|\mu_3|(\bigcup_{n=1}^\infty U_n) < |\mu_3|(E) + 3\varepsilon$, whence $|\mu_3|$ is outer regular, hence regular and so μ_3 is regular. □

148. Let $\sum(x, \varepsilon)$ denote the boundary, $\partial(B(x, \varepsilon))$, i.e., $\sum(x, \varepsilon) = B(x, \varepsilon) \backslash B(x, \varepsilon)^0$. If $\varepsilon_1 \neq \varepsilon_2$, then $\sum(x, \varepsilon_1) \cap \sum(x, \varepsilon_2) = \varnothing$, and so at most countably many $\sum(x, \varepsilon)$ have positive measure. Hence, for each x, there is a sequence $\{\varepsilon_n(x)\}_{n=1}^\infty$ such that $\varepsilon_n(x) \downarrow 0$ and $\mu(\sum(x, \varepsilon_n(x))) = 0$. For each m in $\mathbb{N}$ let $\varepsilon_{n_m}(x)$ be less than $1/m$. For each m, $\bigcup_{x \in X} B(x, \varepsilon_{n_m}(x))^0 = X$ and so there is a finite set $\{x_{mp}\}_{p=1}^{P_m}$ such that $\bigcup_{p=1}^{P_m} B(x_{mp}, \varepsilon_{n_m}(x_{mp}))^0 = X$. Let $\{U_n\}_{n=1}^\infty$ be the countable set $\{B(x_{mp}, \varepsilon_{n_m}(x_{mp}))^0: m = 1, 2, \ldots, p = 1, 2, \ldots, P_m\}$. It will be shown that $\{U_n\}_{n=1}^\infty$ is a basis of the kind specified.

If V is open and $x \in V$, then for some positive ε, $B(x, \varepsilon)^0 \subset V$. If $2/m < \varepsilon$, then for some p, $x \in B(x_{mp}, \varepsilon_{n_m}(x_{mp}))^0$. Furthermore, if $y \in B(x_{mp}, \varepsilon_{n_m}(x_{mp}))^0$, then $d(y, x) \leqq d(y, x_{mp}) + d(x_{mp}, x) < 2\varepsilon_{n_m}(x_{mp}) < 2/m < \varepsilon$, i.e., $B(x_{mp}, \varepsilon_{n_m}(x_{mp}))^0 \subset B(x, \varepsilon)^0 \subset V$. Thus $\{U_n\}_{n=1}^\infty$ is a basis for the topology of X. Since each U_n is some $B(x_{mp}, \varepsilon_{n_m}(x_{mp}))^0$, $\mu(\partial U_n) = 0$, as required. □

149. If $\varepsilon > 0$, there is (Problem 141) a compact set K such that $K \subset U$, $\mu_0(K) > \mu_0(U) - \varepsilon$ and there is in $C(X, \mathbb{C})$ an f_K such that $f_K(K) = 1$, $f_K(X \backslash U) = 0$, and $0 \leqq f_K \leqq 1$. Thus $\mu_0(U) \geqq \int_X f_K(x)\, d\mu_0(x) \geqq \mu_0(U) - \varepsilon$ and

$\mu_n(U) \geqq \int_X f_K(x)\, d\mu_n(x)$. Hence,

$$\liminf_{n=\infty} \mu_n(U) \geqq \lim_{n\to\infty} \int_X f_K(x)\, d\mu_n(x) = \int_X f_K(x)\, d\mu_0(x) \geqq \mu_0(U) - \varepsilon$$

whence $\liminf_{n=\infty} \mu_n(U) \geqq \mu_0(U)$.

For any compact subset K of U, $\int_X f_K(x)\, d\mu_n(x) \geqq \mu_n(K)$ and so

$$\limsup_{n=\infty} \mu_n(K) \leqq \int_X f_K(x)\, d\mu_0(x) \leqq \mu_0(U).$$

Hence if V is open and $V \supset \bar{U}$, then $\mu_0(V) \geqq \limsup_{n=\infty} \mu_n(\bar{U})$; if W is open and $W \supset \partial U$, then $\mu_0(W) \geqq \limsup_{n=\infty} \mu_n(\partial U)$. Since μ_0 is regular, $0 = \mu_0(\partial U) \geqq \limsup_{n=\infty} \mu_n(\partial U) \geqq 0$, i.e., $\mu_n(\partial U) = 0$ for all n. Thus $\mu_n(\bar{U}) = \mu_n(U)$ for all n and $\mu_0(V) \geqq \limsup_{n=\infty} \mu_n(U)$. The regularity of μ_0 implies $\mu_0(\bar{U}) = \mu_0(U) \geqq \limsup_{n=\infty} \mu_n(U)$. The result follows. □

150. If $\varepsilon > 0$, there is for each x in X an open set U_x containing x and such that $\sup\{|f(y_1) - f(y_2)| : y_1, y_2 \in U_x\} < \varepsilon$ and (Problem 148) such that $\mu_0(\partial U_x) = 0$. There is a finite set $\{U_{x_n}\}_{n=1}^N$ such that $\bigcup_{n=1}^N U_{x_n} = X$. If $A_n = U_{x_n} \backslash \bigcup_{m=1}^{n-1} U_{x_m}$ and $V_n = U_{x_n} \backslash \bigcup_{m=1}^{n-1} \bar{U}_{x_m}$, then $X = \bigcup_{n=1}^N A_n = (\bigcup_{n=1}^N V_n) \cup (\bigcup_{m=1}^{n-1} (\bar{U}_{x_m} \backslash \bigcup_{m=1}^{n-1} U_{x_m}))$. If in the last member the first union is denoted U and the second (outer) union is denoted F then U is the union of the disjoint open sets V_n, $F \cap U = \varnothing$ and $\mu_0(F) = 0$. Thus

$$\lim_{n\to\infty} \mu_n(U) + \lim_{n\to\infty} \mu_n(F) = \lim_{n\to\infty} \mu_n(X) = \mu_0(X) = \mu_0(U) + \mu_0(F)$$

$$= \lim_{n\to\infty} \mu_n(U) + \mu_0(F)$$

whence $\lim_{n\to\infty} \mu_n(F) = \mu_0(F) = 0$. Hence, if $y_k \in V_k$, $k = 1, 2, \ldots, N$, then

$$\left| \int_X f(x)\, d\mu_n(x) - \int_X f(x)\, d\mu_0(x) \right|$$

$$\leqq \left| \int_U f(x)\, d\mu_n(x) - \int_U f(x)\, d\mu_0(x) \right| + \left| \int_F f(x)\, d\mu_n(x) \right|$$

$$\leqq \sum_{k=1}^N \left| \int_{V_k} f(x)\, d\mu_n(x) - f(y_k)\mu_n(V_k) \right|$$

$$+ \left| \sum_{k=1}^N f(y_k)(\mu_n(V_k) - \mu_0(V_k)) \right|$$

$$+ \left| \sum_{k=1}^N \left(f(y_k)\mu_0(V_k) - \int_{V_k} f(x)\, d\mu_0(x) \right) \right|$$

$$+ \left| \int_F f(x)\, d\mu_n(x) \right|.$$

According to the choice and construction of the y_k and V_k and the equality $\lim_{n\to\infty} \mu_n(F)=0$, it follows that

$$\limsup_{n=\infty} \left| \int_X f(x)\, d\mu_n(x) - \int_X f(x)\, d\mu_0(x) \right| \leqq 2\varepsilon\mu_0(X).$$

Since ε is an arbitrary positive number, the result follows. □

151. For n in $\mathbb{C}$, μ_n may be regarded as a continuous linear functional on $C(X, \mathbb{C})$. According to the uniform boundedness principle, there is an M such that for all n in $\mathbb{N}$, $\mu_n(X) \leqq M$. Hence $\mu_0(X) \leqq M$ and the argument of Solution 149 may be used. □

7. General Measure Theory

152. The finite additivity of μ is clear. Let a_n be $(-1)^{n+1}/n$. Then $\sum_{n=1}^{\infty} a_n$ converges whereas $\sum_{n=1}^{\infty} |a_n| = \infty$. Thus for some permutation $\{n_k\}_{k=1}^{\infty}$ of $\mathbb{N}$, $\sum_{k=1}^{\infty} a_{n_k} = 2\sum_{n=1}^{\infty} a_n \neq 0$, and μ in this case is not countably additive. $\square$

153. If M is $\{E: E \in \sigma\mathsf{R}(\mathsf{R}), \mu_1(E) = \mu_2(E)\}$ then the finiteness condition *re* μ_1 and μ_2 implies that M is a monotone set of sets and since it contains R, $\mathsf{M} \supset \sigma\mathsf{R}(\mathsf{R})$ (see Problem 2).

On the other hand if R is the ring generated by $\{A: A = [a, b) \cap \mathbb{Q}\}$ then for all A in R, $\text{card}(A) = \text{card}(\mathbb{N})$ unless $A = \emptyset$. Hence if μ_1 is counting measure and $\mu_2 = 2\mu_1$ on $\sigma\mathsf{R}(\mathsf{R})$ then $\mu_1 = \mu_2$ on R. If $r \in \mathbb{Q}$ then $\{r\} = \bigcap_{n=1}^{\infty} [r, r+1/n) \cap \mathbb{Q}$ and so $\{r\} \in \sigma\mathsf{R}(\mathsf{R})$. In particular $\mu_1\{1\} = 1 < \mu_2\{1\} = 2$. $\square$

154. Since $\chi_{X \setminus A} = 1 - \chi_A$, if $A \in \mathscr{F}$ then $X \setminus A \in \mathscr{F}$. Since $\chi_{A \cup B} = \chi_A + \chi_B - \chi_A \cdot \chi_B$, if $A, B \in \mathscr{F}$ then $A \cup B \in \mathscr{F}$. If $\{A_n\}_{n=1}^{\infty} \subset \mathscr{F}$, then $\chi_{\cup_n A_n} = \sum_n \chi_{(A_n \setminus \cup_{m=1}^{n-1} A_m)}$ and so $\mathscr{F}$ is a σ-algebra. $\square$

155. If K is compact there is a sequence $\{U_n\}_{n=1}^{\infty}$ of open sets containing K and a sequence $\{f_n\}_{n=1}^{\infty}$ of continuous functions such that $U_n \supset U_{n+1}$, $\bigcap_n U_n = K$, $f_n(K) = 1$, $f_n(X \setminus U_n) = 0$, $0 \leqq f_n \leqq 1$. Thus $\chi_K = \lim_{n\to\infty} f_n$ and so $\sigma\mathsf{R}(\mathsf{K}(X)) \subset \mathscr{F}$. $\square$

156. If $E \in \mathsf{S}$ and if $\{E_n\}_{n=1}^{\infty}$ is a partition of E and $\nu \in \mathscr{M}$, then $\sum_n |\mu(E_n)| \leqq \sum_n \nu(E_n) = \nu(E)$. Hence $|\mu|(E) \leqq \nu(E)$ and finally $|\mu|(E) \leqq \inf_{\nu \in \mathscr{M}} \nu(E)$. On the other hand, if $E \in \mathsf{S}$ then $|\mu|(E) \geqq \mu(E), -\mu(E)$ and so $|\mu| \in \mathscr{M}$, whence the result follows. $\square$

157. Decompose μ into real and imaginary parts and these into positive and negative parts, *viz.*, $\mu = \mu_1 + i\mu_2 = \mu_1^+ - \mu_1^- + i(\mu_2^+ - \mu_2^-)$. First note that if $A \in \mathsf{S}$ and $A \subset E$, then $E = A \cup (E \backslash A)$ and $|\mu|(E) \geqq |\mu(A)|$ and so $M(E) = M \leqq |\mu|(E)$. On the other hand, if $\{E_n\}_{n=1}^{\infty}$ is a partition of E then $\sum_n |\mu(E_n)| \leqq \sum_n (\mu_1^+(E_n) + \mu_1^-(E_n) + \mu_2^+(E_n) + \mu_2^-(E_n))$ and so $|\mu|(E) \leqq \mu_1^+(E) + \mu_1^-(E) + \mu_2^+ + \mu_2^-(E)$. Thus if $P_i^{\pm}$ are Hahn decompositions for $\mu_i^{\pm}$, $i = 1, 2$, $|\mu|(E) \leqq \mu_1(E \cap P_1^+) + \mu_1(E \cap P_1^-) + \mu_2(E \cap P_2^+) + \mu_2(E \cap P_2^-) \leqq 4M$, by virtue of the earlier conclusions. □

158. For a positive ε there is an $n(\varepsilon)$ such that $\mu(\bigcup_{k=n(\varepsilon)}^{\infty} E_k) < \varepsilon$ if $n \geqq n(\varepsilon)$. Since $\{\bigcup_{k=n}^{\infty} E_k\}_{n=1}^{\infty}$ is a decreasing sequence it follows that $\mu(\bigcup_{k=n}^{\infty} E_k) \downarrow 0$ and thus $\mu(\bigcap_{n=1}^{\infty} \bigcup_{k=n}^{\infty} E_k) = \mu$ (lim $\sup_{n=\infty} E_n) = 0$. □

159. The map $\nu: \mathsf{S} \ni E \mapsto \nu(E) = \mu(f^{-1}(E))$ is a finite measure such that $\nu \ll \mu$. Hence for some integrable and nonnegative h, $\nu(E) = \int_E h(x)\, d\mu(x)$ and so for every bounded measurable function g, $\int_X g(x)\, d\nu(x) = \int_X g(x) h(x)\, d\mu(x)$. On the other hand, since $\chi_{f^{-1}(E)}(x) = \chi_E(f(x))$, it follows that for E in S, $\int_X \chi_E(x)\, d\nu(x) = \nu(E) = \mu(f^{-1}(E)) = \int_X \chi_{f^{-1}(E)}(x)\, d\mu(x) = \int_X \chi_E(f(x))\, d\mu(x)$. The approximation properties of simple functions and the dominated convergence theorem then lead to the desired conclusion. □

160. Since the sequence is contained in $L^1(X, \mu)$ and is a null sequence therein, it follows that $\|f_n - f_m\|_1 \to 0$ as $n, m \to \infty$. Hence there is in $L^1(X, \mu)$ a g that is the limit of the f_n, and indeed $g = 0$, $g = f_0$ a.e. and so $f_0 \in L^1(X, \mu)$, and $\int_X f_0(x)\, d\mu(x) = 0$. □

161. For n in $\mathbb{N}$ let g_n be $f_n \wedge f_0$. Then $g_n \to f_0$ a.e. as $n \to \infty$. Furthermore, $|g_n| \leqq f_0$ for all n and so $\int_E g_n(x)\, d\mu(x) \to \int_E f_0(x)\, d\mu(x)$ as $n \to \infty$. If $\varepsilon > 0$ and lim $\sup_{n=\infty} \int_E f_n(x)\, d\mu(x) \geqq \int_E f_0(x)\, d\mu(x) + \varepsilon$, by passage as needed to subsequences, it may be assumed that $\int_E f_n(x)\, d\mu(x) \geqq \int_E f_0(x)\, d\mu(x) + \varepsilon/2$ for all n. Then $\int_X f_n(x)\, d\mu(x) = \int_E f_n(x)\, d\mu(x) + \int_{X \backslash E} f_n(x)\, d\mu(x) \geqq \int_E f_0(x)\, d\mu(x) + \varepsilon/2 + \int_{X \backslash E} g_n(x)\, d\mu(x)$. Passage to the limit yields a contradiction. □

162. For positive ε let E_n be $\{x : |f_n(x)| \geqq \varepsilon\}$. If

$$A_n = \int_X (|f_n(x)|/(1 + |f_n(x)|))\, d\mu(x),$$

then $A_n \leqq \int_{E_n} (|f_n(x)|/(1 + |f_n(x)|))\, d\mu(x) + \varepsilon\mu(X) \leqq \mu(E_n) + \varepsilon\mu(X)$. It follows that $\lim_{n \to \infty} A_n = 0$ if $\mu(E_n) \to 0$ as $n \to \infty$.

Conversely, if $A_n \to 0$ as $n \to \infty$ and $\mu(E_n) \not\to 0$ as $n \to \infty$, there are positive δ, ε, and a sequence $\{n_k\}_{k=1}^{\infty}$ such that $n_k \to \infty$ as $k \to \infty$ and $\mu\{x : |f_{n_k}(x)| \geqq \varepsilon\} \geqq \delta$. If $B_k = \{x : |f_{n_k}(x)| \geqq \varepsilon\}$, then

$$\int_X \frac{|f_{n_k}(x)|}{(1 + |f_{n_k}(x)|)}\, d\mu(x) \geqq \int_{B_k} \frac{|f_{n_k}(x)|}{(1 + |f_{n_k}(x)|)}\, d\mu(x) \geqq \frac{\delta\varepsilon}{(1 + \varepsilon) > 0},$$

a contradiction, since $A_n \to 0$ as $n \to \infty$. □

163. Egorov's theorem implies that if $\varepsilon > 0$ there is in S an E such that $\mu(X \backslash E) < \varepsilon\mu(X)$ and $\sum_n a_n f_n$ converges uniformly on E. Hence $|a_n f_n|^2 \to 0$

uniformly on E as $n\to\infty$ and $\int_E |f_n(x)|^2\,d\mu(x)=1-\int_{X\setminus E}|f_n(x)|^2\,d\mu(x)\geqq 1-M^2\varepsilon\mu(X)$. These inequalities imply that for all small enough ε and large enough n, i.e., if $\varepsilon<1/2M^2\mu(X)$ and n_0 is so large that $|a_nf_n|^2<\varepsilon/2\mu(E)$ on E whenever $n>n_0$, $|a_n|^2<\varepsilon$ if $n>n_0$. The result follows. □

164. Let A be a measurable subset of E and of finite measure. Egorov's theorem implies there is a sequence $\{A_k\}_{k=1}^\infty$ of measurable subsets of A such that $\mu(A\setminus A_k)<1/k$ and for each k, $f_n\to f$ uniformly on A_k as $n\to\infty$. Bessel's inequality implies that $\int_X \chi_{A_k}(x)f_n(x)\,d\mu(x)\to 0$ for each k as $n\to\infty$. The uniformity of convergence of the f_n to f on each A_k implies $\int_X\chi_{A_k}(x)f_n(x)\,d\mu(x)\to\int_X\chi_{A_k}(x)f(x)\,d\mu(x)$ as $n\to\infty$. It follows that $\int_A f(x)\,d\mu(x)=0$, and since A is an arbitrary measurable set of finite measure, $f=0$, a.e. □

165. If $a_n=\|f\|_{n+1}^{n+1}$, $b_n=\|f\|_n^n$, then since $a_n\leqq\|f\|_\infty b_n$, it follows that $\limsup_{n=\infty}(a_n/b_n)\leqq\|f\|_\infty$. Hölder's inequality implies

$$\|f\|_n\leqq\|f\|_{n+1}\mu(X)^{1/n(n+1)}$$

and so $a_n/b_n\geqq\|f\|_n/\mu(X)^{1/n}$ whence $\liminf_{n=\infty} a_n/b_n\geqq\|f\|_\infty$ and the result follows. □

166. The following will be shown in order: i) if $L^1(X,\mu)\subset L^\infty(X,\mu)$, then $\inf\{\mu(E)$: E measurable and of positive measure$\}>0$; ii) there is at least one atom in X; iii) there are only finitely many atoms in X.

Ad i). Otherwise there is a sequence $\{K_n\}_{n=1}^\infty$ of measurable sets and a sequence $\{k_n\}_{n=1}^\infty$ of positive integers such that $k_{n+1}>2^{k_n}$ and $2^{-k_{n+1}}<\mu(K_n)<2^{-k_n}$. If $L_n=\bigcup_{m=n}^\infty K_m$ then $L_1\supset L_2\supset,\ldots,\mu(L_n)>0$ while $\mu(L_n)\downarrow 0$. If $M_n=L_n\setminus L_{n+1}$, then the M_n are pairwise disjoint and $0<\mu(M_n)<2^{-k_n+1}<2^{-n}$. Thus $\sum_n 2^{n/2}\chi_{M_n}\in L^1(X,\mu)\setminus L^\infty(X,\mu)$, a contradiction.

Ad ii). If there is no atom, then for each n there is in S an M_n such that $0<\mu(M_n)<1/n$, in contradiction of i).

Ad iii). If $\{A_n\}_{n=1}^\infty$ is an infinite sequence of pairwise different atoms and if $\mu(A_n)=a_n(>0)$ either $\sum_n a_n=\infty$, in which case $(x\mapsto 1)\in L^\infty(X,\mu)\setminus L^1(X,\mu)$, or $\sum_n a_n<\infty$, in which case for some sequence $\{n_k\}_{k=1}^\infty$, $\sum_k\sqrt{a_{n_k}}<\infty$ and $(x\mapsto\sum_k\chi_{A_{n_k}}/\sqrt{a_{n_k}})\in L^1(X,\mu)\setminus L^\infty(X,\mu)$. The inevitable contradictions yield the result.

If $A_1,A_2,\ldots,A_n$ are finitely many atoms in X then according to i), $\mu(X\setminus\bigcup_{n=1}^N A_n)=0$. Hence if $f\in L^1(X,\mu)$ then f is constant a.e. on each A_n, say $f=c_n$ a.e. on A_n and $f=\sum_n c_n\chi_{A_n}$ a.e. which shows that $L^1(X,\mu)$ is finite-dimensional. □

167. If $L^1(X,\mu)$ is finite-dimensional, let $\{\chi_{A_n}\}_{n=1}^N$ be a (necessarily finite) maximal linearly independent set of characteristic functions of measurable sets. Then $\mu(X\setminus\bigcup_n A_n)=0$ and if $B_n=A_n\setminus\bigcup_{m=1}^{n-1}A_m$, then the B_n are pairwise disjoint and their union is X. For each n, $\mu(B_n)>0$ since otherwise χ_{A_n} is linearly dependent on $\{\chi_{A_m}\}_{m=1}^{n-1}$. Furthermore, each B_n is an atom since otherwise it is decomposable into two disjoint nonnull measurable subsets that would yield $N+1$ linearly independent characteristic functions. Hence $L^1(X, \mu) = L^\infty(X, \mu)$. A similar argument may be used if it is assumed that $L^\infty(X, \mu)$ is finite-dimensional. □

168. (Counterexample.) If $X = [-\pi, \pi]$, $\mathsf{S} = \sigma\mathsf{R}(\mathsf{K}([-\pi, \pi]))$, $\mu = \lambda$, and f_n is the map $x \mapsto \sin nx$, then $\{f_n\}_{n=1}^\infty$ satisfies the hypothesis. For any sequence $\{n_k\}_{k=1}^\infty$, however, according to Problem 164

$$\mu\left\{x : \lim_k \sin n_k x \text{ exists}\right\} = 0.$$

(Compare with Problems 160 and 161.) □

169. If $f \in L^\infty(X, \mu)$, $\|f\|_\infty = M$, and $E_n = \{x : |f(x)| > M - 1/n\}$ then $\mu(E_n) > 0$. If, for each n, there is in $L^2(X, \mu)$ an h_n such that $h_n = 0$ off E_n and $\|h_n\|_2 = 1$, then $\|T_f\| \geqq M - 1/n$ and since $\|T_f\| \leqq \|f\|_\infty = M$, it follows that $\|T_f\| = \|f\|_\infty$. Hence if $\|T_f\| \neq \|f\|_\infty$, i.e., $\|T_f\| < \|f\|_\infty$, and for some n_0, if it is not true that $h_{n_0} = 0$ a.e. and if, at the same time, $h_{n_0} = 0$ off E_{n_0}, then $\|h_{n_0}\|_2 = \infty$. Hence E_{n_0} contains no measurable subset of positive finite measure, i.e., E_{n_0} is an infinite atom. It has been shown that if $\|T_f\| \neq \|f\|_\infty$ then X contains an infinite atom.

Conversely, if E is an infinite atom, then for any nonzero h in $L^2(X, \mu)$, $\chi_E h = 0$ a.e. and so $\|X_E h\|_2 = 0$, i.e., $\|T_{\chi_E}\| = 0 \neq \|\chi_E\|_\infty$. □

170. The informal reasoning, to be made rigorous below, is that T_f is surjective iff for all h in $L^2(X, \mu)$, $h/f \in L^2(X, \mu)$. Since $1/f$ is undefined where $f = 0$, the argument as just given is incomplete.

First note that if there is a measurable set E of positive finite measure and if $f^{-1}(0) \doteq E$ then T_f cannot be surjective. Indeed, $\chi_E \in L^2(X, \mu)$ and if $fg = \chi_E$ then $g = 0$ a.e. off E and thus $fg = 0$ a.e., a contradiction. On the other hand, if E is an atom and $\mu(E) = \infty$, then every h in $L^2(X, \mu)$ is zero a.e. on E. Thus if $f \in L^\infty(X, \mu)$ and $1/f$ is essentially bounded off E and arbitrary on E, T_f is surjective.

The remarks above lead to the following criterion, to be proved next: T_f is surjective iff for every σ-finite set E, $\|1/f\chi_E\|_\infty < \infty$. (The inequality is to be interpreted in the sense that $\mu((f\chi_E)^{-1}(0)) = 0$ and $1/f\chi_E = 0$ on $(f\chi_E)^{-1}(0)$.)

The proof depends on a lemma: if k is measurable and $\|k\|_\infty = \infty$ and there are no infinite atoms, there is in $L^2(X, \mu)$ a g such that $kg \notin L^2(X, \mu)$. Indeed, if $E_n = \{x : |k(x)| > n\}$, then $E_n \supset E_{n+1}$, $\mu(E_n) > 0$. If $\mu(E_n \backslash E_{n+1}) = 0$ for all but finitely many n, then $|k| = \infty$ on a measurable set E of positive measure, and if F is measurable, a subset of E, and of positive finite measure, then $\chi_F \in L^2(X, \mu)$ and $k\chi_F \notin L^2(X, \mu)$. Thus it may be assumed (*via* subsequencing) that $\mu(E_n \backslash E_{n+1}) > 0$ for all n and then, by hypothesis, there is for each n an F_n, contained in $E_n \backslash E_{n+1}$, measurable, and of finite positive measure. Choose a_n so that $a_n^2 \mu(F_n) = 1/n^2$. If $g = \sum_n a_n \chi_{F_n}$, then $g \in L^2(X, \mu)$ and $kg \geqq 1/\sqrt{\mu(F_n)}$ on F_n, whence $kg \notin L^2(X, \mu)$.

If, therefore, T_f is surjective, E is a measurable set that is σ-finite, and $\|1/f\chi_E\|_\infty = \infty$, according to the lemma there is in $L^2(E, \mu)$ a g such that $g/f\chi_E \notin L^2(E, \mu)$. Extend g from E to X so that $g = 0$ off E and thereby $g \in L^2(X, \mu)$. Hence T_f is not surjective.

Conversely, if $\|1/f\chi_E\|_\infty < \infty$ for every σ-finite measurable set E, then for h in $L^2(X, \mu)$, $\{x: h(x) \neq 0\}$ is σ-finite and $h/f \in L^2(X, \mu)$, whence T_f is surjective. □

171. Let E be $\{x: f(x) \neq 0\}$. Then E is σ-finite, i.e., E is the countable union of disjoint measurable sets E_n of finite measure. Hence $\int_X f(x)\,d\mu(x) = \lim_{n\to\infty} \int_{\cup_{m=1}^n E_m} f(x)\,d\mu(x) \leqq a$. □

172. If $X = \mathbb{R}$, $\mathsf{S} = \{E: E \subset \mathbb{R}, \operatorname{card}(E) \wedge \operatorname{card}(\mathbb{R}\backslash E) \leqq \operatorname{card}(\mathbb{N})\}$, μ is

$$E \mapsto \begin{cases} 0, & \text{if } \operatorname{card}(E) \leqq \operatorname{card}(\mathbb{N}) \\ \infty, & \text{otherwise} \end{cases},$$

and $f = 1$, then for all sets E of finite measure, $\int_E f(x)\,d\mu(x) = 0$, whereas $\int_X f(x)\,d\mu(x) = \infty$. □

173. In the measure situation $(\mathbb{R}, \sigma\mathsf{R}(\mathsf{K}(\mathbb{R})), \lambda)$, let f_n be $\chi_{(-n^2, n^2)}/n$. Then $f_n \to 0$ uniformly as $n \to \infty$ and $\int_{\mathbb{R}} f_n(x)\,dx = 2n$. □

174. Since $\int_X (1 - f_n(x))\,d\mu(x) = \int_E (1 - f_n(x))\,d\mu(x)$, the dominated convergence theorem permits the conclusion: $\int_X (1 - f_n(x))\,d\mu(x) \to 0$ as $n \to \infty$. □

175. If $X = \mathbb{N}$, $\mathsf{S} = 2^{\mathbb{N}}$, $\mu(n) = 1/n$, $E_n = \{n^2, (n+1)^2, \ldots\}$, and

$$f_n(x) = \begin{cases} 1 - 1/n, & \text{if } 1 \leqq x \leqq n \\ 1, & \text{if } x \in E_n \\ 0, & \text{otherwise} \end{cases},$$

then f_n and $1 - f_n$ are integrable and

$$\int_X (1 - f_n(x))\,d\mu(x) = \textstyle\sum_{k=1}^n (n-1)/kn + \sum_{m=n}^\infty 1/m^2 \to \infty \qquad \text{as } n \to \infty.$$

176. Note that $E_n \supset E_{n+1}$ and that by "Abel summation" $\|f\|_1 = \sum_n \int_{E_n \backslash E_{n+1}} f(x)\,d\mu(x) \geqq \sum_n n(\mu(E_n) - \mu(E_{n+1})) = \sum_n \mu(E_n)$. Hence for each k in $\mathbb{N}$ there is an n_k such that $\sum_{n=n_k+1}^\infty \mu(E_n) < 1/k$ and $n_k < n_{k+1}$. Therefore $1/k > \sum_{n=n_k+1}^{n_k+m} \mu(E_n) \geqq m\mu(E_{n_k+m})$. Since $m/(n_k+m) \to 1$ as $m \to \infty$, it follows that $\limsup_{m=\infty} (n_k+m)\mu(E_{n_k+m}) = \limsup_{n=\infty} n\mu(E_n) = \limsup_{m=\infty} (m/n_k+m)(n_k+m)\mu(E_{n_k+m}) \leqq 1/k$ and the result follows. □

177. The following chain of self-explanatory inequalities implies the result:

$$\int_X |f(x)|^p\,d\mu(x) = \int_{\{x:|f(x)|<\varepsilon\}} |f(x)|^p\,d\mu(x) + \int_{\{x:|f(x)|\geqq\varepsilon\}} |f(x)|^p\,d\mu(x)$$
$$\geqq \varepsilon^p \mu\{x: |f(x)| \geqq \varepsilon\}.$$ □

178. If f as described exists and if $\{E_n\}_{n=1}^\infty$ is a partition of X, then $f \geqq 0$ and the Hölder inequality implies that if $\mu_1(E_n) \neq 0$ then $(\mu_2(E_n))^p/(\mu_1(E_n))^{p-1} \leqq \int_{E_n} (f(x))^p \, d\mu_1(x)$ and so $\sum_n (\mu_2(E_n))^p/(\mu_1(E_n))^{p-1} \leqq \|f\|_p^p < \infty$. Thus if $a = \|f\|_p^p$ the result follows.

Conversely, if $\sum_n (\mu_2(E_n))^p/(\mu_1(E_n))^{p-1} \leqq a$ for all partitions $\{E_n\}_{n=1}^\infty$ of X, then $\mu_1(E) > 0$ whenever $\mu_2(E) > 0$ and so $\mu_2 \ll \mu_1$. Hence for some nonnegative integrable f and all measurable sets E, $\mu_2(E) = \int_E f(x) \, d\mu_1(x)$. If $\{b_n\}_{n=1}^\infty$ is a strictly increasing sequence in $[0, \infty)$ and such that $\lim_{n\to\infty} b_n = \infty$ and if $E_n = \{x : b_n \leqq f(x) < b_{n+1}\}$, then $\{E_n\}_{n=1}^\infty$ is a partition of X, $\mu_2(E_n) \geqq b_n\mu_1(E_n)$, $(\mu_2(E_n))^p/(\mu_1(E_n))^{p-1} \geqq b_n^p\mu_1(E_n)$, $\sum_n b_n^p \mu_1(E_n) \leqq a$, and so $\|f\|_p^p \leqq a$. □

179. In the notation used at the end of Solution 178, $\sum_n b_n^p \mu(E_n)$ can, for proper choice of the sequence $\{b_n\}_{n=1}^\infty$, be brought arbitrarily close to $\|f\|_p^p$. On the other hand, $E_n = F_{b_n} \setminus F_{b_{n+1}}$ and $F_{b_{n+1}} \subset F_{b_n}$ whence $\sum_n b_n^p \mu(E_n) = \sum_n b_n^p (\mu(F_{b_n}) - \mu(F_{b_{n+1}}))$. For sufficiently large N, $\sum_{n=N+1}^\infty b_n^p \mu(E_n)$ is small and $\sum_{n=1}^N b_n^p \mu(E_n) = \sum_{n=1}^{N-1} (b_{n+1}^p - b_n^p)\mu(F_{b_n}) + b_1^p \mu(F_{b_1}) - b_N^p \mu(F_{b_{N+1}})$. Since b_1 may be chosen to be zero and since (Problem 176) $b_N^p \mu(F_{b_{N+1}}) \to 0$ as $N \to \infty$, it follows that $\|f\|_p^p$ is approximable arbitrarily well by $\sum_n (b_{n+1}^p - b_n^p)\mu(F_{b_n})$. Finally, the mean value theorem implies that for some θ_n in (b_n, b_{n+1}), $b_{n+1}^p - b_n^p = p\theta_n^{p-1}(b_{n+1} - b_n)$ and so $\|f\|_p^p$ is approximable by $p\sum_n \theta_n^{p-1} \mu(F_{b_n})(b_{n+1} - b_n)$, i.e., $\|f\|_p^p = p\int_0^\infty t^{p-1} \mu(F_t) \, dt$. □

180. According to the Lebesgue–Radon–Nikodým (LRN) theorem there is in $L^1(X, \mu_3)$ an f such that $\mu_{2a}(E) = \int_E f(x) \, d\mu_3(x)$ for all measurable sets E. Since $\mu_{1s} \perp \mu_3$, μ_{1s} and μ_3 live on disjoint sets, and so μ_{2a} and μ_{1s} live on disjoint sets. □

181. i) If, *via* the LRN theorem, $\mu_{ia}, \mu_{is}, i = 1, 2$, are such that $\mu_i = \mu_{ja} + \mu_{js}$, $\mu_{js} \perp \mu_i$, $\mu_{js} \perp \mu_{ja}$, $i \neq j$, then μ_{ia} lives on A_{ia}, μ_{is} lives on A_{is}, $i = 1, 2$, and $A_{ia} \cap A_{is} = \emptyset$, $i, j = 1, 2$, $A_{is} \cap A_{js} = \emptyset$, $i \neq j$. Let E be $A_{1a} \cap A_{2a}$. Then $\mu_i = \mu_{iE} + \mu_{i(X\setminus E)}$, $\mu_{i(X\setminus E)} \perp \mu_{iE}$, $i = 1, 2$, and $\mu_{i(X\setminus E)} \perp \mu_{j(X\setminus E)}$, $i \neq j$. Furthermore if $d\mu_{ia}/d\mu_{ja} = f_{ij}$, $i \neq j$, then $E = \{x : f_{12}(x) f_{21}(x) \neq 0\}$ (E is defined modulo a null set).

If $\mu_{1E}(A) = 0$, then $\mu_1(A \cap E) = \int_{A\cap E} f_{12}(x) \, d\mu_2(x) + \mu_{1s}(A \cap E) = 0$ whence $\mu_{1s}(A \cap E) = 0$ and $\int_{A\cap E} f_{12}(x) \, d\mu_2(x) = 0$. Since $f_{12} \neq 0$ a.e. on E it follows that $\mu_2(A \cap E) = 0$. Thus $\mu_{2E}(A) = 0$, $\mu_{2E} \ll \mu_{1E}$ and, by a symmetrical argument, $\mu_{1E} \ll \mu_{2E}$.

ii) The following relations obtain either by hypothesis or by direct deductions therefrom: $\mu_i = \mu_{iF} + \mu_{i(X\setminus F)}$, $i = 1, 2$, $\mu_{iF} \ll \mu_{jF}$, $\mu_{iF} \ll \mu_j$, $\mu_{i(X\setminus F)} \perp \mu_{j(X\setminus F)}$, $\mu_{i(X\setminus F)} \perp \mu_{jF}$, $\mu_{i(X\setminus F)} \perp \mu_j$, $i \neq j$. Thus μ_{iF}, $\mu_{i(X\setminus F)}$ are the unique LRN components of μ_i, $i = 1, 2$, and so

$$\mu_{iE} = \mu_{iF},\ \mu_{i(X\setminus E)} = \mu_{i(X\setminus F)},\ i = 1, 2.$$

It follows that

$$\mu_1(E\backslash F)=\mu_{1E}(E\backslash F)+\mu_{1(X\backslash E)}(E\backslash F)=\mu_{1F}(E\backslash F)+\mu_{1(X\backslash E)}(E\backslash F)=0+0.$$

A symmetrical argument shows $\mu_1(F\backslash E)=0$ and so $\mu_1(E\Delta F)=0$. Similarly $\mu_2(E\Delta F)=0$, from which the result follows. □

182. If $\mu_1\perp\mu_2$, then μ_1 and μ_2 live on disjoint (measurable) sets A_1 and A_2. If $a_1, a_2\in\mathbb{C}$ and if $\{E_n\}_{n=1}^{\infty}$ is a partition of the measurable set E, then $\sum_n|a_1\mu_1(E_n)+a_2\mu_2(E_n)|=\sum_n|a_1\mu_1(E_n\cap A_1)+a_2\mu_2(E_n\cap A_2)|\leqq(|a_1||\mu_1|+|a_2||\mu_2|)(E)$ and so $|a_1\mu_1+a_2\mu_2|(E)\leqq(|a_1||\mu_1|+|a_2||\mu_2|)(E)$. However, if $\{E_{ni}\}_{n=1}^{\infty}$ is a partition of $E\cap A_i$, $i=1,2$, then $\{E_{ni}\}_{n=1,}^{\infty}{}_{i=1}^{2}$ together with $E\backslash(A_1\cup A_2)$ is a partition of E. Thus

$$|a_1\mu_1+a_2\mu_2|(E)\geqq\sum_n|a_1\mu_1(E_{n1})+a_2\mu_2(E_{n1})|+\sum_n|a_1\mu_1(E_{n2})+a_2\mu_2(E_{n2})|$$
$$+|a_1\mu_1(E\backslash(A_1\cup A_2))+a_2\mu_2(E\backslash(A_1\cup A_2))|.$$

Since μ_i lives on A_i, $i=1,2$, the last term is zero and the first two terms reduce to $\sum_n|a_i||\mu_i(E_{ni})|$, $i=1,2$; finally $|a_1\mu_1+a_2\mu_2|(E)\geqq(|a_1||\mu_1|+|a_2||\mu_2|)(E)$ as required.

Conversely, the argument used in Solution 181 i) may be repeated *mutatis mutandis* to produce μ_{ia}, μ_{is} such that $\mu_i=\mu_{ia}+\mu_{is}$, $\mu_{is}\perp\mu_{ia}$, $i=1,2$, $\mu_{ia}\ll|\mu_j|$, $\mu_{is}\perp|\mu_j|$, $i\neq j$. If μ_{ia} lives on A_i, $i=1,2$, let E be $A_1\cap A_2$. As in Solution 181 i), $\mu_i=\mu_{iE}+\mu_{i(X\backslash E)}$, $i=1,2$, $\mu_{iE}\ll\mu_{jE}$, $\mu_{i(X\backslash E)}\perp\mu_{j(X\backslash E)}$, $i\neq j$.

In the notation introduced in Conventions, if A is measurable, $\mu_{iE}(A)=\int_A(d\mu_{iE}(x)/d)\mu_j|)\ d|\mu_j|(x)$, $i\neq j$. If $f_{ij}=d\mu_{iE}/d|\mu_j|$, $i\neq j$, then $(\mu_{1E}\pm\mu_{2E})(E)=\int_E f_{12}(x)\ d|\mu_2|(x)\pm\int_E f_{21}(x)\ d|\mu_1|(x)=\int_E(f_{12}(x)f_{21}(x)\pm f_{21}(x))\ d|\mu_1(x)$. Hence

$$(|\mu_{1E}\pm\mu_{2E}|)(E)=\int_E|f_{12}(x)f_{21}(x)\pm f_{21}(x)|\ d|\mu_1|(x)=(|\mu_{1E}|+|\mu_{2E}|)(E)$$
$$=\int_E(|f_{12}(x)||f_{21}(x)|+|f_{21}(x)|)\ d|\mu_1|(x).$$

Since the relations above are true for all measurable sets A it follows that $|f_{12}||f_{21}|+|f_{21}|=|f_{12}f_{21}\pm f_{21}|$ a.e. $(|\mu_1|)$ on E.

Since $\mu_{iE}\ll\mu_{jE}$, $i\neq j$, it follows that if A is a measurable set then $|\mu_1|(A)$ and $|\mu_2|(A)$ are both zero or neither is zero. If $|\mu_1|(E)=0$ then automatically $\mu_1\perp\mu_2$. If $|\mu_1|(E)\neq0$, let B be $\{x: f_{21}(x)=0,\ x\in E\}$, let C be $\{x: |f_{12}(x)|+1=|f_{12}(x)+1|,\ x\in E\}$, and let D be $\{x: |f_{12}(x)|+1=|f_{12}(x)-1|,\ x\in E\}$. Then $|\mu_1|(E\backslash(B\cup C))=|\mu_1|(E\backslash(B\cup D))=0$. Hence $|\mu_2|(B)=0$ and so $|\mu_1|(B)=0$, whence $E\doteq C$ and $E\doteq D(|\mu_1|)$. But if $a\in\mathbb{C}$ and $|a|+1=|a+1|$, then $a\geqq0$ and if $|a|+1=|a-1|$, then $a\leqq0$. Hence, since $E\doteq C\cap D(|\mu_1|)$, it follows that $f_{12}=0$ on E a.e. $(|\mu_1|)$. In sum, $\mu_{iE}=0$, $\mu_i=\mu_{i(X\backslash E)}$, $i=1,2$, and $\mu_1\perp\mu_2$, as required. □

183. i) For (γ, k, n) in $\Gamma\times\mathbb{Z}\times\mathbb{N}$ let $E(\gamma, k, n)$ be $\{x: k/2^n \leqq f_\gamma(x)\}$, a measurable set. For (k, n) in $\mathbb{Z}\times\mathbb{N}$ let $\delta(k, n)$ be $\sup_\gamma \mu(E(\gamma, k, n))$. If $f_\gamma \vee f_{\gamma'}$ is denoted $f_{\gamma\vee\gamma'}$, it follows that $E(\gamma\vee\gamma', k, n) = E(\gamma, k, n)\cup E(\gamma', k, n)$. Thus if $\lim_{p\to\infty} \mu(E(\gamma_{knp}, k, n)) = \delta(k, n)$ then $\bigcup_{p=1}^\infty E(\gamma_{knp}, k, n) = \lim_{p\to\infty} E(\gamma_{kn1}\vee\cdots\vee\gamma_{knp}, k, n)$, denoted $E(k, n)$, is a measurable set such that $\mu(E(k, n)) = \delta(k, n)$.

If $\gamma\in\Gamma$ and $\mu(E(\gamma, k, n)\backslash E(k, n)) = \varepsilon$, a nonnegative number, and if $\varepsilon > 0$ let p be such that $\mu(E(\gamma_{kn1}\vee\cdots\vee\gamma_{knp}, k, n) > \delta(k, n)n - \varepsilon/2$. Thus $E(\gamma_{kn1}\vee\cdots\vee\gamma_{knp}\vee\gamma, k, n) = (E(\gamma, k, n)\backslash E(\gamma_{kn1}\vee\cdots\vee\gamma_{knp}, k, n))\cup E(\gamma_{kn1}\vee\cdots\vee\gamma_{knp}, k, n)$ and thus $\delta(k, n) \geqq \mu(E(\gamma_{kn1}\vee\cdots\vee\gamma_{knp}\vee\gamma, k, n)) = \mu(E(\gamma, k, n)\backslash E(\gamma_{kn1}\vee\cdots\vee\gamma_{knp}, k, n)) + \mu(E\gamma_{kn1}\vee\cdots\vee\gamma_{knp}, k, n)) \geqq \varepsilon + \delta(k, n) - \varepsilon/2 = \delta(k, n) + \varepsilon/2$ and hence $\varepsilon = 0$.

For (γ, n) in $\Gamma\times\mathbb{N}$, $\bigcup_{k=-\infty}^\infty E(\gamma, k, n) = X$ and so $\bigcup_{k=-\infty}^\infty E(k, n) = X$. For n fixed, let $A(k, n)$ be $E(k, n)\backslash\bigcup_{l=k+1}^\infty E(l, n)$, k in $\mathbb{Z}$. If $k\neq k'$, $A(k, n)\cap A(k', n) = \varnothing$ and so the map $g_n: x\mapsto (k+1)/2^n$ if $x\in A(k, n)$ is properly defined, since $\bigcup_k A(k, n) = X$. If $\gamma\in\Gamma$,

$$\{x: f_\gamma(x)\geqq g_n(x)\} = \bigcup_k \{x: f_\gamma(x) \geqq g_n(x)\}\cap A(k, n)$$
$$\subset \bigcup_k (E(\gamma, k+1, n)\backslash\bigcup_{l=k+1}^\infty E(l, n).$$

According to an earlier observation, $\mu(E(\gamma, k+1, n)\backslash E(k+1, n)) = 0$ and hence $\mu(E(\gamma, k+1, n)\backslash\bigcup_{l=k+1}^\infty E(l, n)) = 0$. Consequently,

$$\mu\{x: f_\gamma(x)\geqq g_n(x)\} = 0.$$

Since $E(\gamma_{knp}, k, n) = E(\gamma_{knp}, 2k, n+1)$, it differs from $E(\gamma_{knp}, k, n)\cap E(2k, n+1)$ by a null set; similarly, $E(\gamma_{2k,n+1,q}, 2k, n+1) = E(\gamma_{2k,n+1,q}, k, n)$ and so differs from $E(\gamma_{2k,n+1,q}, 2k, n+1)\cap E(k, n)$ by a null set. Hence

$$E(k, n) = \bigcup_{p=1}^\infty E(\gamma_{knp}, 2k, n+1)$$
$$\doteq \bigcup_{q=1}^\infty \bigcup_{p=1}^\infty E(\gamma_{knp}, 2k, n+1)\cap E(\gamma_{2k,n+1,q}, 2k, n+1)$$
$$\doteq \bigcup_{q=1}^\infty E(\gamma_{2k,n+1,q}, 2k, n+1)\cap E(k, n) \doteq E(2k, n+1)\cap E(k, n).$$

Similarly, $E(2k, n+1)\doteq E(k, n)\cap E(2k, n+1)$ and so $E(2k, n+1)\doteq E(k, n)$. Thus

$$A(k, n) = E(k, n)\backslash\bigcup_{l=k+1}^\infty E(l, n) \doteq E(2k, n+1)\backslash\bigcup_{l=k+1}^\infty E(2l, n+1)$$
$$= (E(2k, n+1)\backslash\bigcup_{l=2k+1}^\infty E(l, n+1))$$
$$\cup(\bigcup_{l=2k+1}^\infty E(l, n+1)\backslash\bigcup_{l=k+1}^\infty E(2l, k+1)).$$

Since

$$\bigcup_{l=2k+1}^\infty (E(l, n+1)\backslash\bigcup_{l=k+1}^\infty E(2l, n+1)$$
$$= E(2k+1, n+1)\backslash\bigcup_{l=k+1}^\infty E(2l, n+1)$$
$$= A(2k+1, n+1),$$

it follows that $A(k,n)\doteq A(2k,n+1)\cup A(2k+1,n+1)$. Thus $g_n\geqq g_{n+1}$ a.e., the measurable function $\lim_{n\to\infty} g_n$, denoted g, exists a.e., and for all γ, $f_\gamma\leqq g$ a.e. (Where g is not yet defined it may be set equal to zero.)

ii) If h is measurable and if for all γ, $f_\gamma\leqq h$, let H be $\{x: h(x)<g(x)\}$. It will be shown that $\mu(H)=0$. First note that

$$H=\bigcup_{m=1}^{\infty}\{x: h(x)\leqq g(x)-1/2^m\}\subset\bigcup_m\{x: h(x)\leqq g_{m+1}(x)-1/2^m\}.$$

If $\mu(H)>0$, for some m, $\mu(x: h(x)\leqq g_{m+1}(x)-1/2^m\}>0$. However, $\{x: h(x)\leqq g_{m+1}(x)-1/2^m\}=\bigcup_{k=-\infty}^{\infty}\{x: h(x)\leqq g_{m+1}(x)-1/2^m\}\cap A(k,m+1)$. Hence for some k, $\{x: h(x)\leqq g_{m+1}(x)-1/2^m\}\cap A(k,m+1)$, denoted A, is not a null set. In A and off a null set in A, $h(x)\leqq g_{m+1}(x)-1/2^m=(k+1)/2^{m+1}-2/2^{m+1}=(k-1)/2^{m+1}$. Furthermore

$$\begin{aligned}A&=\{x: h(x)\leqq g_{m+1}(x)-1/2^m\}\cap(E(k,m+1)\setminus\bigcup_{l=k+1}^{\infty}E(l,m+1))\\&=\bigcup_{p,q=1}^{\infty}\{x: h(x)\leqq g_{m+1}(x)-1/2^m\}\\&\quad\cap(E(\gamma_{k,m+1,p},k,m+1)\setminus E(\gamma_{l,m+1,q},l,m+1))\end{aligned}$$

and so at least one of the summands is not a null set. On this set, corresponding to indices p, q, l, $k/2^{m+1}\leqq f_{\gamma k,m+1,p}\leqq g_{m+1}-1/2^m=(k-1)/2^{m+1}$, a.e., a contradiction. Hence $\mu(H)=0$ and $h\geqq g$ a.e., as required. □

184. There is a sequence $\{\{a_{mn}\}_{m=1}^{\infty}\}_{n=1}^{\infty}$ of sequences such that for each m, $0=a_{m1}<a_{m2}<\cdots$, such that $\sum_n a_{mn}\mu\{x: a_{mn}\leqq f(x)<a_{m,n+1}\}=\sum_n(a_{m,n+1}-a_{mn})\mu\{x: a_{mn}\leqq f(x)\}$, and such that the left member approaches

185. Since $\mu(X)=1$ and $f\in L^2(X,\mu)$, $f\in L^1(X,\mu)$. If $m=\int_X f(x)\,d\mu(x)$, the calculation proceeds directly to the conclusion. □

186. Again, the calculation is direct. □

187. Let X be the two-point space $\{0,1\}$. If $\mathsf{S}=2^X$, $\mu\{0\}=1-\mu\{1\}=x$ and $f=\chi_{\{0\}}$, the results of Problems 185 and 186 may be applied directly. The inequality is used in S. Bernstein's proof of the Weierstrass approximation theorem. □

8. Measures in $\mathbb{R}^n$

188. i) The following general result will be established: if $\mathcal{M} \subset 2^X$ and $\operatorname{card}(\mathcal{M}) \geqq 2$, then $\operatorname{card}(\sigma\mathsf{R}(\mathcal{M})) \leqq (\operatorname{card}(\mathcal{M}))^{\operatorname{card}(\mathbb{N})}$. The proof uses transfinite induction. Let $\mathcal{M}_0$ be $\mathcal{M}$. If $0 < \beta < \Omega$ (Ω is the ordinal number of the (well-ordered) set of ordinal numbers of well-ordered countable sets), let $\mathcal{M}_\beta$ be the set of all countable unions of differences of sets drawn from $\bigcup_{0 \leqq \alpha < \beta} \mathcal{M}_\alpha$. It will be shown that $\mathsf{S} = \bigcup_{0 \leqq \alpha < \Omega} \mathcal{M}_\alpha = \sigma\mathsf{R}(\mathcal{M})$.

If A and $B \in \mathsf{S}$, then for some α, A and $B \in \mathcal{M}_\alpha$ and so $A \backslash B \in \mathcal{M}_{\alpha+1}$. If $\{A_n\}_{n=1}^\infty \subset \mathsf{S}$, there is in $[0, \Omega)$ an α such that $\{A_n\}_{n=1}^\infty \subset \mathcal{M}_\alpha$ and so $\bigcup_n A_n \in \mathcal{M}_{\alpha+1}$, whence S is a σ-ring. Since $\mathcal{M}_0 = \mathcal{M} \subset \mathsf{S}$ it follows that $\sigma\mathsf{R}(\mathcal{M}) \subset \mathsf{S}$. Transfinite induction shows that for all α, $\mathcal{M}_\alpha \subset \sigma\mathsf{R}(\mathcal{M})$, whence $\mathsf{S} = \sigma\mathsf{R}(\mathcal{M})$.

Since $\operatorname{card}(\mathcal{M}) \geqq 2$, it follows that $\operatorname{card}(\mathcal{M}_0) \leqq \operatorname{card}(\mathcal{M}))^{\operatorname{card}(\mathbb{N})}$. If $\beta \in [0, \Omega)$ and if for all α in $[0, \beta)$, $\operatorname{card}(\mathcal{M}_\alpha) \leqq (\operatorname{card}(\mathcal{M}))^{\operatorname{card}(\mathbb{N})}$, then $\operatorname{card}(\bigcup_{0 \leqq \alpha < \beta} \mathcal{M}_\alpha) \leqq (\operatorname{card}(\mathcal{M}))^{\operatorname{card}(\mathbb{N})}$. $\operatorname{card}(\mathbb{N}) = \operatorname{card}(\mathcal{M}))^{\operatorname{card}(\mathbb{N})}$ and so $\operatorname{card}(\mathcal{M}_\beta) \leqq (\operatorname{card}(\mathcal{M}))^{\operatorname{card}(\mathbb{N}))2} = (\operatorname{card}(\mathcal{M}))^{\operatorname{card}(\mathbb{N})}$, from which the result follows, because $\operatorname{card}([0, \Omega)) \leqq \operatorname{card}(\mathbb{R})$, and $(\operatorname{card}(\mathcal{M}))^{\operatorname{card}(\mathbb{N})} \geqq \operatorname{card}(\mathbb{R})$.

In particular, if $\mathcal{M} = \{(a, b) : 0 \leqq a < b \leqq 1,\ a, b \in \mathbb{Q}\}$, then $\sigma\mathsf{R}(\mathcal{M}) = \mathsf{S}_\beta(I)$ and so $\operatorname{card}(\mathsf{S}_\beta(I)) \leqq \operatorname{card}(\mathbb{R})$. Since $\mathsf{S}_\beta(I) \ni \{x\}$ for all x in I, it follows that $\operatorname{card}(\mathsf{S}_\beta(I)) = \operatorname{card}(\mathbb{R})$.

ii) Let $\mathcal{S}$ denote the set in question. Since $\mathcal{S} \subset ([0, \infty])^{\mathsf{S}_\beta(I)}$, it follows that $\operatorname{card}(\mathcal{S}) \leqq (\operatorname{card}(\mathbb{R})^{\operatorname{card}(\mathsf{S}_\beta(I))} = \operatorname{card}(\mathbb{R}))^{\operatorname{card}(\mathbb{R})} = \operatorname{card}(2^{\mathbb{R}})$. Moreover if $\mathcal{B} = \{(A, B) \colon A, B \in \mathsf{S}_\beta(I),\ A \cap B = \varnothing,\ \operatorname{card}(A) = \operatorname{card}(B) = \operatorname{card}(\mathbb{R})\}$, then $\operatorname{card}(\mathcal{B}) = \operatorname{card}(\mathbb{R})$. For each pair (A, B) in $\mathcal{B}$ and each pair (s, t) in $(0, \infty)^2$ there is a nonatomic measure μ that lives on $A \cup B$ and is such that $\mu(A) = s$, $\mu(B) = t$. Thus $\operatorname{card}(\mathcal{S}) \geqq \operatorname{card}(((0, \infty)^2)^{\mathcal{B}}) = (\operatorname{card}(\mathbb{R}))^{\operatorname{card}(\mathbb{R})} = \operatorname{card}(2^{\mathbb{R}})$ and so $\operatorname{card}(\mathcal{S}) = \operatorname{card}(2^{\mathbb{R}})$. □

189. The construction of *the* Cantor set (see Problem 12) on $[0, 1]$ may be imitated on any finite interval $[a, b]$ as follows. Let $\{\{a_{km}\}_{m=1}^{2k}\}_{k=0}^\infty$ be a double sequence such that for all k, m, $a_{km}>0$, for k fixed a_{km} is a constant function c_k of m, $c_k>c_{k+1}$, and $\sum_k 2^k c_k = c \leqq b-a$. At the midpoint $\frac{1}{2}(a+b)$ of $[a, b]$ center an open interval I_0 of length c_0, at the midpoints of the two intervals remaining, center open intervals I_{01}, I_{11} of length $c_1, \ldots$, etc. Then $[a, b]\backslash\bigcup_{n=1}^{\infty}\{I_{\varepsilon_1\varepsilon_2\cdots\varepsilon_n}: \varepsilon_k = 0 \text{ or } 1\}$ is a perfect nowhere dense subset F of $[a, b]$ and $\lambda(F)=(b-a)-c \geqq 0$. (In the present situation, $[a, b]=[0, 7/9]$, $c_0 = 56/90$, $c_1 = 56/900, \ldots$, and $E=F$.) The generalized Cantor function g_F for the set F is defined as follows: for t in $[a, b]$ let $\mathscr{I}_t$ be the set of all intervals $I_{\varepsilon_1\varepsilon_2\cdots\varepsilon_n}$ such that the right endpoint of each does not exceed t. Then $g_F(t)=\sup\{\sum_{k=1}^{n}\varepsilon_k 2^{-k}: I_{\varepsilon_1\varepsilon_2\cdots\varepsilon_n}\in\mathscr{I}_t\}$. Then g_F is a continuous monotone increasing function on $[a, b]$, $g_F(a)=0=1-g_F(b)$.

For every $[c, d]$ contained in $[a, b]$ let $\mu([c, d])$ be $g_F(d)-g_F(c)$. Then μ serves to define a Stieltjes measure on $\mathsf{S}_\beta(I)$ and $\operatorname{supp}(\mu)=F$ since $\mu(I_{\varepsilon_1\varepsilon_2\cdots\varepsilon_n})=0$ for all $I_{\varepsilon_1\varepsilon_2\cdots\varepsilon_n}$ and if U is an open set meeting F then $\mu(U)>0$. □

190. Let S be a countable infinite set $\{x_n\}_{n=1}^{\infty}$ contained in $[0, 1]$, e.g., let S be $\{S^{-n}\}_{n=1}^{\infty}$. For any Borel set A let $\mu(A)$ be $\sum_{x_n\in A} 2\cdot 3^{-n}$. Then $\mu(\mathsf{S}_\beta)=$ *the* Cantor set C. □

191. According to Jensen's inequality (Problem 102), since $x\mapsto e^x$ is a convex function, $\exp(\int_I f(x)\, d\mu(x) \leqq \int_I e^{f(x)}\, d\mu(x)$. Furthermore (see Problem 102), $e^a(x-a)+e^a \leqq e^x$ and equality obtains iff $x=a$. Thus if $a=\int_I f(x)\, d\mu(x)$ and if it is not true that $f=a$ a.e. (μ), then $e^a(f(x)-a)+e^a<\exp(f(x))$ on a set of positive measure, and integration leads to the inequality: $\exp(\int_I f(x)\, d\mu(x))<\int_I e^{f(x)}\, d\mu(x)$. □

192. If F is a closed set of I and if $0\notin F$ there is in $C(I, [0, \infty))$ an h such that $h(F)=1$, $h(0)=0$ and $0\leqq h \leqq 1$. Then $0 \geqq \int_I h(x)\, d\mu(x) \geqq \mu(F)$, whence $\operatorname{supp}(\mu)=\{0\}$. It follows that if $f\in C(I, \mathbb{C})$ then $\int_I f(x)\, d\mu(x)=\mu(0)f(0)=c.f(0)$. □

193. Let A be $\{f: f\in C(I, \mathbb{R}), f(x)=f(1-x), f(0)=f(1)=0\}$. Then, according to the Stone–Weierstrass theorem the linear span of $\{\sin^k \pi t\}_{k=1}^{\infty}$ is dense in A relative to the $\|\cdot\cdot\cdot\|_\infty$-induced topology of A. Hence if $E\in \mathsf{S}_\beta([0, \frac{1}{2}])$, $\mu(E)=-\mu(1-E)$. □

194. The linear span of $\{\cos^k \pi t\}_{k=1}^{\infty}$ is dense (see Problem 193) in $\{f: f\in C(I, \mathbb{C}), f(\frac{1}{2})=0\}$. Consequently μ lives on $\{\frac{1}{2}\}$. □

195. Let B be the $\mathbb{R}$-linear span of $\{\cos^k \pi t\}_{k=1}^{N}$. Then B is a closed linear subspace of $C(I, \mathbb{R})$. The map $B\ni\sum_{k=1}^{N} c_k \cos^k \pi t\mapsto\sum_{k=1}^{N} c_k a_k$, denoted L, is a continuous linear functional on B and, *via* the Hahn–Banach theorem, L may be extended without increase of norm to some linear functional L' on $C(I, \mathbb{R})$. According to the Riesz representation theorem there is on I a finite (possibly signed) Borel measure μ such that for all f in $C(I, \mathbb{R})$, $L'(f)=\int_I f(x)\, d\mu(x)$ and the result follows. □

196. For f in $\mathscr{F}$ the map $f^*: x \mapsto \int_0^x f(t)\,dt$ is in $C(I, \mathbb{C})$. Hence $|f^*(x)| \leqq \int_I |f(t)|\,dt = \int_{I \setminus G(f,K(1))} |f(t)|\,dt + \int_{G(f,K(1))} |f(t)|\,dt \leqq K(1)+1$. Hence $\{f^*\}_{f \in \mathscr{F}}$ is a uniformly bounded set in $C(I, \mathbb{C})$.

If $x \leqq y$ and $\varepsilon > 0$ then $|f^*(x) - f^*(y)| \leqq \int_{[x,y] \setminus G(f,K(\varepsilon/2))} |f(t)|\,dt + \int_{[x,y] \cap G(f,K(\varepsilon/2))} |f(t)\,dt \leqq (y-x)K(\varepsilon/2) + \varepsilon/2$. If $y - x \leqq \varepsilon/2K(\varepsilon/2)$ then $|f^*(x) - f^*(y)| \leqq \varepsilon$. In sum, $\{f^*\}_{f \in \mathscr{F}}$ is a uniformly bounded equicontinuous set and the Arzelà-Ascoli theorem implies the required result. □

197. If $\mu \in P$ and if there are in S_β sets A_1 and A_2 such that $A_1 \cap A_2 = \varnothing$, $A_1 \cup A_2 = I$, and $\mu(A_1)\mu(A_2) > 0$, then the maps $\mu_i: \mathsf{S}_\beta \ni E \mapsto \mu(E \cap A_i)/\mu(A_i)$, $i = 1, 2$, are also in P. Furthermore $\mu = \mu(A_1)\mu_1 + \mu(A_2)\mu_2$ and $\mu \neq \mu_1, \mu_2$. In other words μ is an extreme point of P iff for all E in S_β, $\mu(E) = 0$ or 1. □

198. Let A be the linear span of the sequence $\{x \mapsto e^{nx}\}_{n=0}^\infty$. Then the Stone–Weierstrass theorem implies that A is dense in $C(I, \mathbb{R})$. On A the linear functional $L: \sum_{n=0}^N a_n e^{nx} \mapsto \sum_{n=0}^N a_n t_n$ is nonnegative.

It will be shown that if $\{g_m: x \to \sum_{n=0}^{N_m} a_{mn} e^{nx}\}_{m=1}^\infty$ is a sequence in A and if $g_m \downarrow 0$ then $L(g_m) \downarrow 0$. Indeed, since L is nonnegative, $L(g_m) \geqq L(g_{m+1})$. If $L(g_m) \downarrow a > 0$, then, since by Dini's theorem, $\|g_m\|_\infty \downarrow 0$, there is for each k in $\mathbb{N}$ an m_k such that $g_m < a/k$ if $m > m_k$. Thus $a \leqq L(g_m) < aL(1)/k$, i.e., $L(1) \geqq k$. Since k is arbitrary in $\mathbb{N}$, $L(1)$ is not defined and the contradiction implies $a = 0$.

Next it will be shown that if $\{g_m\}_{m=1}^\infty \subset A$ and $\|g_m\|_\infty \to 0$ as $m \to \infty$ then $L(g_m) \to 0$. To this end let G_m be $g_m + (m+1)\|g_m\|_\infty/m$. Then $\|G_m\|_\infty \to 0$ as $m \to \infty$. It may be assumed that all g_m are not zero. If $L(G_m) \nrightarrow 0$, by passage to a subsequence it may be assumed that for some positive a, $L(G_m) \geqq a$, since all G_m are strictly positive. Again, because all G_m are strictly positive and the G_m converge uniformly to zero, by passage to a subsequence it may be assumed that $G_m > G_{m+1}$. The preceding paragraph assures that $L(G_m) \downarrow 0$, a contradiction. Thus $L(g_m) = L(G_m) - ((m+1)\|g_m\|_\infty/m)L(1) \to 0$ as $m \to \infty$. Hence L is continuous and the result now follows from the Riesz representation theorem. □

199. Direct calculation shows that if μ_1 exists and $p \leqq q \leqq r \leqq s$ then $\mu_1([p,q] \times [r,s]) = (q-p)(s-r)(1 + (q+p)(s+r))$ and $\mu_2([p,q] \times [r,s]) = (q-p)(s-r)(1 - (p^2 + pq + q^2)(r^2 + rs + s^2))$. Testing these formulae when $p = 0$, $q = r = \frac{1}{2}$, and $s = 1$ leads to a contradiction in each instance. □

200. Integration by parts shows that if $h = 0$ off $[a, b]$ then, $F(x)$ denoting $\int_a^x f(t)\,dt$, $-\int_{\mathbb{R}} f(t)h(t)\,dt = -[F(t)h(t)|_a^b - \int_a^b F(t)h'(t)\,dt] = \int_a^b F(t)h'(t)\,dt = \int_a^b h'(t)\,d\mu(t)$. Hence if $E \in \mathsf{S}_\beta$, then $\mu(E) = \int_E F(t)\,dt$ and so $\mu \ll \lambda$ and $d\mu/d\lambda = F$. □

201. i) Since $\mathbb{R} \setminus (x+E) = x + (\mathbb{R} \setminus E)$ and $E \Delta (x+E) = (\mathbb{R} \setminus E) \Delta (x + (\mathbb{R} \setminus E))$, the result is clear.

ii) If E' denotes $\mathbb{R}\backslash E$ then by hypothesis and i), for all x, $\lambda((x+E)\cap E')=\lambda((x+E')\cap E)=0$. If $\lambda(E).\lambda(E')>0$ there is in E resp. E' a point p resp. p' such that for small positive a, $\lambda(E\cap(p-a,p+a))$ resp. $\lambda(E'\cap(p'-a,p'+a))$ is near $2a$ (metric density theorem). If $x=p-p'$ then $\lambda((x+E')\cap(p-a,p+a))$ and $\lambda(E\cap(p-a,p+a))$ are near $2a$ and so $\lambda((x+E')\cap E\cap(p-a,p+a))>0$ and a contradiction results. □

202. i) According to Fejér's theorem, the hypothesis implies that if f is continuous and periodic with period 2π, then $\int_{\mathbb{R}} f(t)\,d\mu(t)=0$. Hence if A is Borel measurable and for all n in $\mathbb{Z}$, $A+2\pi n=A$ (A is "periodic" with "period" 2π) then A is the union of the disjoint sets $A\cap[2\pi m, 2\pi(m+1))$, denoted A_m, m in $\mathbb{Z}$, $A_m+2\pi=A_{m+1}$, and so $\mu(\bigcup_{n\in\mathbb{Z}} A+2\pi n)=\mu(A)=\sum_{m\in\mathbb{Z}}\mu(A_m)=0$. If A is a Borel measurable set then $\bigcup_{n\in\mathbb{Z}}(A+2\pi n)$, denoted B, is periodic with period 2π, and according to the last result, $\mu(\bigcup_{n\in\mathbb{Z}}(B+2\pi n))=0=\mu(\bigcup_{n\in\mathbb{Z}}(A+2\pi n))$.

ii) If $\{a_m\}_{m=0}^{\infty}$ is a sequence in $\mathbb{R}$, $\sum_{m=0}^{\infty}|a_m|=\infty$, and $\sum_{m=0}^{\infty}a_m=0$, then for any Borel set E let $\mu(E)$ be 0 if $E\cap(2\pi\mathbb{Z})=\varnothing$ and otherwise let $\mu(E)$ be $\sum_{2\pi m\in E}a_m$. (For any Borel set E, $\mu(E)=\sum_{2\pi m\in E}a_m$, if the following convention is adopted: $\sum_{m\in\varnothing}a_m=0$.) Then for m in $\mathbb{Z}$, $\int_{\mathbb{R}} e^{-itm}\,d\mu(t)=\sum_{n=0}^{\infty}a_n=0$. Yet, if $E=\{2k\pi\}_{k=0}^{\infty}$, then $\mu(E)=0$, $\mu(E+2\pi)=-a_0$, $\mu(E+2.2\pi)=-a_0-a_1$, etc.; $\mu(E-2p\pi)=0$, $p=1,2,\ldots$. Thus if $N>0$, $\sum_{n=-\infty}^{N}\mu(E+2n\pi)=-Na_0-(N-1)a_1-\cdots-a_{N-1}=-N(a_0+\cdots a_{N-1})+a_1+2a_2+\cdots+(N-1)a_{N-1}$.

In particular if $a_0=1$ and $a_n=(-1)^n(2n+1)/n(n+1)$ then $\sum_{n=0}^{\infty}a_n=0$, $\sum_{n=0}^{\infty}|a_n|=\infty$, $-N(a_0+a_1+\cdots+a_{N-1})=(-1)^N$, and $\sum_{n=1}^{N-1}na_n=S_{N-1}=\sum_{n=1}^{N-1}(-1)^n(2n+1)/(n+1)$. It follows that $S_{2M}\uparrow\bar{S}>\log 3/2$ and $S_{2M+1}\downarrow\underline{S}<-3/2-\log 3/4$. Hence $-N(a_0+a_1+\cdots+a_{N-1})+\sum_{n=1}^{N-1}na_n\not\to 0$ as $N\to\infty$. □

203. i) Extend f so that $f=0$ off I. There is a Borel set E on which μ lives and such that $\lambda(E)=0$. It may be assumed that 0, 1 are in E. To show that $\int_I f(x-y)\,d\mu(y)$ exists it suffices to note first that for $\lambda-$ almost every x, $f(x-y)$ is a Borel measurable function of y and so the integral exists and is finite a.e. (λ) on $\mathbb{R}$ iff $\int_I|f(x-y)|\,d\mu(y)<\infty$ a.e. (λ) on $\mathbb{R}$.

Let E_n be $\{z: |f(z)|\leqq n,\ z\in I\}$. Then each E_n is Borel measurable and $\lambda(E_n)\uparrow 1$ as $n\to\infty$. If $E_n+E=\{x: x=z+y, z\in E_n, y\in E\}$ then, since $0,1\in E$, $[0,2]\supset E_n+E\supset E_n\cup(E_n+\{1\})$, $E_n\cap(E_n+\{1\})\doteq\varnothing(\lambda)$. Hence, $\lambda_*(\lambda^*)$ denoting inner (outer) Lebesgue measure, it follows that $2\geqq\lambda^*(E_n+E)$ and $\lambda_*(E_n+E)\uparrow 2$ and so $2\geqq\lambda^*(\bigcup_{n=1}^{\infty}(E_n+E))\geqq\lambda_*(\bigcup_{n=1}^{\infty}(E_n+E))\geqq 2$, i.e., $\bigcup_{n=1}^{\infty}(E_n+E)$ is Lebesgue measurable and its measure is 2.

If $x\in\bigcup_{n=1}^{\infty}(E_n+E)$, say $x\in E_{n_0}+E$, then for y in E, $x-y\in E_{n_0}$, $|f(x-y)|\leqq n_0$ and so $\int_I|f(x-y)|\,d\mu(y)<\infty$ a.e. (λ) in $[0,2]$. If $x\notin\mathbb{R}\backslash[0,2]$ and $y\in E$ then $x-y\notin[0,1]$, $f(x-y)=0$, and $\int_I|f(x-y)|\,d\mu(y)=0$. Consequently $\int_I|f(x-y)|\,d\mu(y)<\infty$ a.e. (λ).

ii) The Fubini theorem implies [23] that if $f,g\in L^1(\mathbb{R},\lambda)$ then $\int_{\mathbb{R}} f(x-y)g(y)\,dy$ exists for almost every x. The function so defined is denoted $f*g$; $f*g=g*f$ and $f*g\in L^1(\mathbb{R},\lambda)$.

The Lebesgue–Radon–Nikodým decomposition of μ, $\mu = \mu_a + \mu_s$, where $\mu_a \ll \lambda$, $\mu_s \perp \mu_a$, $\mu_s \perp \lambda$, permits the reduction of the problem, in view of i), to the study of $\int_I |f(x-y)|\, d\mu_a(y)$. If $d\mu_a/d\lambda = g$ and if f and g are extended so that each is zero off I, then each is in $L^1(\mathbb{R}, \lambda)$ and then $\int_I |f(x-y)|\, d\mu_a(y) = \int_{\mathbb{R}} |f(x-y)| g(y)\, dy = |f| * g(x)$, which exists and is finite a.e. (λ) (and is actually in $L^1(\mathbb{R}, \lambda)$). □

204. i) Since each μ_n is complex, $|\mu_n|(\mathbb{R}) = a_n < \infty$. Hence there are positive numbers b_n, n in $\mathbb{N}$, such that $\sum_{n=1}^{\infty} a_n b_n < \infty$, e.g., $b_n = 1/2^n a_n$. If $\nu = \sum_n b_n |\mu_n|$ then for n in $\mathbb{N}$, $|\mu_n| \ll \nu$.

ii) For n in $\mathbb{N}$ the maps

$$f_n : x \mapsto \begin{cases} 1, & \text{if } x \in [n, n+1) \\ 0, & \text{if } x \notin [n, n+1) \end{cases}$$

permit the definition of measures $\mu_n : \mathsf{S}_\beta \ni E \mapsto \int_E f_n(x)\, dx$. Each may be regarded as complex (and hence finite). If $\nu \ll \mu_n$ for all n, then $\nu((-\infty, n]) = 0$ for all n and so $\nu = 0$. □ [13]

205. Consider the set S of all sequences $\{y_n + E : y_n \in \mathbb{R}\}_{n=1}^{\infty}$ and let a be $\sup\{\mu(\bigcup_{n=1}^{\infty} (y_n + E)) : \{y_n + E\}_{n=1}^{\infty} \in S\}$. Thus $a < \infty$ and there is in $\mathbb{R}$ a sequence $\{y_{nm}\}_{n,m=1}^{\infty}$ such that $\mu(\bigcup_{n=1}^{\infty} (y_{nm} + E)) > a - 1/m$. If $\{x_k\}_{k=1}^{\infty}$ is an enumeration of the y_{mn} then $\mu(\bigcup_{k=1}^{\infty} (x_k + E)) = a$. Let G be $\mathbb{R} \backslash \bigcup_{k=1}^{\infty} (x_k + E)$ and let ν_1 be μ_G (for each Borel set A, $\nu_1(A) = \mu(A \cap G)$) and let ν_2 be $\mu - \nu_1$. Then for x in $\mathbb{R}$, $\nu_1(x + E) = \mu((x + E) \backslash \bigcup_{k=1}^{\infty} (x_k + E))$, and if $\nu_1(x + E)$ is positive it follows that $\mu((x + E) \cup (\bigcup_{k=1}^{\infty} (x_k + E))) > a$, a contradiction.

On the other hand, $\nu_2(G) = 0$ by definition. □

206. Let $\{x_n\}_{n=1}^{\infty}$ be a dense subset of F. For f in $C_0(\mathbb{R}, \mathbb{C})$ let $L_n(f)$ be $f(x_n)$. Then $\sum_n L_n/2^n$, denoted L, is in $(C_0(\mathbb{R}, \mathbb{C}))^*$ and $\|L\| \leqq 1$. If μ is the Borel measure (provided by the Riesz representation theorem) such that for f in $C_0(\mathbb{R}, \mathbb{C})$, $L(f) = \int_{\mathbb{R}} f(x)\, d\mu(x)$, then it follows in the next few lines that $\operatorname{supp}(\mu) = F$.

Indeed, if U is an open subset of $\mathbb{R} \backslash F$ and if K is a compact subset of U, then there is in $C_0(\mathbb{R}, \mathbb{C})$ an f such that $f(K) = 1$, $f(F) = 0$, whence $L(f) = 0$ and hence $\mu(K) = 0$. It follows that $\operatorname{supp}(\mu) \subset F$. If V is open and $V \cap F \neq \emptyset$, there is in $V \cap F$ some x_{n_0} and in $C_0(\mathbb{R}, \mathbb{C})$ a g such that $g(x_{n_0}) = 1$, $0 \leqq g \leqq 1$, and $g = 0$ off V. Hence $\mu(V) \geqq L(g) \geqq g(x_{n_0})/2^{n_0} > 0$. Hence $\operatorname{supp}(\nu) = F$. □

207. It may be assumed that $a < \mu(\mathbb{R})$. Since $\sum_{n=-\infty}^{\infty} \mu([n, n+1)) = \mu(\mathbb{R})$ for some M, N in $\mathbb{N}$, $\sum_{n=-M}^{N} \mu([n, n+1)) > a$. Then $t \mapsto \mu([tM, t(N+1))$, denoted f, is continuous because μ is nonatomic. Since $f(0) = 0$ and $f(1) > a$, there is in $[0, 1)$ a t_a such that $f(t_a) = a$, as required. □

208. In Figure 8 it can be seen that if (a_1, a_2) is close to (b_1, b_2) then $|\mu(Q(a_1, a_2)) - \mu(Q(b_1, b_2))|$ is the sum of the μ-measures of six (some possibly empty) nonoverlapping rectangles with sides that are horizontal or vertical. Since $\mathbb{R}^2$ is metric, compact sets have finite measure, and horizontal and vertical lines are null sets, it follows that i) μ confined to a rectangle is regular and ii) if $\{R_n\}_{n=1}^{\infty}$ is a sequence of rectangles with horizontal and vertical sides, $R_n \supset R_{n+1}$, and $\lambda_2(R_n)\downarrow 0$, then $\mu(R_n)\downarrow 0$.

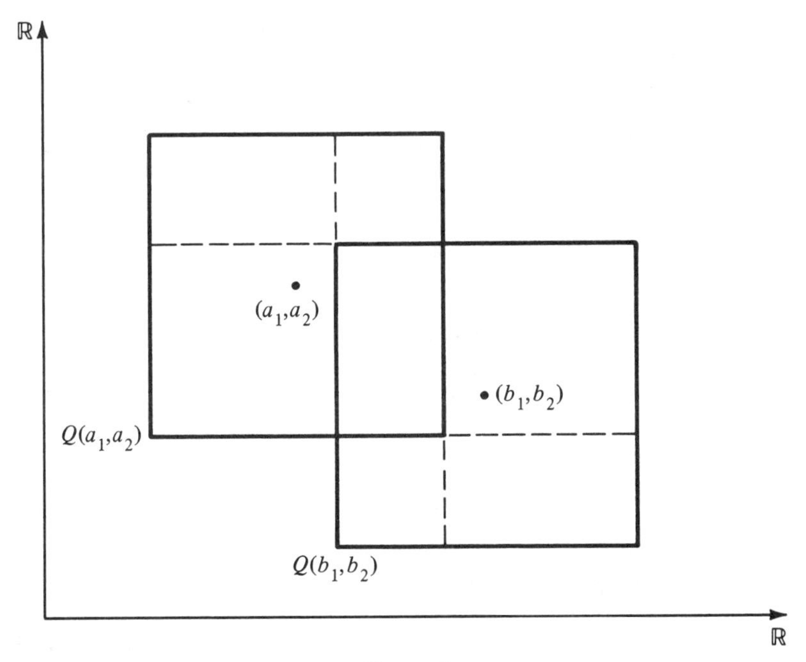

Figure 8

As (b_1, b_2) approaches (a_1, a_2) the six rectangles noted in the first sentence generate six sets of rectangles, each set partially ordered by inclusion, and each partially ordered set containing a countable cofinal sequence descending to a point or a line segment. In consequence, f is continuous. □

209. For t in $(0, \infty)$ let S_t be $\{x_1, x_2): |x_1|+|x_2|\leqq t\}$. Then $\mu(\partial S_t)=0$ and $\mu((0,0))=0$. The argument in Solution 208 may be repeated *mutatis mutandis* to show that $f: t\mapsto \mu(S_t \cap E)$ is continuous. Since E is bounded, for large t, $f(t)=\mu(E)$. Since $f(0)=0$, it follows that for some t_a, $f(t_a)=a$. The required set F is $S_{t_a}\cap E$. □

210. i) The hypothesis implies that if K is compact then $\mu_n(K)<\infty$, $n=0,1,2,\ldots$, (see Problems 149 and 150, and 151). Since $\mathbb{R}^n = \bigcup_{k=1}^{\infty} B(0,k)$ and $B(0,k)$ is compact for each k, it follows that $\mathbb{R}^n$ is σ-finite with respect to each μ_m. It will be shown that these facts permit the conclusion that each μ_m is regular.

Indeed, if U is open there is a sequence $\{K_p\}_{p=1}^{\infty}$ of compact sets such that $K_p \subset K_{p+1}$ and $\bigcup_p K_p = U$. Then $\mu_m(U) = \sum_p \mu_m(K_{p+1} \backslash K_p) = \sup_p \mu_m(K_p)$, i.e., U is inner regular with respect to each μ_m. As in the solution of problem 141, it can be concluded that each μ_m is regular since the set of regular sets is a σ-ring and contains all open sets.

If $\mu_0(U) = \infty$, there is, for each N, a compact set K_N contained in U and such that $\mu_0(K_N) > N$. For large m, $\mu_m(K_N) > N$ and so for large m, $\mu_m(U) > N$, whence $\liminf_{m=\infty} \mu_m(U) = \infty \geqq \mu_0(U)$.

If $\mu_0(U) < \infty$, there is for each N a compact set K_N contained in U and such that $\mu_0(K_N) > \mu_0(U) - 1/N$. For large m, $\mu_m(K_N) > \mu_0(K_N) - 1/N$ and so $\mu_m(K_N) > \mu_0(U) - 2/N$. Hence for large m, $\mu_m(U) > \mu_0(U) - 2/N$ and $\liminf_{m=\infty} \mu_m(U) \geqq \mu_0(U) - 2/N$. The inequality obtains for all N in $\mathbb{N}$ and the result follows.

ii) The regularity of all the μ_m permits repetition of the argument given in Solution 149 and the conclusion follows. □

211. As a compact metric space, K is the continuous image of *the* Cantor set C [15]: for some f in $C(C, \mathbb{R}^N)$, $f(C) = K$. If, on C, ν is the measure derived from the Cantor function (see Problem 189), the map $\mu: \mathsf{S}_\beta(\mathbb{R}^n) \ni E \mapsto \nu(f^{-1}(E) \cap C)$ defines a Borel measure on $\mathbb{R}^n$. If $U \cap K = \emptyset$ then $\mu(U) = 0$. If V is open and $V \cap K \neq \emptyset$, extend f to a continuous function, again denoted f, in $C(\mathbb{R}, \mathbb{R}^n)$. (This extension exists according to the Tietze extension theorem; the extension may be constructed "explicitly" by linear extension of f to $\mathbb{R} \backslash C$.) Then $f^{-1}(V)$ is an open set W, $W \cap C \neq \emptyset$ and so $\nu(W \cap C) > 0$, whence $\text{supp}(\mu) = K$. □

212. If U_n is open in $\mathbb{R}^n$ and U_m is open in $\mathbb{R}^m$ then $U_m \times U_n$ is open in $\mathbb{R}^{m+n}$, whence $\mathsf{S}_\beta(\mathbb{R}^m) \times \mathsf{S}_\beta(\mathbb{R}^n) \subset \mathsf{S}_\beta(\mathbb{R}^{m+n})$. On the other hand, if U_{m+n} is open in $\mathbb{R}^{m+n}$, U_{m+n} is a (countable) union of products of "half-open" rectangles in $\mathbb{R}^m$ and $\mathbb{R}^n$ and so $\mathsf{S}_\beta(\mathbb{R}^{m+n}) \subset \mathsf{S}_\beta(\mathbb{R}^m) \times \mathsf{S}_\beta(\mathbb{R}^n)$. □

213. If $A = p^{-1}((-\infty, 0)) \cap K \neq \emptyset$ there is in $C_{00}(\mathbb{R}^n, \mathbb{R})$ an f such that $0 \leqq f \leqq 1$, $f = 1$ somewhere on A and $f = 0$ on $\mathbb{R} \backslash A$. Then $\int_{\mathbb{R}^n} f(x)p(x)\, d\mu(x) = -c < 0$. Furthermore $f^{1/2}$ is definable as an element of $C_{00}(\mathbb{R}^n, \mathbb{R})$ and may, *via* the Stone–Weierstrass theorem, be approximated by a polynomial q such that $\|(f - q^2)p|_K\|_\infty < c/2\mu(K)$, in which case $\int_{\mathbb{R}^n} (f(x) - (q(x))^2)p(x)\, d\mu(x) = -c - \int_{\mathbb{R}^n} (q(x))^2 p(x)\, d\mu(x) > -c/2$ and thus $\int_{\mathbb{R}^n} (q(x))^2 p(x)\, d\mu(x) < -c/2 < 0$, a contradiction. □

214. (See Problem 198.) The Stone–Weierstrass theorem implies that the linear span A of $\{x \mapsto e^{-nx} : n \text{ in } \mathbb{N}\}$ is dense in $C_0([0, \infty), \mathbb{R})$ and so, since $|\mu|$ is bounded, $\int_0^\infty f(x)\, d\mu(x) = 0$ for all f in $C_0([0, \infty), \mathbb{R})$. Thus for all compact sets K, $\mu(K) = 0$ and finally $\mu = 0$. □

215. For n in $\mathbb{N}$, the sequence $\{[k/2^n, (k+1)/2^n)\}_{k=0}^{\infty}$ is a partition of $[0, \infty)$ and $\sum_{k=0}^{\infty} (1 - \mu([0, k/2^n])).2^{-n} \to \int_0^\infty (1 - \mu([0, x))\, dx$ as $n \to \infty$.

However, (Abel summation)

$$\textstyle\sum_{k=0}^{K}(1-\mu([0,k/2^n)).2^{-n}=\sum_{k=1}^{K-1}\mu([(k-1)/2^n,k/2^n))k2^{-n}$$
$$+K\mu([K/2^n,\infty)).2^{-n}.$$

Thus

$$\sum_{k=0}^{\infty}(1-\mu([0,k/2^n)).2^{-n}$$
$$=\lim_{K\to\infty}\left(\sum_{k=1}^{K-1}\mu([(k-1)/2^n,k/2^n))k2^{-n}+K\mu([K/2^n,\infty))2^{-n}\right).$$

The result of Problem 176 applied here to the map $f:x\mapsto x$ shows $K\mu([K/2^n,\infty)).2^{-n}\to 0$ as $K\to\infty$. Consequently,

$$\sum_{k=1}^{\infty}\mu([(k-1)/2^n,k/2^n))k.2^{-n}\to\int_0^{\infty}(1-\mu([0,x))\,dx$$

and, on the other hand, $\sum_{k=1}^{\infty}\mu([(k-1)/2^n,\ k/2^n))k2^{-n}\to\int_0^{\infty}x\,d\mu(x)$ as $n\to\infty$. □

216. The Abel summation used in Solution 179 may be applied again. Since $p=1$ the result follows. □

9. Lebesgue Measure in $\mathbb{R}^n$

217. It may be assumed that $a_1 \leqq a_2 \leqq \cdots \leqq a_N$, $0 \leqq b_1$ (and, of course, $a_n \leqq b_n$). Then $b_1 \geqq a_2$ since otherwise $b_1 < a_2$ and $(b_1, a_2) \subset I \setminus \bigcup_{n=1}^{N} [a_n, b_n]$. Similar reasoning shows $b_{n-1} \geqq a_n$, $n = 2, 3, \ldots, N$ and since some $b_n \geqq 1$, the result follows. □

218. Since $\lambda_2(L) = 0$ for any line L, if $\{L_n\}_{n=1}^{\infty}$ is a sequence of lines, their union is a null set (λ_2), whereas $\lambda_2(\mathbb{R}^2) = \infty$. □

219. Since $\lambda(\bigcup_n (r_n - 1/n^2, r_n + 1/n^2)) \leqq 2 \sum_n n^{-2} = \pi^2/3$, it follows that $\mathbb{R} \setminus \bigcup_n (r_n - 1/n^2, r_n + 1/n^2) \neq \emptyset$.

Let A be $\mathbb{N} \setminus \{n^2\}_{n=2}^{\infty}$ and enumerate A as $a_1, a_2, \ldots$, so that $a_k < a_{k+1}$. In $\mathbb{Q}$ choose a sequence $\{t_k\}_{k=1}^{\infty}$ such that for all k, $|t_k - 1| < 1/a_k$. If $\mathbb{Q} \setminus \{t_k\}_{k=1}^{\infty}$ is enumerated as $\{r_p\}_{p=1}^{\infty}$, enumerate $\mathbb{Q}$ according to the scheme: $s_1 = t_1$, $s_2 = t_2$, $s_3 = t_3$, $s_4 = r_1, \ldots$; more precisely, let s_p be $r_{p^{1/2}-1}$ if $p \notin A$ and for p in A let s_p successively pick up the t_k. Since $\bigcup_{p \in A} (s_p - 1/p, s_p + 1/p) \subset [-1, 2]$ and $1 + \sum_{p \notin A} \lambda((s_p - 1/p, s_p + 1/p)) \leqq \pi^2/3$, it follows again that $\mathbb{R} \setminus \bigcup_m (s_m - 1/m, s_m + 1/m) \neq \emptyset$. □

220. If $\lambda(\{x_2 : \lambda(E_{x_2}) = 1\}) > \frac{1}{2}$, then $\lambda_2(E) = \int_I (\int_I \lambda(E_{x_2})\, dx_2)\, dx_1 = \int_I (\int_{\{x_2 : \lambda(E_{x_2}) = 1\}} 1\, dx_2 + \int_{\{x_2 : \lambda(E_{x_2}) \neq 1\}} \lambda(E_{x_2})\, dx_2)\, dx_1 > \frac{1}{2}$, whereas $\lambda_2(E) = \int_I (\int_I \lambda(E_{x_1})\, dx_1)\, dx_2 \leqq \frac{1}{2}$. Thus the implication is valid. □

221. Since $(x_1, x_2) \in E$ iff $x_1 - x_2 \notin \mathbb{Q}$, it follows that $A_1 \times A_2 \subset E$ iff $(A_1 - A_2) \cap \mathbb{Q} = \emptyset$. However, if $\lambda_2(A_1 \times A_2) > 0$, then A_1 resp. A_2 contains a measurable subset B_1 resp. B_2 such that $\lambda(B_1) + \lambda(B_2) < \infty$ and $\lambda_2(B_1 \times B_2) > 0$. In other words it may be assumed from the start that $\lambda(A_1) + \lambda(A_2) < \infty$. Then $x_1 \mapsto \int_{\mathbb{R}} \chi_{A_1}(x_1 + x_2) \chi_{A_2}(x_2)\, dx_2$ is a continuous map f vanishing off $A_1 - A_2$ and not identically zero, e.g., $\int_{\mathbb{R}} f(x_1)\, dx_1 =$

$\lambda_1(A_1)\lambda_1(A_2)=\lambda_2(A_1\times A_2)>0$. Hence f is nonzero on a nonempty open subset U of A_1-A_2 and $U\cap\mathbb{Q}\neq\emptyset$, a contradiction. □

222. For m in $\mathbb{Z}$, n in $\mathbb{N}$, and x_1 in $[(m-1)/n, m/n]$ define $f_n\colon(x_1,x_2)\mapsto n(m/n-x_1)f((m-1)/n, x_2)+n(x_1-(m-1)/n)f(m/n, x_2)$ ($f_n(x_1,x_2)$ is a convex (linear) combination of $f((m-1)/n, x_2)$ and $f(m/n, x_2)$). Hence $f_n(x_1,x_2)$ is between $f((m-1)/n, x_2)$ and $f(m/n, x_2)$ on $[(m-1)/n, m/n]$. For each (x_1,x_2) there is a sequence $\{[(m_k-1)/n_k, m_k/n_k]\}_{k=1}^{\infty}$ such that $\bigcap_k[(m_k-1)/n_k, m_k/n_k]=x_1$ and so $f_n(x_1,x_2)\to f(x_1,x_2)$ as $n\to\infty$ because $f^{x_2}\in C(\mathbb{R},\mathbb{R})$. Furthermore, for all k, $f(x_1,x_2)$ is between $f((m_k-1)/n_k, x_2)$ and $f(m_k/n_k, x_2)$. Since f^{x_1} is Borel measurable for all x_1 it follows that f is also Borel measurable. □

223. In analogy with the solution of Problem 222 for each n in $\mathbb{N}$ choose a sequence $\{a_{nm}\}_{m=-\infty}^{\infty}$ contained in E and such that for all m, $1/2n<a_{n,m+1}-a_{nm}<1/n$. This time, if $a_{nm}\leqq x_1\leqq a_{n,m+1}$ let f_n be the map $(x_1,x_2)\mapsto(a_{n,m+1}-x_1)/(a_{n,m+1}-a_{nm}))f(a_{nm},x_2)+((x_1-a_{nm})/(a_{n,m+1}-a_{nm}))f(a_{n,m+1}m, x_2)$ (again a convex (linear) combination!). As in the earlier situation, $f(x_1,x_2)$ is between $f(a_{nm},x_2)$ and $f(a_{n,m+1},x_2)$ and, as before, for each (x_1,x_2), $f_n(x_1,x_2)\to f(x_1,x_2)$ as $n\to\infty$, whence f is Lebesgue measurable. □

224. Use the device in the solution of Problem 222, this time in its original form. The maps $h_n: x_2\mapsto f_n(g(x_2),x_2)$ are then Lebesgue measurable and so h is also Lebesgue measurable since $h_n\to h$ as $n\to\infty$. □

225. Tonelli's theorem shows that $\int_{[-1,1]^2}|f(x_1,x_2)|\,d\lambda_2(x_1,x_2)=\int_{[-1,1]}|x_2|(\int_{[-1,1]}(|x_1|/(x_1^2+x_2^2))\,dx_1)\,dx_2$. For $x_2\neq 0$,

$$\int_{[-1,1]}\frac{|x_1|}{x_1^2+x_2^2}\,dx_1=2\log\frac{1+x_2^2}{x_2^2}$$

and hence both the double and the iterated integrals are finite and so $f\in L^1([-1,1]^2,\lambda_2)$. □

226. Since f is continuous on $I^2\backslash(0,0)$ and $|f|\leqq 1$ it follows that f is in $L^1(I^2,\lambda_2)$ and so the two integrals are equal to the integral of f over I^2. Since $f(x_1,x_2)=-f(x_2,x_1)$, $\int_{I^2}f(x_1,x_2)\,d\lambda_2(x_1,x_2)=0$. □

227. Let f be the map $(x_1,x_2)\mapsto x_2^{|x_1-\frac{1}{2}|^{\frac{1}{2}}-1}$. Then

$$\int_0^1\left(\int_0^1 f(x_1,x_2)\,dx_2\right)dx_1=\int_0^1\left\{\begin{matrix}|x_1-\frac{1}{2}|^{-1/2}, & \text{if } x_1\neq\frac{1}{2}\\ +\infty, & \text{if } x_1=\frac{1}{2}\end{matrix}\right\}dx_1=2\sqrt{2}.$$

Hence $f\in C((0,1)^2,\mathbb{R})\cap L^1((0,1)^2,\lambda_2)$ and yet $\int_0^1 f(\frac{1}{2},x_2)\,dx_2=\int_0^1 x_2^{-1}\,dx_2=\infty$. □

228. If μ_1 is $E\mapsto\int_E x^{1/2}\,dx$ and if $\mu_2=\lambda$ for $(I,\mathsf{S}_\lambda,\mu_i)$, $i=1,2$, then $d\mu_1/d\mu_2=(x\mapsto x^{1/2})$, which is in $L^\infty(I,\mu_2)$ and $d\mu_2/d\mu_1=(x\mapsto x^{-1/2})$ which is not in $L^\infty(I, \mu_1)$ since if $M>0$, $\mu_1(\{x: (d\mu_2/d\mu_1)(x)>M\}) = \mu_1((0, M^{-2})) = 2M^{-3}/3>0$. □

229. Let f be $d\mu/d\lambda$. Then if $0<x-a<x+a<1$,

$$\frac{\mu(I\cap(x-a, x+a))}{\lambda(I\cap(x-a, x+a))} = (2a)^{-1}\int_{x-a}^{x+a} f(t).$$

As $a\to 0$ the fraction converges to $f(x)$ a.e. on $(0, 1)$. □

230. Since, for any Lebesgue measurable set E in $\mathbb{R}$ the metric density, i.e., $\lim_{a>0, a\to 0}\lambda(E\cap(x-a, x+a))/2a$, exists and is one a.e. in E, it follows that there is no Lebesgue measurable set E with the property described. □

231. For J and J' in $\mathscr{J}$ write JRJ' iff there are $\mathscr{J}$ elements $J_1, J_2, \ldots, J_n$ such that $J\cap J_1, J_1\cap J_2, \ldots, J_{n-1}\cap J_n, J_n\cap J'\neq\emptyset$. (The J_i link successively to join J and J'.) Then R is an equivalence relation and $\bigcup\{J: J$ in one equivalence class$\}$ is connected and so is some kind of interval (open, half-open, or closed). Since equivalence classes are disjoint and no member of $\mathscr{J}$ is a single point, it follows that there are at most countably many equivalence classes and hence $\bigcup_{J\in\mathscr{J}} J$ is the countable union of intervals and therefore in S_β, *a fortiori* in S_λ. □

232. The equivalence relation R of Solution 231 may be used again. This time each union over an equivalence class is a component of $\bigcup_n J_n$, and every component is such a union. Since components are disjoint it follows that $\sum_n \lambda(J_n)\geqq\lambda(\bigcup_n J_n)=\lambda(\bigcup_n C_n)=\sum_n\lambda(C_n)$. □

233. Use the equivalence relation R and the notation of Solution 232. Then the components C_n are finite in number and pairwise disjoint, whence the problem is reduced to proving the result for each C_n. Furthermore, each C_n is the union of interlocking (linking) intervals, again finite in number, denotable $(a_1, b_1), (a_2, b_2), \ldots, (a_p, b_p)$ and such that $a_1<a_2<b_1<a_3<b_2<a_4<b_3<\cdots<b_{p-1}<b_p$. If $\sum_{k \text{ odd}}(b_k-a_k)<\frac{1}{2}\lambda(C_n)$ and $\sum_{k \text{ even}}(b_k-a_k)<\frac{1}{2}\lambda(C_n)$, then $\sum_{k=1}^p(b_k-a_k)<\lambda(C_n)$. □

234. Since $\mu\ll\lambda$ there is in $L^1(I, \lambda)$ an f such that for all a in I, $\mu([0, a]) = \int_0^a f(x)\,dx$. If f is decomposed into its canonical parts $f\vee 0=f^+$ and $-(f\wedge 0)=f^-$ then $\int_0^a f^+(x)\,dx=\int_0^a f^-(x)\,dx$. Hence for all Lebesgue measurable sets E, $\int_E f^+(x)\,dx=\int_E f^-(x)\,dx$, i.e., $f^+=f^-$ a.e., whence $f=0$ a.e. and $\mu=0$. □

235. Since spheres centered at x are T-invariant and since $T(B(y, r)^0) = B(T(y), r)^0$ it follows that $\lambda(T(B(y, r)^0))=\lambda(B(y, r)^0$. If K is compact there is a sequence $\{\bigcup_{k=1}^{N_n} B_{nk}^0\}_{n=1}^\infty$ of finite unions of open balls such that $K=\bigcap_n(\bigcup_k B_{nk}^0)$ and $\lambda(T(\bigcup_k B_{nk}^0))=\lambda(\bigcup_k B_{nk}^0)$, whence $\lambda(T(K))=\lambda(K)$. Since λ is regular and T is continuous and bijective it follows that $T(\mathsf{S}_\lambda)\subset\mathsf{S}_\lambda$ and that T preserves measure. □

236. Let $\{r_n\}_{n=1}^{\infty}$ be an enumeration of $\mathbb{Q}\cap I$ and let I_{nk} be $(r_n-2^{-n-k}, r_n+2^{-n-k})$, k, n in $\mathbb{N}$. Then $\lambda(\bigcup_n I_{nk})\leqq 2^{-k+1}$ and so $\lambda(\bigcap_k(\bigcup_n I_{nk}))=0$. Since each set $I\backslash\bigcup_n I_{nk}$ is nowhere dense and I is a complete metric space, it follows that $\bigcap_k(\bigcup_n I_{nk})$ is a null set of the second category. □

237. If $f\in L^1(\mathbb{R},\lambda)$, then since $\chi_E\in L^1(\mathbb{R},\lambda)$ and $g=f*\chi_E$ it follows that $g\in L^1(\mathbb{R},\lambda)$. Conversely, if $g\in L^1(\mathbb{R},\lambda)$, then since $f\geqq 0$, $\int_{\mathbb{R}} g(x)\,dx=\int_{\mathbb{R}}\int_{\mathbb{R}}\chi_E(t)f(x-t)\,dt\,dx$ and thus the translation invariance of Lebesgue measure and Tonelli's theorem show that $f\in L^1(\mathbb{R},\lambda)$. □

238. Since $\lambda(E)>0$, then essentially as shown in Solution 221, $E-E$ contains an open set U containing 0. Hence for some positive a, $\{x: |x|<a\}\subset E-E$ and if $|x|<a$ there are in E, e_1, e_2 such that $x=e_1-e_2$, i.e., $x+e_2\in E$, $(x+E)\cap E\neq\emptyset$. □

239. Since λ is regular, there is in E a compact subset K_1 such that $\lambda(K_1)>a$. Choose x in K_1 and let r be such that $B(x, r/2)\supset K_1\cup\{0\}$. Then since $t\mapsto\lambda(K_1\cap B(x,t))$ is a continuous function, since $\lambda(K_1\cap B(x,r))>a$, and $\lambda(K_1\cap B(x,0))=0$, it follows that for some t_0 in $(0,r)$, $\lambda(K_1\cap B(x,t_0))=a$. The set $K_1\cap B(x,t_0)$ may serve as the set K required. □

240. Since $\mu(\mathbb{R})=1$ it follows that $x\mapsto\mu((-\infty,b)+x)=\mu((-\infty,b+x))$ is, for every b a monotone increasing nonnegative function that approaches 1 as $x\to\infty$ (and approaches 0 as $x\to-\infty$). Thus $\int_{\mathbb{R}}\mu((-\infty,b)+x)\,dx=\infty=\lambda((-\infty,b))$ for all b. Furthermore, $\mu([a,b)+x)=\mu([a+x,b+x))\to 0$ if $a\to\infty$ or $b\to-\infty$. Hence, since $\mu([a+x,b+x))\geqq 0$, $\int_{\mathbb{R}}\mu([a,b)+x)\,dx$ exists although *a priori*, its value is ∞. For all n in $\mathbb{N}$, let E_n be $[0,1/n)$. Then for m in $\mathbb{N}$ the sets $\{E_{mn}+k/mn\}_{k=0}^{m-1}$ are disjoint and $E_n=\bigcup_{k=0}^{m-1}(E_{mn}+k/mn)$.

Hence

$$\sum_{p=-\infty}^{\infty}\mu\left(E_n+\frac{p}{mn}\right)\Big/ mn=\sum_{k=0}^{m-1}\sum_{p=-\infty}^{\infty}\mu\left(E_{mn}+\frac{k}{mn}+\frac{p}{mn}\right)\Big/ mn.$$

For each k in question, $\sum_{p=-\infty}^{\infty}\mu(E_{mn}+k/mn+p/mn)/mn=\sum_{p=-\infty}^{\infty}\mu(E_{mn}+p/mn)/mn=1/mn$ and so $\sum_{p=-\infty}^{\infty}\mu(E_n+p/mn)/mn=1/n\geqq\sum_{p=-\infty}^{\infty}\inf\mu(\{E_n+x: x\in[p/mn,(p+1)/mn)\})/mn$. Thus $\lambda(E_n)=1/n=\int_{\mathbb{R}}\mu(E_n+x)\,dx$. The same argument applied to $[a,a+1/n)$ shows $\lambda([a,a+1/n))=\int_{\mathbb{R}}\mu([a,a+1/n)+x)\,dx$. The map $\nu: E\mapsto\int_{\mathbb{R}}\mu(E+x)\,dx$ is a measure defined for all E in S_β. Since for all a in $\mathbb{R}$ and all n in $\mathbb{N}$, $\lambda([a,a+1/n))=\nu([a,a+1/n))$, it follows (see Solution 153) that for all Borel sets E, $\nu(E)=\lambda(E)$ as required. □

241. For each k in $\mathbb{N}$, $\bigcup_{r\in\mathbb{Q}^m}B(r,1/k)^0=\mathbb{R}^m$ and so for each k there is an r_k such that $\lambda_n(T^{-1}(B(r_k,1/k)^0))>0$. If $T(x)\in B(r_k,1/k)^0$ then $\lambda_n(-x+T^{-1}(B(r_k,1/k)^0))>0$ and $\lambda_n(T^{-1}(B(0,1/k)^0))\geqq\lambda_n(-x+T^{-1}(B(r_k,1/k)^0))>0$. Hence $T^{-1}(B(0,1/k)^0)-T^{-1}(B(0,1/k)^0)$

contains an open set containing 0 in $\mathbb{R}^n$. However, $B(0, 1/3k)^0 - B(0, 1/3k)^0 \subset B(0, 1/k)^0 \subset 3B(0, 1/3k)^0$ and since $T^{-1}(B(0, 1/k)^0) \supset T^{-1}(B(0, 1/3k)^0) - T^{-1}(B(0, 1/3k)^0)$, $T^{-1}(B(0, 1/k)^0)$ contains an open set containing 0 in $\mathbb{R}^n$. Correspondingly, if $T(y) = z$, then $T^{-1}(B(z, 1/k)^0) \supset y + T^{-1}(B(0, 1/k)^0)$ which contains an open set containing y. Thus if U is open in $\mathbb{R}^m$ then for some sequence $\{z_p, 1/k_p\}_{p=1}^\infty$ in $\mathbb{Q}^m \times \mathbb{Q}$, $U = \bigcup_p B(z_p, 1/k_p)^0$ and so $T^{-1}(U)$ is open, i.e., T is continuous. Hypothesis i) implies that for n in $\mathbb{Z}$, $T(nx) = nT(x)$ and then for p in $\mathbb{Z}$ and q in $\mathbb{Z}\backslash\{0\}$, $T(px/q) = pT(x)/q$. The continuity of T permits the conclusion that for all t in $\mathbb{R}$, $T(tx) = tT(x)$. □

242. Let E_m be $\{x: a_m = 0\}$ and let E be $\bigcup_{m=1}^\infty E_m$. Each E_m is the union of 10^{m-1} disjoint intervals each of length 10^{-m}. Hence each E_m is measurable, E and $I\backslash E$ are measurable, and $I\backslash E = f^{-1}(k)$. Furthermore $E = E_1 \cup (E_2\backslash E_1) \cup (E_3\backslash(E_1 \cup E_2)) \cup \cdots$ and so $\lambda(E) = \sum_{r=0}^\infty 10^{-1}(9/10)^r = 1$, whence $\lambda(f^{-1}(k)) = 0$.

Since $f^{-1}(1) = (E_1\backslash E_2) \cup (E_1 \cap E_2 \cap E_3\backslash E_4) \cup \cdots$ and since the sets in parentheses are pairwise disjoint it follows that $f^{-1}(1)$ is measurable and that $\lambda(f^{-1}(1)) = (10^{-1} - 10^{-2}) + (10^{-3} - 10^{-4}) + \cdots = 1/11$. Thus f is measurable, nonnegative and $\int_I f(x)\,dx = 1/11$. □

243. i) If $\{x_m\}_{m=1}^\infty \subset A$ and if $x_m \to x$ as $m \to \infty$, then $x = \sum_{n=1}^\infty b_n 10^{-n}$, $b_n = 0, 1, 2, \ldots,$ or 9. If $|x - x_m| < 10^{-p}$ then $b_n = 2$ or 7, $n = 1, 2, \ldots, p-1$, whence $x \in A$.

ii) According to i), A is closed. Since I is connected and A is neither empty nor all I, A cannot be open.

iii) Let X_n be $\{2, 7\}$, n in $\mathbb{N}$. Then $\text{card}(\prod_{n=1}^\infty X_n) = \text{card}(\mathbb{R})$. Since A and $\prod_n X_n$ are in one-one correspondence, $\text{card}(A) = \text{card}(\mathbb{R})$.

iv) Since $A \cap (.28, .7) = \varnothing$, A is not dense in I.

v) Since A is closed i) A is Borel measurable and hence also Lebesgue measurable.

As in Solution 189, it can be shown that A is a Cantor-like set and the calculation of its measure is the evaluation of $1 - .8\sum_{n=0}^\infty (.2)^n$ which is zero. □

244. By induction choose a sequence $\{a_n\}_{n=1}^\infty$ as follows: each a_n is 0 or 7 and these occur alternately in blocks of size 1, 2, $2^2, \ldots,$ *viz.*, 0, 7, 7, 0, 0, 0, 0, 7, 7, 7, 7, 7, 7, 7, 7, ... If $x = \sum_{n=1}^\infty a_n 10^{-n}$ then $\limsup_{n=\infty} A_n(x)/n = 2/3 > \liminf_{n=\infty} A_n(x)/n = 1/3$ and so E is not empty.

For m in $\mathbb{N}$ and x given by $\sum_{n=1}^\infty a_n 10^{-n}$ let $f_m(x)$ be 1 or 0 according as a_m is or is not 7. Then each f_m is Lebesgue measurable and since $A_n = (\sum_{m=1}^n f_m)/n$ it follows that A_n is also Lebesgue measurable. Thus $I\backslash E$, as the set of points where $\lim_{n\to\infty} A_n$ exists, is Lebesgue measurable and so is E.

For x in $[0, 1]$, if $a > 0$ and $x = \sum_{n=1}^\infty a_n 10^{-n}$, let m be such that $10^{-m} < a$. The process applied in showing E is not empty may be used on the decimal places following the mth to provide in E a y such that $|x-y|<a$. Hence E is dense. Since $.3$ is not in E it follows that E is not closed.

For x, a and m as in the preceding paragraph let y be $\sum_{n=1}^{m} a_n 10^{-n} + 7.10^{-m}/9$. Then $y \notin E$ and $|x-y|<a$, whence $I \backslash E$ is dense and so E is not open. □

245. In each interval $[n, n+1)$, n in $\mathbb{Z}$, construct a countable dense union S_n of Cantor-like sets such that $\lambda(S_n) \in (0, 1)$ and $\sum_{n=-\infty}^{\infty} \lambda(S_n) < \infty$. (See Solution 189.) The process described there is carried out countably many times, each time in one of the intervals deleted at a preceding stage of construction. All intervals deleted in one stage are addressed in subsequent stages of the construction.) If $E = \bigcup_{n \in \mathbb{Z}} S_n$, then $\lambda(E) < \infty$ and any nondegenerate interval (a, b) contains some interval deleted in the process of constructing all the S_n. Such an interval contains one of the Cantor-like sets of positive measure and thus $0 < \lambda(E \cap [a, b]) < b-a$. □

246. The set E of Solution 245 serves. □

247. Let E_{11} be $[0, \frac{1}{2})$, E_{12} be $[\frac{1}{2}, 1)$. By induction, if $E_{n1}, \ldots, E_{n2}n$ have been constructed, let $E_{n+1,1}$ be the left half of E_{n1}, $E_{n+1,2}$ the right half of E_{n1}, etc. Then let $E_1 = E_{11}$, $E_2 = E_{21} \cup E_{23}$, $E_3 = E_{31} \cup E_{33} \cup E_{35} \cup E_{37}$, and, in general, $E_n = E_{n1} \cup E_{n3} \cup \cdots \cup E_{n2^n-1}$. It follows that $\lambda(E_n) = \frac{1}{2}$, n in $\mathbb{N}$ but $\lambda(\bigcap_{k=1}^{m} E_{n.}) \leqq 2^{-m}$. □

248. The proof of the Borel–Cantelli lemma (Problem 158) shows that $\lambda(\limsup_{n=\infty} A_n) = 0$. Note that $\limsup_{n=\infty} A_n = \{x : x \text{ belongs to infinitely many } A_n\}$ may be denoted appropriately as G_∞ and therefore $\lambda(G_\infty) = 0$.

Let H_k be $\bigcup\{\bigcap_{p=1}^{k} A_{n_p} : 1 \leqq n_1 < n_2 < \cdots < n_k\}$, k in $\mathbb{N}$. Then $H_k \supset H_{k+1}$ and $H_k = \{x : x \text{ belongs to at least } k \text{ of the } A_n\}$. Thus $G_k = H_k \backslash H_{k+1}$. It follows that each H_k and each G_k is Lebesgue measurable. Note that the G_k are pairwise disjoint.

Since $A_n = (A_n \backslash \bigcup_k (G_k \cap A_n)) \cup (\bigcup_k (G_k \cap A_n))$ and since

$$A_n \backslash \bigcup_k (G_k \cap A_n) \subset G_\infty,$$

it follows that $\lambda(A_n) = \sum_k \lambda(G_k \cap A_n)$ and that $\sum_n \lambda(A_n) = \sum_{k,n} \lambda(G_k \cap A_n)$. (All series considered are absolutely convergent and thus the Fubini–Tonelli theorems justify the double summation in any convenient order.)

Since $G_k \subset \bigcup_{p_1 < p_2 < \cdots < p_k} A_{p_1} \cap A_{p_2} \cap \cdots \cap A_{p_k}$ it follows that $G_k = \bigcup_{p_1 < p_2 < \cdots < p_k} G_k \cap A_{p_1} \cap A_{p_2} \cap \cdots \cap A_{p_k}$. Moreover, the elements of the last union are pairwise disjoint because the members of each element belong to precisely k of the A_n.

Thus

$$\sum_{n=1}^{\infty} \lambda(G_k \cap A_n) = \sum_{n=1}^{\infty} \sum_{\substack{p_1 < p_2 < \cdots < p_{k-1} \\ n \neq p_1, p_2, \ldots, p_{k-1}}} \lambda(G_k \cap A_{p_1} \cap A_{p_2} \cap \cdots \cap A_{p_{k-1}} \cap A_n).$$

Each term in the right member, e.g., $\lambda(G_k \cap A_{r_1} \cap A_{r_2} \cap \cdots \cap A_{r_k})$ occurs k times, namely when $n = r_k$ and $p_1 = r_1$, $p_2 = r_2, \ldots, p_{k-1} = r_{k-1}$, when $n = r_{k-1}$, $p_1 = r_1$, $p_2 = r_2, \ldots, p_{k-1} = r_k$, etc. Consequently,

$$\sum_{n=1}^{\infty} \lambda(G_k \cap A_n) = k\lambda(G_k) \text{ and } \sum_k k\lambda(G_k) = \sum_n \lambda(A_n). \qquad \square$$

10. Lebesgue Measurable Functions

249. In Solution 239 it is shown that if E is Lebesgue measurable and $0<a<\lambda(E)$ then there is a (compact) subset K of E and for which $\lambda(K)=a$. Hence if $m\in\mathbb{N}$ and $\lambda(E)=m$ then E is the union of m pairwise disjoint sets, each of measure 1. Similarly, if $n\in\mathbb{N}$ and $\lambda(E)=n\sqrt{2}$ then E is the union of n pairwise disjoint sets each of measure $\sqrt{2}$. Hence if $\lambda(E)=m+n\sqrt{2}$ then $\int_E f(x)\,dx=0$.

Since $0<\sqrt{2}-1<1$, if $a>0$ there is in $\mathbb{N}$ a k such that $0<(\sqrt{2}-1)^k<a$ and $(\sqrt{2}-1)^k=m+n\sqrt{2}$, m, n in $\mathbb{Z}$. Consequently $\{m+n\sqrt{2}: m, n$ in $\mathbb{Z}\}$ is dense in $\mathbb{R}$.

If $\lambda(E)=m+n\sqrt{2}$, two cases arise: i) both m and n are nonnegative; ii) $mn<0$. If i) obtains, E is the union of two disjoint sets E_m and $E_{n\sqrt{2}}$ such that $\lambda(E_m)=m$ and $\lambda(E_{n\sqrt{2}})=n\sqrt{2}$. If ii) obtains and $m>0$ there are two sets E_m and $E_{n\sqrt{2}}$ such that $E_m\supset E_{n\sqrt{2}}$ and $E=E_m\backslash E_{n\sqrt{2}}$; if $m<0$ then $E_{n\sqrt{2}}\supset E_m$ and $E=E_{n\sqrt{2}}\backslash E_m$. (Under ii, $\lambda(E_m)=\pm m$, $\lambda(E_{n\sqrt{2}})=\pm n\sqrt{2}$.) Hence for every E such that $\lambda(E)=m+n\sqrt{2}$, $\int_E f(x)\,dx=0$. In particular, for all intervals J of lengths constituting a dense set in $(0,\infty)$, $\int_J f(x)\,dx=0$ and so $f=0$ a.e. □

250. The formula for the sum of a (finite) geometric series shows that Fejér's kernel $F_N: y\mapsto(2\pi(N+1))^{-1}(\sin(N+1)y/2)^2/(\sin\frac{1}{2}y)^2$ satisfies $\sigma_N(f)(x)=\int_{-\pi}^{\pi}F_N(x-y)f(y)\,dy$. If $E=\{x:|f(x)|=a<1,\ x$ in $[-\pi,\pi]\}$ is not a null set then $|\sigma_N(f)(x_0)|\leqq\int_E+\int_{[-\pi,\pi]\backslash E}F_N(x_0-y)|f(y)|\,dy<\int_E+\int_{[-\pi,\pi]\backslash E}F_N(x_0-y)\,dy=1$.

Thus $|f(x)|=1$ a.e. If $A=f^{-1}(1)$ and $B=f^{-1}(-1)$ and if $\lambda(A)\lambda(B)>0$ then $|\sigma_N(f)(x_0)|=|\int_A-\int_B F_N(x_0-y)dy|<1$. Thus either $f=1$ a.e. or $f=-1$ a.e. If f is complex-valued the method used in Solution 308 yields that for some real θ, $f=e^{i\theta}$ a.e. □

REMARK. The Fejér kernels are examples in a general class $\mathscr{K}$ of kernels K satisfying: i) $K \geqq 0$; ii) $\int_{-\pi}^{\pi} K(y)\,dy = 1$. In the context above if $|\int_{-\pi}^{\pi} K(x_0 - y)f(y)\,dy| = 1$ then again $f = 1$ a.e. or $f = -1$ a.e. If, to boot: iii) for each open set U containing 0 and each positive b there is in $\mathscr{K}$ a $K_U(b)$ such that off U, $K_U(b) < b$, the class $\mathscr{K}$ enjoys an additional property: if $g \in C([-\pi, \pi], \mathbb{C})$ and $g(-\pi) = g(\pi)$ and if $b > 0$ there is a U such that if $V \subset U$ then $\sup_x |g(x) - \int_{-\pi}^{\pi} K_V(x-y)g(y)\,dy| < b$. The set of Fejér kernels (for all N in $\mathbb{N}$) is a class $\mathscr{K}$ in which iii) holds.

251. Let $g: \mathbb{Q} \cap (1/4, 3/4) \mapsto \mathbb{Q} \cap [1/4, 3/4]$ be a bijection and extend g^{-1} to I according to the rule

$$G: x \mapsto \begin{cases} x, & \text{if } x \notin \mathbb{Q} \cap [1/4, 3/4] \\ g^{-1}(x), & \text{otherwise} \end{cases}.$$

Then G is a bijection and if $f = G^{-1}$ then f satisfies all requirements posed. □

252. Revise mildly the construction described in Solution 245. This time insure that for each interval $[n, n+1)$ the sum of the measures of the Cantor-like sets constructed in it is one. If A is a Cantor-like set in the interval $[a, b]$ there is a corresponding Cantor-like (monotone increasing) function g_A that maps A onto $[0, 1]$. Then on (a, b), $x \mapsto \tan \frac{1}{2}\pi g_A(x)$ maps $A \cap (a, b)$ onto $(0, \infty)$. Thus if, for every Cantor-like set A constructed, $G|_A$ is g_A and G is zero otherwise, then on every interval (a, b), $\int_a^b G(x)\,dx = \infty$. □

253. For k in $\mathbb{Z}$ and n in $\mathbb{N}$ let E_{kn} be $f^{-1}([k \,.\, 2^{-n}, (k+1) \,.\, 2^{-n}))$ and let f_n be $x \mapsto \sum_{k=-\infty}^{\infty} k \,.\, 2^{-n} \,.\, \chi_{E_{kn}}$. Since $0 \leqq f - f_n \leqq 2^{-n}$, the result follows. (Note: $f_n \uparrow f$.) □

254. If U is an open subset of $\mathbb{R}$, then $(g \circ f)^{-1}(U) = f^{-1}(g^{-1}(U))$. Since g is Lebesgue measurable, $g^{-1}(U)$ is Lebesgue measurable; there are two null sets (λ) N_1 and N_2 and a Borel set A such that $g^{-1}(U) = (A \backslash N_1) \cup N_2$, $A \supset N_1$, and $A \cap N_2 = \varnothing$. Thus $f^{-1}(g^{-1}(U)) = (f^{-1}(A) \backslash f^{-1}(N_1)) \cup f^{-1}(N_2)$. Because f is continuous and A is a Borel set, $f^{-1}(A)$ is a Borel set; the hypothesis implies $f^{-1}(N_1)$ and $f^{-1}(N_2)$ are Lebesgue measurable from which the result follows. □ (See [9] for a counterexample to the conclusion if the hypothesis: *if* $\lambda(N) = 0$ *then* $f^{-1}(N)$ *is Lebesgue measurable*, is dropped.)

255. According to the metric density theorem [24], if x is outside a null set N in A and $\lim_{d>0, d\to 0} \lambda^*(A \cap (x-d, x+d))/2d = 1$. Since $\lambda^*(A) > 0$, $A \backslash N \neq \varnothing$ and for some positive d, $\lambda^*(A \cap (x-d, x+d))/2d > \theta$ and the result follows. □

256. If $x_{2n} \to x_2$ as $n \to \infty$ then for all x_1, $f(x_1, x_{2n}) \to f(x_1, x_2)$ as $n \to \infty$. The dominated convergence theorem implies that $\int_I f(x_1, x_{2n})\,dx_1 \to \int_I f(x_1, x_2)\,dx_1$ as $n \to \infty$, i.e., that $h(x_{2n}) \to h(x_2)$ as $n \to \infty$, whence h is continuous. □

257. Let g be

$$x \mapsto \begin{cases} f(x) & \text{if } |f(x)| \leqq \|f\|_\infty \\ 0 & \text{otherwise} \end{cases}.$$

Then $f = g$ a.e. and $\sup_x |g(x)| = \|f\|_\infty$. □

258. The map $a \mapsto \lambda\{x : f(x) \geqq a\}$ is monotone decreasing and, since all sets considered are contained in I, $1 \geqq \lambda\{x : f(x) \geqq a\} \geqq 0$. Since $\lim_{a\to-\infty} \lambda\{x : f(x) \geqq a\} = 1 = 1 - \lim_{a\to\infty} \lambda\{x : f(x) \geqq a\}$, let a_0 be $\sup\{a : \lambda\{x : f(x) \geqq a\} \geqq \frac{1}{2}\}$. Then if $a > a_0$, $\lambda\{x : f(x) \geqq a\} < \frac{1}{2}$. Furthermore, $\{x : f(x) \geqq a_0\} = \bigcap_{n=1}^{\infty} \{x : f(x) > a_0 - 1/n\}$ and so $\lambda\{x : f(x) \geqq a_0\} = \lim_{n\to\infty} \lambda\{x : f(x) > a_0 - 1/n\} \geqq \frac{1}{2}$. If $a_1 < a_0$ and if $\lambda\{x : f(x) \geqq a_1\} \geqq \frac{1}{2}$ then for a_2 in (a_1, a_0), $\lambda\{x : f(x) \geqq a_1\} \geqq \frac{1}{2}$ and so only a_0 satisfies all requirements posed. □

259. i) Since $g_C = \frac{1}{2}$ on the middle (deleted) interval (the length of which is 1/3), etc., and since the values of g_C on *the* Cantor set are irrelevant to the value of $\int_I g_C(t)\,dt$ (because $\lambda(C) = 0$) it follows that the integral is $\sum_{n=1}^{\infty} 2^{n-2}(1/3)^n = 3/8$.

ii) Since g_C is monotone and bounded, l_{g_C} is finite. Furthermore if $M = \sum_{k=0}^{m} 2^k$ and if $\{t_n\}_{n=1}^{M}$ are the endpoints, arranged in increasing order, of the intervals deleted up to and including the mth stage of the construction of C, let f_M be the piecewise linear function such that $f_M(t_n) = g_C(t_n)$, $n = 1, 2, \ldots, M$. Then $l_{g_C} = \lim_{M\to\infty}$ (sum of the lengths of the horizontal line segments in the graph of $f_M +$ sum of the lengths of the sloped line segments in the graph of f_M). The first sum approaches one from below, the second approaches one from above and thus $l_{g_C} = 2$. □

260. If $(a, b) \subset I$ there are in $[a, b]$ points c, d such that $f([a, b]) = [f(c), f(d)]$. Hence $\lambda(f((a, b))) = \lambda(f([a, b])) = |f(d) - f(c)|$. According to the law of the mean, $|f(d) - f(c)| \leqq M|d - c|$, whence $\lambda(f((a, b))) \leqq M(b - a)$. (Note: The following regarding f and f' are not needed in the subsequent discussion. They are included here for their intrinsic interest. i) The function f is in Lip (1) on I and hence f is absolutely continuous. ii) Since $f' = \lim_{n\to\infty} \Delta_{1/n} f$, f' is Borel measurable. Since $|f'| \leqq M$, $f' \in L^1(I, \lambda)$ and for all x in I, $f(x) = f(0) + \int_0^x f'(t)\,dt$.)

If N is a null set (λ) and if $a > 0$ there is a sequence $\{(a_n, b_n)\}_{n=1}^{\infty}$ of pair-wise disjoint (open) intervals such that $N \subset \bigcup_n (a_n, b_n)$ and $\sum_n (b_n - a_n) < a/M$. Thus $f(N) \subset \bigcup_n f((a_n, b_n)$ and $\lambda(f(N)) \leqq M \sum_n (b_n - a_n) < a$, i.e., $f(N)$ is a null set.

Since λ is regular there is a sequence $\{K_n\}_{n=1}^{\infty}$ of compact sets and a null set $N(\lambda)$ such that $E = \bigcup_n K_n \cup N$, whence $f(E) = \bigcup_n f(K_n) \cup f(N)$. Since $f(N)$ is a null set (λ) and each $f(K_n)$ is compact it follows that $f(E)$ is Lebesgue measurable.

Finally, if $a > 0$ and $E \subset \bigcup_{n=1}^{\infty} (a_n, b_n)$ and $\sum_n (b_n - a_n) < \lambda(E) + a/M$, then $\lambda(f(E)) \leqq M\lambda(E) + a$ and the result follows. □

261. Make the substitutions: $x=e^z$, $y=e^u$, $f(e^z)=F(z)$, $g(e^u)=G(u)$. Then $F, G\in L^1(\mathbb{R},\lambda)$ and $\int_{(0,\infty)} y^{-1}f(xy)g(y^{-1})\,dy = F*G(z)$, from which the result follows. □

262. The Stone–Weierstrass theorem implies that for g in $C(I,\mathbb{C})$, $\int_I g(t)f(t)\,dt=0$ and the result follows by approximation. □

263. For n in $\mathbb{N}$, since $|f_n(x)|\leqq e^{-x^2}$ for all x, it follows that $f_n\in L^1(\mathbb{R},\lambda)$. If $a>0$ let A be positive and such that $\int_{\mathbb{R}\setminus[-A,A]} e^{-x^2}\,dx<a/6$. Let b be positive and such that $b\,.\,2A<a/3$. Then there is in $\mathbb{N}$ an N such that if $m, n>N$ then $\lambda\{x:|f_m(x)-f_n(x)|\geqq b\}<a\int_{\mathbb{R}} e^{-x^2}/6\,dx$. If $E_{mn}=\{x:|f_m(x)-f_n(x)|\geqq b\}$, then $\int_{\mathbb{R}}|f_m(x)-f_n(x)|\,dx=\int_{\mathbb{R}\setminus[-A,A]}+\int_{[-A,A]\cap E_{mn}}+\int_{[-A,A]\setminus E_{mn}}|f_m(x)-f_n(x)|\,dx<a/3+a/3+a/3=a$. It follows that $\{f_n\}_{n=1}^\infty$ is a Cauchy sequence in $L^1(\mathbb{R},\lambda)$. If g is the limit then there is a subsequence $\{f_{n_k}\}_{k=1}^\infty$ converging a.e. to g, hence converging in measure to g and so $f_0=g$ a.e. and the result follows. □

264. For some n in $\mathbb{N}$, $\lambda(E\cap[-n,n])>0$ and so it may be assumed that $0<\lambda(E)<\infty$. (If $x, y\in E\cap[-n,n]$ then $\frac{1}{2}(x+y)\in E\cap[-n,n]$.) Thus $\chi_E\in L^1(\mathbb{R},\lambda)$ and $f:x\mapsto\int_{\mathbb{R}}\chi_E(2x-y)\chi_E(y)\,dy$ is continuous, nonnegative and vanishes off $\frac{1}{2}(E+E)$. By hypothesis $\frac{1}{2}(E+E)\subset E$. Furthermore $f(\frac{1}{2}x)=\chi_E*\chi_E(x)$. If $f=0$ then $\hat{\chi}_E^2=0$ while χ_E is a nonzero element of $L^1(\mathbb{R},\lambda)$. The fundamental injectivity of the Fourier transform does not permit such a situation and so $f>0$ on some open set U, which, since $f=0$ off $\frac{1}{2}(E+E)$, is contained in E and the result follows. □

265. Every subset of *the* Cantor set C is in S_λ whence $\text{card}(\mathsf{S}_\lambda)\geqq 2^{\text{card}(C)}=2^{\text{card}(\mathbb{R})}$ and the result follows. □

266. Since $\bigcup_{a\in E} B(a,r_a)\supset E$, the Lindelöf (covering) theorem implies that there is in E a countable set $\{a_n\}_{n=1}^\infty$ such that $\bigcup_n B(a_n,r_{a_n})\supset E$, i.e., $E=\bigcup_n B(a_n,r_{a_n})\cap E$. Since each constituent of the union is Lebesgue measurable, so is E. □

267. According to Problem 265, $\text{card}(\mathsf{S}_\lambda)=2^{\text{card}(\mathbb{R})}$ whereas according to Problem 188, $\text{card}(\mathsf{S}_\beta)=\text{card}(\mathbb{R})$ $(<2^{\text{card}(\mathbb{R})})$. If $E\in\mathsf{S}_\lambda\setminus\mathsf{S}_\beta$ nothing need be proved. If $E\in\mathsf{S}_\beta$ then (see Problem 516) $\text{card}(E)=\text{card}(\mathbb{N})$ or $\text{card}(E)=\text{card}(\mathbb{R})$. By hypothesis, $\text{card}(E)>\text{card}(\mathbb{N})$ whence $\text{card}(E)=\text{card}(\mathbb{R})$. Since $\lambda(E)=0$ every element of 2^E is Lebesgue measurable. Since $\text{card}(2^E)>\text{card}(\mathbb{R})=\text{card}(\mathsf{S}_\beta)$ there must be some F contained in E and such that F is not Borel measurable. □

268. If E is Lebesgue measurable there are null sets (λ) N_1 and N_2 and in S_β an A such that $E=(A\setminus N_1)\cup N_2$. The hypotheses imply that N_1 and N_2 are null sets (μ). Since $\mathsf{S}_\beta\subset\mathsf{S}_\mu$ it follows that $E\in\mathsf{S}_\mu$, i.e., $\mathsf{S}_\lambda\subset\mathsf{S}_\mu$. □

269. The following is a counterexample to the assertion. Let

$$f_n \text{ be } x \mapsto \begin{cases} x & \text{if } 0 \leqq x < 1 \\ 2 & \text{if } x = 1 \end{cases},$$

$n = 1, 2, \ldots$. If $f_0(x) = x$ for x in $[0, 1]$ then $f_n \to f_0$ in measure as $n \to \infty$ but $f_n(1) = 2 \not\to f_0(1) = 1$ as $n \to \infty$. □

270. The set S is the (classical) example of a set that is a subset of I and is not in $\mathsf{S}_\lambda(I)$. The sets S_k are pairwise disjoint: if $k \neq k'$ and if $x \in S_k \cap S_{k'}$ then for s, s' in S, $x = s + r_k = s' + r_{k'}$, $s - s' = r_{k'} - r_k$ and so $s = s'$. Furthermore, $\bigcup_k S_k = [0, 1)$ since if $x \in [0, 1)$ there is in S a unique s such that $\theta(x) = \theta(s)$, i.e., $x - s \in \mathbb{Q} \cap [0, 1)$ and $x = s + r_k$ for some k.

Consequently $\lambda^*(S) = \lambda^*(S_k) = M > 0$ (and $\lambda_*(S) = \lambda_*(S_k) = 0$). (The existence of S depends on the use of the axiom of choice.)

If $t \to 0$, $k_t \to \infty$. Hence for x fixed, x belongs to precisely one S_k and hence if t is sufficiently close to zero, $x \notin S_{k_t}$, i.e., $f_t(x) = 0$. In sum $\lim_{t \to 0} f_t(x) = 0$ for all x.

Next note that for all positive t, $\text{card}\{x : f_t(x) \neq 0\} = 1$ or 0 according as $2^{k_t+1}t - 1$ is or is not in S_{k_t}. Hence, each f_t is Lebesgue measurable.

Let J_k be $[2^{-k-1}, 2^{-k})$, k in $\mathbb{N}$. Then as t traverses J_k, $x = 2^{k+1}t - 1$ traverses $[0, 1)$. Hence if $x \in S_k$, let t be $(x+1)/2^{k-1}$, in which case $f_t(x) = 1 > \frac{1}{2}$, i.e., $\bigcup_{t \in J_k} \{x : f_t(x) > \frac{1}{2}\} \supset S_k$, $\lambda^*(\bigcup_{t \in J_k} \{x : f_t(x) > \frac{1}{2}\}) \geqq M$. If $0 < a < M$, if E is Lebesgue measurable, and if $\lambda(E) < a$, then as $t \to 0$, t traverses all but finitely many J_k and f_t fails to approach zero uniformly off E since off E there are points in the sets $\bigcup_{t \in J_k} \{x : f_t(x) > \frac{1}{2}\}$. □

271. For n in $\mathbb{N}$ let f_n be

$$x \mapsto \begin{cases} 1/n, & \text{if } 1/n < x \leqq 1 \\ n, & \text{if } 0 < x \leqq 1/n. \\ 0, & \text{if } x = 0 \end{cases}$$

Then for all x, $f_n(x) \to 0$ as $n \to \infty$ while if $\lambda(E) = 1$, $\max_{x \in E} f_n(x) = n$. □

272. For each n let $\{J_{nm}\}_{m=1}^\infty$ be a sequence of pairwise disjoint open intervals such that $U_n = \bigcup_m J_{nm} \supset E$, $\sum_m \lambda(J_{nm}) < 3^{-n}$, and for each $J_{n+1,m}$ there is a $J_{nm'}$ containing $J_{n+1,m}$. In other words $\{J_{n+1,m}\}_{m=1}^\infty$ is a refinement of $\{J_{nm}\}_{m=1}^\infty$. If $J_{nm} = (a_{nm}, b_{nm})$ let f_{nm} be

$$x \mapsto \begin{cases} 0, & \text{if } x < a_{nm} \\ 3^n(x - a_{nm}), & \text{if } a_{nm} \leqq x < b_{nm} \\ 3^n(b_{nm} - a_{nm}), & \text{if } b_{nm} \leqq x \end{cases}$$

and let f_n be $\sum_m f_{nm}$. Note that $0 \leqq f_n \leqq 1$, f_n is monotone increasing, and $f_n' = 3^n$ on E. If $f = \sum_n f_n/2^n$ then f is monotone increasing; if $x \in E$ for each n there is a unique m_n such that $x \in J_{nm_n}$. If h is so small that $x + h \in J_{nm_n}$ then $\Delta_h f(x) \geqq 3^n$ and so $f'(x) = \infty$. □

273. i) If $\{g_m\}_{m=1}^{\infty} \subset A_f$ and $\|g_m - g_0\|_1 \to 0$ as $m \to \infty$, by passage to a subsequence as needed it may be assumed that $g_m \to g_0$ a.e. as $m \to \infty$. Since $|g_m| \leqq f$ a.e. for all m it follows that $|g_0| \leqq f$ a.e. and so $g_0 \in A_f$, i.e., A_f is closed. (Note that this conclusion may be drawn without the assumption that f is Lebesgue measurable.)

ii) Let E_n be $\{x : f(x) \leqq n\}$. Then E_n is Lebesgue measurable and $\bigcup_n E_n = I$. If $g_n = \chi_{E_n} \cdot f$ then g_n is in A_f and $g_n \to f$ a.e. as $n \to \infty$. If A_f is compact there is a subsequence $\{g_{n_k}\}_{k=1}^{\infty}$ and in A_f a g_0 such that $\|g_{n_k} - g_0\|_1 \to 0$ as $k \to \infty$. Again, *via* subsequences it may be assumed that $g_{n_k} \to g_0$ a.e. as $k \to \infty$. Since $g_0 \in A_f \subset L^1(I, \lambda)$ and since $g_0 = f$ a.e., it follows that $f \in L^1(I, \lambda)$. □

274. Let A belong to $2^{\mathbb{R}} \backslash \mathsf{S}_\lambda$ (A is not Lebesgue measurable) and let E be $A \times \{0\} \cup \{0\} \times A$ (in $\mathbb{R}^2$). Then $\lambda_2(E) = 0$ whereas $E + E = (A + A) \times \{0\} \cup \{0\} \times (A + A) \cup (A \times A)$, which is not in $\mathsf{S}_\lambda(\mathbb{R}^2)$, as the following lines show. The first two constituents of E have (two-dimensional) Lebesgue measure equal to zero. The third is not Lebesgue measurable (not in $\mathsf{S}_\lambda(\mathbb{R}^2)$) because if it were its sections $(A \times A)_{x_1}$ would be in $\mathsf{S}_\lambda(\mathbb{R})$ a.e. and on the other hand, for x_1 in the nonnull set (λ) A, the section is the nonmeasurable set A. □

11. $L^1(X, \mu)$

275. If for each p in $\mathbb{N}\{y_m(p)\}_{m=1}^{\infty} = \{\delta_{pm}\}_{m=1}^{\infty}$ then $\sum_{m=1}^{\infty} x_{nm}y_m(p) = x_{np}$. Hence $\lim_{n\to\infty} x_{np} = x_{0p}$ for all p in $\mathbb{N}$.

Let z_n be $\{x_{nm} - x_{0m}\}_{m=1}^{\infty} = \{z_{nm}\}_{m=1}^{\infty}$. If $\|z_n\|_1 \nrightarrow 0$ as $n \nrightarrow \infty$, then by passage to subsequences as needed, it may be assumed that for some positive a and all n in $\mathbb{N}$, $\|z_n\|_1 \geqq a$. Thus let m_1 be such that $\sum_{m=1}^{m_1} |z_{1m}| > \frac{1}{2}a$, $\sum_{m=m_1+1}^{\infty} |z_{1m}| < a/5$. Then let n_1 be such that $\sum_{m=1}^{m_1} |z_{n_1m}| < a/5$ and let m_2 be greater than m_1 and such that $\sum_{m=m_1+1}^{m_2} |z_{n_1m}| > \frac{1}{2}a$ and $\sum_{m=m_2+1}^{\infty} |z_{n_1m}| < a/5$. Proceed inductively and produce two sequences $n_1, n_2, \ldots,$ $m_1, m_2, \ldots,$ such that $n_1 < n_2 < \ldots,$ $m_1 < m_2 \ldots,$ $\sum_{m=1}^{m_k} |z_{n_km}| < a/5$, $\sum_{m=m_k+1}^{m_{k+1}} |z_{n_km}| > a/2$, and $\sum_{m=m_{k+1}+1}^{\infty} |z_{n_km}| < a/5$. If $y_m = \operatorname{sgn} z_{n_km}$, $m_k + 1 \leqq m \leqq m_{k+1}$, then $\{y_m\}_{m=1}^{\infty} \in l^{\infty}(\mathbb{N})$ and $|\sum_{m=1}^{\infty} z_{n_km}y_m| = |\sum_{m=1}^{m_k} + \sum_{m=m_k+1}^{m_{k+1}} + \sum_{m=m_{k+1}+1}^{\infty} z_{n_km}y_m| \geqq -a/5 + a/2 - a/5 = a/10$. On the other hand, by hypothesis, $\lim_{k\to\infty} \sum_{m=1}^{\infty} z_{n_km}y_m = 0$ and the contradiction implies the result.

Note that for each p, $\{y_m(p)\}_{m=1}^{\infty} \in l^1(\mathbb{N})$ and for any $\{u_m\}_{m=1}^{\infty}$ in $C_0(\mathbb{N}, \mathbb{C})$, $\lim_{p\to\infty} \sum_{m=1}^{\infty} y_m(p)u_m = 0$. On the other hand there is in $l^1(\mathbb{N})$ no $\{y_m(0)\}_{m=1}^{\infty}$ such that $\|\{y_m(p)\}_{m=1}^{\infty} - \{y_m(0)\}_{m=1}^{\infty}\|_1 \to 0$ as $p \to \infty$. However, whereas $\{y_m(p)\}_{m=1}^{\infty} \to 0$ in the weak* topology $(\sigma(l^1(\mathbb{N}), C_0(\mathbb{N}, \mathbb{C})))$, $\{y_m(p)\}_{m=1}^{\infty} \nrightarrow 0$ in the weak topology $(\sigma(l^1(\mathbb{N}), l^{\infty}(\mathbb{N})))$ as $n \to \infty$. □

276. The use of bridging functions (see [9]) lying in $C^{\infty}([0, \infty), \mathbb{C})$ permits the construction, for any interval $[a, b]$ $(a < b)$ of $[0, \infty)$ and any c in $(0, b-a)$, of a nonnegative function f_{abc} in $C^{\infty}([0, \infty), \mathbb{C})$, such that

$$f_{abc}(x) = \begin{cases} 0, & \text{if } x \leqq a \\ 1, & \text{if } x \geqq b \end{cases}$$

and such that $\int_a^b f_{abc}(x)\,dx = c$. (Indeed there are two points p_c, q_c such that $a \leqq p_c < q_c \leqq b$ and such that f_{abc} is

$$x \mapsto \begin{cases} \exp(-(x-p_c)^{-2} + (q_c - p_c)^{-2}), & \text{if } p_c < x < q_c \\ 0, & \text{if } x \leqq p_c \\ 1, & \text{if } x \geqq q_c \end{cases} \quad .)$$

For n in $\mathbb{N}$ let g_n be $x \mapsto f_{2n-1,2n,2^{-n}}x) \cdot f_{2n-1,2n,2^{-n}}(4n-x)$ and let g be $\sum_n g_n$. Then $\int_{[0,\infty)} g(x)\,dx = 2$ and $\sum_{n=1}^{\infty} g(n) = \sum_{n \text{ even}} 1 = \infty$. □

277. Since

$$\begin{aligned} \int_0^\infty x^{-1}|\sin x|\,dx &= \sum_{n=0}^{\infty} \int_{n\pi}^{(n+1)\pi} x^{-1}|\sin x|\,dx \\ &\geqq \sum_{n=0}^{N} \frac{1}{n+1\pi} \int_{n\pi}^{(n+1)\pi} |\sin x|\,dx \\ &= 2 \sum_{n=0}^{N} ((n+1)\pi)^{-1}, \end{aligned}$$

it follows that the integral is infinite and so $x \mapsto x^{-1} \sin x \notin L^1((0,\infty), \lambda)$. □

(NOTE. Nevertheless, $\lim_{R\to\infty} \int_0^R x^{-1} \sin x\,dx$ exists and is the sum of the convergent alternating series $\sum_{n=0}^{\infty} \int_{n\pi}^{(n+1)\pi} x^{-1} \sin x\,dx$.)

278. By hypothesis, $|\int_X g(x)\,d\mu_n(x)| \leqq \|g\|_1$. Thus $\{\mu_n\}_{n=1}^{\infty}$, viewed as a subset of $(L^1(X, \mu_0))^* = L^\infty(X, \mu_0)$, is contained in the unit ball of $L^\infty(X, \mu_0)$. However, the Alaoglu theorem insures that the unit ball of the dual space of a Banach space is compact in the weak* topology, from which the result follows. □

279. Note that T is the canonical injection of $L^1(X, \mu)$ into its second dual $(L^1(X, \mu))^{**} = (L^\infty(X, \mu))^*$. For each n in $\mathbb{N}$ let h_n be $\chi_{A_n}/\mu(A_n)$. Thus for all n, $\|h_n\|_1 = 1$ and it suffices, by virtue of Eberlein's theorem, to show that every subsequence $\{h_{n_k}\}_{k=1}^{\infty}$ fails to converge in the weak topology of $L^1(X, \mu)$. Indeed, if $h_{n_k} \to h_0$ in the weak topology as $k \to \infty$, then for each n and any g in $L^\infty(X, \mu)$, $\int_X h_{n_k}(x) g(x) \chi_{A_n}(x)\,d\mu(x) \to \int_X h_0(x) g(x) \chi_{A_n}(x)\,d\mu(x)$ as $k \to \infty$. Hence $h_0 = 0$ on $\bigcup_n A_n$ and so $h_0 = 0$. On the other hand if $g = \chi_{\bigcup_n A_n}$, then $\int_X h_{n_k}(x) g(x)\,d\mu(x) = (\mu(A_{n_k}))^{-1} \int_{A_{n_k}} d\mu(x) = 1$, and so $h_{n_k} \not\to 0$ in the weak topology as $k \to \infty$. The contradiction implies the result. □

280. If $f \in L^1(I, \lambda)$ and $f \geqq 0$ then $\int_I f(x)\,dx \geqq \sum_{n=1}^{\infty} (n-1)\lambda(E_n) = \sum_n n\lambda(E_n) - \sum_n \lambda(E_n) = \sum_n n\lambda(E_n) - 1$, whence $\sum_n n\lambda(E_n) < \infty$.

Conversely, if $\sum_n n\lambda(E_n) < \infty$ then $\int_X f(x)\,dx \leqq \sum_n n\lambda(E_n) < \infty$. □

281. If $x \neq 0$ then f is the product of differentiable functions and so is differentiable at x. If $x = 0$, $|\Delta_h f| \leqq |h|$ and so $f'(0)$ exists and is zero. In sum, f' is

$$x \mapsto \begin{cases} -2x^{-1}\cos(x^{-2}) + 2x\sin(x^{-2}), & \text{if } x \neq 0 \\ 0, & \text{if } x = 0 \end{cases}.$$

If $2/\pi < x^2 < \infty$, $|\sin(x^{-2})| \geqq 2/\pi x^2$ and so $|2x \sin x^{-2}| \geqq 4/\pi x$ while $|-2x^{-1}\cos x^{-2}| \leqq 2/|x|$. Consequently $|f'(x)| \geqq (2-4/\pi)|1/x|$ whence $f' \notin L^1(\mathbb{R}, \lambda)$. □

282. If $f = 2\chi_{[0,1/3)} + \chi_{[1/3,2/3)} + 1.5\chi_{[2/3,1]}$ and if

$$F(x) = \begin{cases} 0, & \text{if } x = 0 \\ x^{-1}\displaystyle\int_0^x f(t)\,dt, & \text{if } x \neq 0 \end{cases}$$

then $F \geqq f$ a.e. and yet f is not monotone decreasing. □

283. There is a sequence $\{J_n\}_{n=1}^{\infty}$ of pairwise disjoint nondegenerate closed intervals such that $\int_{J_n} |f(x)|\,dx = a_n > 0$. Then $\sum_n a_n < \infty$ and by passage to a subsequence as needed it may be assumed that $\sum_n a_n^{1/2} < \infty$. If $g = \sum_n a_n^{-1/2}\chi_{J_n}$ then g behaves in accordance with the requirements. □

284. Let E be $f^{-1}(1)$. If $A \subset I \backslash E$ then $A = \bigcup_{m=1}^{\infty} \{x : |f(x) - 1| \geqq 1/m\}$. If $\lambda(A) = 0$ then $f = 1$ a.e. If $\lambda(A) > 0$ then for some m, $\lambda\{x : |f(x) - 1| \geqq 1/m\} > 0$, i.e., i) $\lambda\{x : f(x) \geqq 1 + 1/m\} > 0$ or ii) $\lambda\{x : f(x) < 1 - 1/m\} > 0$. If i) obtains then $\int_I (f(x))^n\,dx \to \infty$ as $n \to \infty$ in denial of the hypothesis. Hence $f < 1$ on A and so $f^n \to 0$ on A as $n \to \infty$. The dominated convergence theorem implies $\int_A (f(x))^n\,dx \to 0$ as $n \to \infty$. In the presence of the hypothesis, $f = 0$ a.e. on A and so $f = \chi_E$ a.e.

If the hypothesis that $f \geqq 0$ is dropped then, e.g., for any c in $\mathbb{C}$ and any k in $\mathbb{Z}\backslash\{0\}$, $f : x \mapsto c\,e^{2\pi i k x}$ is such that $\int_I (f(x))^n\,dx = 0$ for all n in $\mathbb{N}$. If f is assumed to be $\mathbb{R}$-valued, then $f^2 \geqq 0$ and so for some Lebesgue measurable set E, $f^2 = \chi_E$. Hence, if $E_1 = (f \wedge 0)^{-1}(-1)$ and $E_2 = (f \vee 0)^{-1}(1)$, then $f = \chi_{E \cap E_2} - \chi_{E \cap E_1}$. □

285. It may be assumed that f is $\mathbb{R}$-valued since otherwise, $f = p + iq$, p, q, $\mathbb{R}$-valued, and then $\int_U p(x)\,dx = \int_U q(x)\,dx = 0$ for every open set U such that $\lambda(U) = 1$ and the discussion may be carried out for p and q.

Thus $\int_a^{a+1} f(x)\,dx = 0$ for all a and so for any nonzero b, $\int_a^{a+b} f(x)\,dx = \int_{a+1}^{a+1+b} f(x)\,dx$, $b^{-1}\int_a^{a+b} f(x)\,dx = b^{-1}\int_{a+1}^{a+1+b} f(x)\,dx$ whence $f(a) = f(a+1)$ a.e.

Let N_1 be the set of a such that $f(a) \neq f(a+1)$, N_2 the set of a such that $f(a) = f(a+1) \neq f(a+2)$, etc. Then for all k, $\lambda(N_k) = 0$ and if $a \in \mathbb{R}\backslash\bigcup_k N_k = S$ then $f(a) = f(a+1) = \cdots$, i.e., $f(a) = f(a+1) = \cdots$ a.e. Off a null set N in S, $\lim_{n\to\infty} n\int_a^{a+1/n} f(x)\,dx = f(a)$. If $a \in S\backslash N$ let U be $(a, a+1/n) \cup (a+1, a+1+1/n) \cup \cdots \cup (a+n-1, a+n-1+1/n)$. Then U is an open set, $\lambda(U) = 1$ and $\int_U f(x)\,dx = 0 = n\int_a^{a+1/n} f(x)\,dx \to f(a)$ as $n \to \infty$. Thus $f = 0$ a.e. □

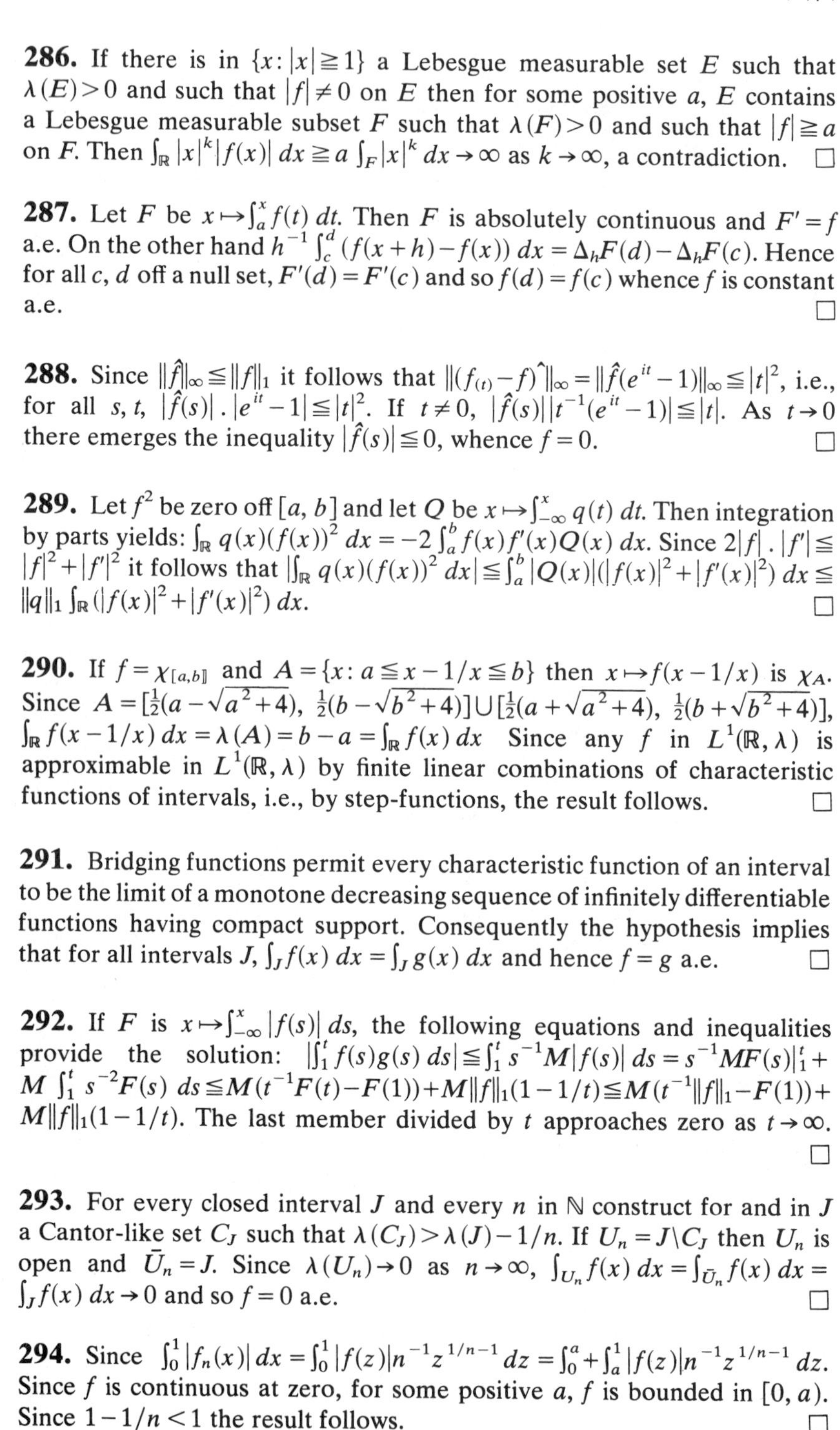

286. If there is in $\{x: |x| \geqq 1\}$ a Lebesgue measurable set E such that $\lambda(E) > 0$ and such that $|f| \neq 0$ on E then for some positive a, E contains a Lebesgue measurable subset F such that $\lambda(F) > 0$ and such that $|f| \geqq a$ on F. Then $\int_{\mathbb{R}} |x|^k |f(x)|\, dx \geqq a \int_F |x|^k\, dx \to \infty$ as $k \to \infty$, a contradiction. □

287. Let F be $x \mapsto \int_a^x f(t)\, dt$. Then F is absolutely continuous and $F' = f$ a.e. On the other hand $h^{-1} \int_c^d (f(x+h) - f(x))\, dx = \Delta_h F(d) - \Delta_h F(c)$. Hence for all c, d off a null set, $F'(d) = F'(c)$ and so $f(d) = f(c)$ whence f is constant a.e. □

288. Since $\|\hat{f}\|_\infty \leqq \|f\|_1$ it follows that $\|(f_{(t)} - f)\hat{}\|_\infty = \|\hat{f}(e^{it} - 1)\|_\infty \leqq |t|^2$, i.e., for all s, t, $|\hat{f}(s)| \,.\, |e^{it} - 1| \leqq |t|^2$. If $t \neq 0$, $|\hat{f}(s)|\,|t^{-1}(e^{it} - 1)| \leqq |t|$. As $t \to 0$ there emerges the inequality $|\hat{f}(s)| \leqq 0$, whence $f = 0$. □

289. Let f^2 be zero off $[a, b]$ and let Q be $x \mapsto \int_{-\infty}^x q(t)\, dt$. Then integration by parts yields: $\int_{\mathbb{R}} q(x)(f(x))^2\, dx = -2\int_a^b f(x)f'(x)Q(x)\, dx$. Since $2|f| \,.\, |f'| \leqq |f|^2 + |f'|^2$ it follows that $|\int_{\mathbb{R}} q(x)(f(x))^2\, dx| \leqq \int_a^b |Q(x)|(|f(x)|^2 + |f'(x)|^2)\, dx \leqq \|q\|_1 \int_{\mathbb{R}} (|f(x)|^2 + |f'(x)|^2)\, dx$. □

290. If $f = \chi_{[a,b]}$ and $A = \{x: a \leqq x - 1/x \leqq b\}$ then $x \mapsto f(x - 1/x)$ is χ_A. Since $A = [\frac{1}{2}(a - \sqrt{a^2+4}),\ \frac{1}{2}(b - \sqrt{b^2+4})] \cup [\frac{1}{2}(a + \sqrt{a^2+4}),\ \frac{1}{2}(b + \sqrt{b^2+4})]$, $\int_{\mathbb{R}} f(x - 1/x)\, dx = \lambda(A) = b - a = \int_{\mathbb{R}} f(x)\, dx$ Since any f in $L^1(\mathbb{R}, \lambda)$ is approximable in $L^1(\mathbb{R}, \lambda)$ by finite linear combinations of characteristic functions of intervals, i.e., by step-functions, the result follows. □

291. Bridging functions permit every characteristic function of an interval to be the limit of a monotone decreasing sequence of infinitely differentiable functions having compact support. Consequently the hypothesis implies that for all intervals J, $\int_J f(x)\, dx = \int_J g(x)\, dx$ and hence $f = g$ a.e. □

292. If F is $x \mapsto \int_{-\infty}^x |f(s)|\, ds$, the following equations and inequalities provide the solution: $|\int_1^t f(s)g(s)\, ds| \leqq \int_1^t s^{-1}M|f(s)|\, ds = s^{-1}MF(s)|_1^t + M\int_1^t s^{-2}F(s)\, ds \leqq M(t^{-1}F(t) - F(1)) + M\|f\|_1(1 - 1/t) \leqq M(t^{-1}\|f\|_1 - F(1)) + M\|f\|_1(1 - 1/t)$. The last member divided by t approaches zero as $t \to \infty$. □

293. For every closed interval J and every n in $\mathbb{N}$ construct for and in J a Cantor-like set C_J such that $\lambda(C_J) > \lambda(J) - 1/n$. If $U_n = J \backslash C_J$ then U_n is open and $\bar{U}_n = J$. Since $\lambda(U_n) \to 0$ as $n \to \infty$, $\int_{U_n} f(x)\, dx = \int_{\bar{U}_n} f(x)\, dx = \int_J f(x)\, dx \to 0$ and so $f = 0$ a.e. □

294. Since $\int_0^1 |f_n(x)|\, dx = \int_0^1 |f(z)| n^{-1} z^{1/n - 1}\, dz = \int_0^a + \int_a^1 |f(z)| n^{-1} z^{1/n-1}\, dz$. Since f is continuous at zero, for some positive a, f is bounded in $[0, a)$. Since $1 - 1/n < 1$ the result follows. □

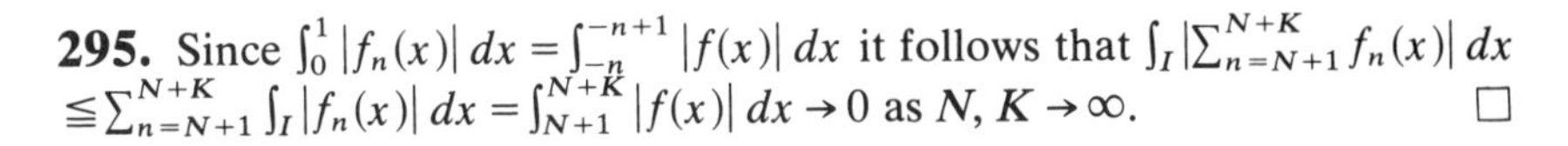

295. Since $\int_0^1 |f_n(x)|\,dx = \int_{-n}^{-n+1} |f(x)|\,dx$ it follows that $\int_I |\sum_{n=N+1}^{N+K} f_n(x)|\,dx \leqq \sum_{n=N+1}^{N+K} \int_I |f_n(x)|\,dx = \int_{N+1}^{N+K} |f(x)|\,dx \to 0$ as $N, K \to \infty$. □

296. Since $\|f_n\|_\infty \leqq M$ it follows that $\|f\|_\infty \leqq M$ and so $f_n g_n$, $fg \in L^1(G, \lambda)$. If $\|f_n g_n - fg\|_1 \nrightarrow 0$ as $n \to \infty$ there is a sequence $\{n_k\}_{k=1}^\infty$ such that $n_k < n_{k+1}$, such that for some positive a, $\|f_{n_k} g_{n_k} - fg\|_1 \geqq a$ and yet such that $f_{n_k} \to f$, $g_{n_k} \to g$ (whence $f_{n_k} g \to fg$) a.e. as $k \to \infty$. Since $|f_{n_k} g - fg| \leqq 2M|g|$ the dominated convergence theorem implies that $\|f_{n_k} g - fg\|_1 \to 0$ as $k \to \infty$. Since $\|f_{n_k} g - f_{n_k} g_{n_k}\|_1 \leqq M\|g - g_{n_k}\|_1$ it follows that $\|f_{n_k} g_{n_k} - fg\|_1 \leqq \|f_{n_k} g_{n_k} - f_{n_k} g\|_1 + \|f_{n_k} g - fg\|_1$ which approaches zero as $k \to \infty$. The contradiction ($\|f_{n_k} g_{n_k} - fg\|_1 \geqq a > 0$ was assumed) implies the result. □

297. For all t in $[0, \infty)$, $1 - e^{-t} \leqq t$ whence $1 - e^{-f_n} \leqq f_n$ and the result follows. □

298. Assume i) and ii) obtain. Since $f_n \to f$ a.e. as $n \to \infty$, f is Lebesgue measurable. If $b > 0$ there is, by virtue of Egorov's theorem, in $A_{b/2}$ a Lebesgue measurable set B such that for all n $\int_{A_{b/2} \setminus B} |f_n(x)|\,dx < b/2^2$ and $|f_n - f_m| \to 0$ uniformly on B as $n, m \to \infty$. Thus $\|f_n - f_m\|_1 \leqq \int_{\mathbb{R} \setminus A_{b/2}} + \int_{A_{b/2} \setminus B} + \int_B |f_n(x) - f_m(x)|\,dx \leqq b/2 + b/2^2 + \int_B |f_n(x) - f_m(x)|\,dx$. Since convergence is uniform on B it follows that for large n, m, the third term is less than $b/2^3$ and so $\{f_n\}_{n=1}^\infty$ is a Cauchy sequence in $L^1(\mathbb{R}, \lambda)$. If f_n converges to g in $L^1(\mathbb{R}, \lambda)$ as n gets large, then, *via* subsequences as needed, $f_n \to g$ a.e. as $n \to \infty$ and so $f = g$ a.e. and $f_n \to f$ in $L^1(\mathbb{R}, \lambda)$ as $n \to \infty$.

Conversely if $\|f_n - f\|_1 \to 0$ as $n \to 0$ and if $a > 0$ there is a Lebesgue measurable set E_a such that $\int_{\mathbb{R} \setminus E_a} |f(x)|\,dx < a/2$ and there is an n_0 such that if $n \geqq n_0$, $\int_{\mathbb{R}} |f_n(x) - f(x)|\,dx < a/2^2$. Hence $\int_{\mathbb{R} \setminus E_a} |f_n(x)|\,dx < a/2 + a/2^2$ if $n \geqq n_0$. There is a Lebesgue measurable set A_a containing E_a, of finite positive measure, and such that $\int_{\mathbb{R} \setminus A_a} |f_n(x)|\,dx < a/2^3$ if $n \leqq n_0$ and thus i) obtains.

Since $f_n \to f$ in $L^1(\mathbb{R}, \lambda)$ as $n \to \infty$, for some M, $\|f_n\|_1 \leqq M < \infty$ for all n. If B is a Lebesgue measurable set, let T_B be $L^1(\mathbb{R}, \lambda) \ni g \mapsto \int_B g(x)\,dx$. Then $T_B(g) \to 0$ as $\lambda(B) \to 0$ and so, according to the uniform boundedness principle, for some K, $\|T_B\| \leqq K < \infty$. Hence $|T_B(|f_n|)| \leqq |T_B(|f|)| + K\|f_n - f\|_1$. For b positive there is an n_0 such that $\|f_n - f\|_1 < b/K$ if $n \geqq n_0$ and so $\lim_{\lambda(B) \to 0} \sup_{n \geqq n_0} |T_B(|f_n|)| \leqq \lim_{\lambda(B) \to 0} |T_B(|f|)| + b = b$ and $\lim_{\lambda(B) \to 0} \sup_{n < n_0} |T_B(|f_n|)| = 0$. Since b is arbitrary it follows that $\lim_{\lambda(B) \to 0} \sup_n |T_B(|f_n|)| = 0$ and hence ii) holds. □

299. Since all coefficients of the series are nonnegative, the partial sums $S_n(|x|)$ for any x in $(-1, 1)$ constitute a monotone increasing sequence. Furthermore if $x \in (-1, 0)$ the signs of the terms of the alternate and the terms decrease to zero in absolute value. Thus for all x in $(-1, 1)$, $0 \leqq S_n(x) \leqq S_n(|x|) \leqq (1 - |x|)^{-1/2}$. Since $x \mapsto (1 - |x|)^{-1/2}$ is in $L^1((-1, 1), \lambda)$, the dominated convergence theorem is applicable and the result follows. □

300. Let $\sup_{x \geqq K} x/f(x)$ be $F(K)$. Since

$$\int_I |g_n(x)|\,dx = \int_{\{x:|g_n(x)|\geqq K)} + \int_{\{x:|g_n(x)|<K\}} |g_n(x)|\,dx$$

$$\leqq F(K) \textstyle\int_{\{x:|g_n(x)|\geqq K\}} f(|g_n(x)|)\,dx + K \leqq F(K)M + K < \infty$$

it follows that $\{g_n\}_{n=1}^{\infty} \subset L^1(I, \lambda)$. Furthermore if $a > 0$ there is a K so that $F(K) < a/8(M+1)$. Then

$$\|g_n - g_m\|_1 \leqq \int_{\{x:\,|g_n(x)-g_m(x)|/2 \leqq K\}}$$

$$+ \int_{\{x:\,|g_n(x)-g_m(x)|/2 > K\}} 2\,.\,|g_n(x) - g_m(x)|/2\,.\,dx.$$

Since $\int_I f(|g_n(x)|)\,dx \leqq M$ it follows that $\int_I f(|g_n(x)| \vee |g_m(x)|)\,dx = \int_{\{x:\,g_n(x)|\geqq g_m(x)|\}} f(|g_n(x)|)\,dx + \int_{\{x:\,|g_n(x)|<|g_m(x)|\}} f(|g_m(x)|)\,dx \leqq 2M$. Since $|g_n(x) - g_m(x)|/2 \leqq |g_n(x)| \vee |g_m(x)|$ and f is monotone increasing it follows that $f(\frac{1}{2}|g_n(x) - g_m(x)|) \leqq f(|g_n(x)| \vee |g_m(x)|)$ and so for all n, m the integrand in the second integral displayed above does not exceed $F(K) f(\frac{1}{2}|g_n(x) - g_m(x)|)$ and the second integral displayed above does not exceed $2\,.\,2MF(K) < a/2$. Owing to bounded convergence, as $n, m \to \infty$ the first integral displayed above approaches zero. Hence for some n_0, if $n, m > n_0$, $\|g_n - g_m\|_1 < a$. As in earlier discussions it follows that if $\|g_n - h\|_1 \to 0$ as $n \to \infty$, then $g = h$ a.e. and the result follows. □

301. If $\{f_n\}_{n=1}^{\infty} \subset A$ and $\|f_n - f\|_1 \to 0$ as $n \to \infty$ then a subsequence converges a.e. to f and since for all n, $|f_n| \geqq 1$ a.e. it follows that $|f| \geqq 1$ a.e. Hence A is closed in the norm-induced topology of $L^1(I, \lambda)$. (The same proof applies for $L^1(\mathbb{R}, \lambda)$.) □

302. Let r_0 be $x \mapsto 1$, r_1 be $x \mapsto \operatorname{sgn}(\sin 2\pi x), \ldots r_n$ be $x \mapsto r_1(2^{n-1}x)$, $n \geqq 2$. (These are known as the Rademacher functions.) If $m = \sum_{k=1}^{K} 2^{n_k}$ let W_m be $f_{n_1} \cdots f_{n_K}$. Then $\{W_m\}_{m=1}^{\infty}$ is the sequence of Walsh functions and they constitute a complete orthonormal set in $L^2(I, \lambda) \subset L^1(I, \lambda)$ (see Problem 492). Clearly $A \supset \{W_m\}_{m=1}^{\infty}$. If $g \in L^{\infty}(I, \lambda)$ then $g \in L^2(I, \lambda)$ and so $\int_I g(x) W_m(x)\,dx \to 0$ as $m \to 0$. (This conclusion is also valid for the sequence $\{r_n\}_{n=0}^{\infty}$. In each instance Bessel's inequality yields the result.) Hence $W_m \to 0$ in the weak topology of $L^1(I, \lambda)$ as $m \to \infty$ and so zero is in the weak closure of A but zero is not in A whence A is not weakly closed. □

303. Since $\int_I |\cos \pi f(x)|^n\,dx = \int_S + \int_{I\setminus S} |\cos \pi f(x)|^n\,dx$, $|\cos \pi f(x)|^n = 1$ iff $x \in S$, and for x in $I\setminus S$, $|\cos \pi f(x)|^n \to 0$ as $n \to \infty$, the result follows upon an application of the bounded convergence theorem. □

304. If $\|f_n\|_\infty \leqq M$ and $h_n = f_n * g$ then

$$|h_n(x+y) - h_n(x)| \leqq M\|g_{(x+y)} - g_{(x)}\|_1.$$

Since $\|g_{(x+y)} - g_{(x)}\|_1 \to 0$ uniformly in x as $y \to 0$, $\{h_n\}_{n=1}^\infty$ is a uniformly bounded equicontinuous sequence. Hence, according to the Arzelà–Ascoli theorem $\{h_n\}_{n=1}^\infty$ contains a uniformly convergent subsequence.

If f_n is $x \mapsto \cos nx$ the Riemann–Lebesgue lemma assures that since $|h_n(x)| \leqq |\int_{\mathbb{T}} g(y) \cos ny\, dy| \,.\, |\cos nx| + |\int_{\mathbb{T}} g(y) \sin ny\, dy| \,.\, |\sin nx|$, $h_n \to 0$ as $n \to \infty$. □

305. Since $|f| = 0$ a.e. off $[-n, n]$, $\hat{f}(t) = \int_{-n}^{n} e^{-itx} f(x)\, dx$. Hence

$$(\hat{f}(t+s) - \hat{f}(t))/s = \int_{-n}^{n} e^{-itx} \,.\, s^{-1}(e^{-isx} - 1) f(x)\, dx.$$

The dominated convergence theorem permits passage of s to zero and shows $\hat{f}'$ exists and $\hat{f}'(t) = \int_{-n}^{n} (-ix) e^{-itx} f(x)\, dx$. Similar calculations show that $\hat{f}^{(k)}$ exists for all k in $\mathbb{N}$ and $\hat{f}^{(k)}(t) = \int_{-n}^{n} (-ix)^k e^{-itx} f(x)\, dx$. The Riemann–Lebesgue lemma implies the result. □

306. The hypothesis implies $\hat{f}^2 = \hat{f}$. Hence for each t, $\hat{f}(t) = 0$ or 1. Since $\hat{f}$ is continuous, either $\hat{f} = 0$ or $\hat{f} = 1$. The Riemann–Lebesgue lemma excludes the latter possibility and the injectivity of the Fourier transform implies $f = 0$. □

307. Viewed geometrically, $\sum_{n=1}^{N} a_n z_n$ lies inside or on the polygon determined by the vertices z_n lying on $\mathbb{T}$. This polygon is contained in $\{z : |z| \leqq 1\}$ and only the vertices lie on $\mathbb{T}$. Any vertex corresponds to the situation in which only one a_n is not zero (and that a_n is one).

Viewed analytically, the theorem is clearly true if $N = 1$. If $N = 2$, $a_1 \,.\, a_2 > 0$, and $z_j = e^{it_j}$, $j = 1, 2$, then

$$|a_1 z_1 + a_2 z_2|^2 = a_1^2 + a_2^2 + 2a_1 a \cos(t_1 - t_2) \leqq a_1^2 + a_2^2 + 2a_1 a_2 = (a_1 + a_2)^2 = 1.$$

Equality obtains iff $\cos(t_1 - t_2) = 1$ in which case $z_1 = z_2$, a contradiction. Thus $|a_1 z_1 + a_2 z_2| \leqq 1$ and equality obtains iff $a_1 = 1 = 1 - a_2$ or $a_1 = 0 = 1 - a_2$.

If the result holds for $N = k$ and if $0 < a_{k+1} < 1$, let a be $\sum_{n=1}^{k} a_n$, which is positive. Then $\sum_{n=1}^{k+1} a_n z_n = a(\sum_{n=1}^{k} (a_n/a) z_n) + a_{k+1} z_{k+1}$. Since $\sum_{n=1}^{k} (a_n/a) = 1$ it follows from the inductive hypothesis that if $z = \sum_{n=1}^{k} (a_n/a) z_n$, then $|z| \leqq 1$ and the argument for the case in which $N = 2$ shows $|az + a_{k+1} z_{k+1}| < 1$. The result follows. □

308. Let g be $f \,.\, \chi_E$. Then $\|g\|_1 = 1$ and hence $|\hat{g}| \leqq 1$. If $\hat{g}(1) = e^{ia}$ then $\int_{\mathbb{R}} g(x) e^{-i(x+a)}\, dx = 1 = \int_{\mathbb{R}} g(y - a) e^{-iy}\, dy = (g_{(-a)})\hat{}\,(1)$. Since $\|g_{(-a)}\|_1 = 1$ the problem is reduced to showing that if $f \geqq 0$ and $\|f\|_1 = 1$ then $\hat{f}(1) \neq 1$.

If $\hat{f}(1) = 1$ then $\int_{\mathbb{R}} f(x) \cos x\, dx = 1$. If $E_n = \{x : \cos x \leqq 1 - 1/n\}$ then $1 = \int_{\mathbb{R}} f(x) \cos x\, dx = \int_{E_n} + \int_{\mathbb{R}\setminus E_n} f(x) \cos x\, dx \leqq (1 - 1/n) \int_{E_n} f(x)\, dx + \int_{\mathbb{R}\setminus E_n} f(x)\, dx$ and so $\int_{E_n} f(x)\, dx \leqq (1 - 1n) \int_{E_n} f(x)\, dx$, whence $\int_{E_n} f(x)\, dx = 0$ and $f = 0$ a.e. on E_n. Since $\mathbb{R} = \bigcup_{n=2}^{\infty} E_n \cup 2\pi\mathbb{Z}$ it follows that $f = 0$ a.e. However $\|f\|_1 = 1$ and a contradiction results. □

(Note how Problem 308 is a generalization of Problem 307.)

309. Since

$$\hat{f}(y) = \sum_{n \in \mathbb{Z}} \int_{2n\pi/y}^{2(n+1)\pi/y} f(x)\, e^{-ixy}\, dx = \sum_{n \in \mathbb{Z}} r_n e^{ia_n},$$

if $\int_{2n\pi/y}^{2(n+1)\pi/y} f(x)\, dx = F_n$ then $r_n \leqq F_n$ and, indeed, according to Problem 308, for some n_0, $r_{n_0} < F_{n_0}$. Thus $|\hat{f}(y)| \leqq \sum_{n \in \mathbb{Z}} r_n < \sum_{n \in \mathbb{Z}} F_n = \int_{\mathbb{R}} f(x)\, dx = \|f\|_1$. □

310. Let $[a, b]$ be disjoint from $\text{supp}(\hat{f})$ and let G be in $C^\infty(\mathbb{R}, \mathbb{C})$, $G \neq 0$ and $\text{supp}(G) \subset [a, b]$. Then $\hat{G} \in L^1(\mathbb{R}, \lambda) \cap C_0(\mathbb{R}, \mathbb{C})$, $\hat{\hat{G}} = G$ and $(\hat{G} * f)\hat{} = G \,.\, \hat{f} = 0$, whence $\hat{G} * f = 0$ and thus g may be chosen to be $\hat{G}$. □

311. By hypothesis $\hat{f}(t) = \int_{-n}^{n} f(x)\, e^{-itx}\, dx$ and, as shown in Problem 305, $\hat{f}^{(k)}$ exists for all k in $\mathbb{N}$ and $\hat{f}^{(k)}(t) = \int_{-n}^{n} (-ix)^k f(x)\, e^{-itx}\, dx$. Thus for all t, $|\hat{f}^{(k)}(t)| \leqq 2n^{k+1}\|f\|_1$. Hence if $\hat{f}(b) \neq 0$, let $\sum_{k=0}^{N} \hat{f}^{(k)}(b)(t-b)^k/k! + R_{N+1}(t)$ be the finite Taylor series for $\hat{f}(t)$. Since $|R_{N+1}(t)| \leqq \hat{f}^{(N+1)}(a)| \,.\, |t-b|^{N+1}/(N+1)!$, a between b and t, and since $|\hat{f}^{(N+1)}(a)| \leqq 2n^{N+2}\|f\|_1$ it follows that $|R_{N+1}(t)| \leqq 2n^{N+2}|t-b|^{N+1}\|f\|_1/(N+1)!$ which approaches zero for each t as N approaches infinity. In a word, $\hat{f}$ is analytic, and since $\text{supp}(\hat{f})$ is compact, $\hat{f} = f = 0$. □

312. For positive a and f in $L^1(\mathbb{R}, \lambda)$, $g_U * f(x) = \int_U f(x-y) g_U(y)\, dy$, $|g_U * f(x) - f(x)| = |\int_U (f(x-y) - f(x)) g_U(y)\, dy|$, and

$$\|g_U * f - f\|_1 \leqq \int_U \left(\int_{\mathbb{R}} |f(x-y) - f(x)|\, dx \right) g_U(y)\, dy.$$

Since $\|f_{(-y)} - f\|_1 \to 0$ as $y \to 0$, there is a U such that $\int_{\mathbb{R}} |f(x-y) - f(x)|\, dx < a$ if $y \in U$ and, since $\|g_U\|_1 = 1$ it follows that if $V \subset U$ (i.e., if $V \geqq U$) $\|g_V * f - f\|_1 < a$ and the result follows. □

313. If $f \in L^1(\mathbb{R}, \lambda)$, $\|\hat{f}\|_\infty \leqq \|f\|_1$ and so $\|(g_\gamma * \hat{f}) - \hat{f}\|_\infty = \|\hat{f}(\hat{g}_\gamma - 1)\|_\infty \leqq \|g_\gamma * f - f\|_1$. Since $\lim_\gamma \|g_\gamma * f - f\|_1 = 0$ it follows that $\lim_\gamma (\hat{g}_\gamma - 1) = 0$ (pointwise convergence). □

314. The uniform boundedness principle implies that for some M and all γ, $\|g_\gamma\| \leqq M < \infty$. If $n = 2$ and $f \in A$ then $\|g_\gamma \circ g_\gamma \circ f - f\| \leqq \|g_\gamma \circ g_\gamma \circ f - g_\gamma \circ f\| + \|g_\gamma \circ f - f\| \leqq \|g_\gamma\| \,.\, \|g_\gamma \circ f - f\| + \|g_\gamma \circ f - f\|$ and thus for $n = 2$ the result follows. The result for arbitrary n is proved by induction. □

315. The equations $(\int_{\mathbb{R}} e^{-x^2}\, dx)^2 = \int_{\mathbb{R}^2} e^{-(x^2+y^2)}\, dx\, dy = \int_0^{2\pi} (\int_0^\infty r e^{-r^2}\, dr) \times d\theta = \pi$ show that $\int_{\mathbb{R}} e^{-x^2}\, dx = \pi^{1/2}$. Thus $k_t = (\pi t)^{-1/2}$. If $f \in L^1(\mathbb{R}, \lambda)$ then

$$\int_{\mathbb{R}} \left| \int_{\mathbb{R}} (f(x-y) g_t(y) - f(x) g_t(y))\, dy \right| dx$$

$$\leqq \int_{\mathbb{R}} \left(\int_{\mathbb{R}} |f(x-y) - f(x)|\, dx \right) g_t(y)\, dy$$

$$= \int_{-a}^{a} + \int_{\mathbb{R} \setminus [-a,a]} \left(\int_{\mathbb{R}} |f(x-y) - f(x)|\, dx \right) g_t(y)\, dy.$$

Since $\|f_{(-y)} - f\|_1 \to 0$ as $y \to 0$, if $b > 0$, for small enough a and all t, the first integral in the last member is less than $\frac{1}{2}b$. Then, since $g_t \downarrow 0$ on $\mathbb{R} \setminus [-a, a]$

as $t \to 0$ and since the inner integral in the second term of the last member is not more than $2\|f\|_1$, the monotone convergence theorem yields the desired conclusion. □

316. If f is $x \mapsto e^{-x^2}$ then $\hat{f}(t) = \int_{\mathbb{R}} e^{-(itx+x^2)}\,dx$ and the dominated convergence theorem implies $\hat{f}'$ exists and is $t \mapsto \int_{\mathbb{R}} (-ix)\, e^{-(itx+x^2)}\,dx$. Since $\int_{\mathbb{R}} (-ix)\, e^{-(itx+x^2)}\,dx = -i\, e^{-t^2/4} \int_{\mathbb{R}} (x - it/2)\, e^{-(x+it/2)^2}\,dx - \frac{1}{2}t \int_{\mathbb{R}} e^{-(itx+x^2)}\,dx$, and since the first integral is zero it follows that $\hat{f}'(t) = -\frac{1}{2}t\hat{f}(t)$. Since $(e^{-t^2/4})' = -\frac{1}{2}t\, e^{-t^2/4}$ it follows that for some constant c, $\hat{f}(t) = c\, e^{-t^2/4}$. Because $\hat{f}(0) = \pi^{1/2}$, $\hat{f}$ is $t \mapsto \pi^{1/2} e^{-t^2/4}$. In particular for all t, $\hat{f}(t) \neq 0$. Wiener's Tauberian theorem implies that the linear span of $\{f_{(-y)} \colon y \in \mathbb{R}\}$ is dense in $L^1(\mathbb{R}, \lambda)$. Hence for any g in $L^1(\mathbb{R}, \lambda)$, $\int_{\mathbb{R}} g(x)k(x)\,dx = 0$. Since $(L^1(\mathbb{R}, \lambda))^* = L^\infty(\mathbb{R}, \lambda)$ it follows that $k = 0$ a.e. □

NOTE. If the assumption that k is in $L^\infty(\mathbb{R}, \lambda)$ is replaced by the assumption that k is in $L^1(\mathbb{R}, \mathrm{d})$ the result follows more easily. Indeed, if g is $x \mapsto e^{-x^2}$ then $g * f = 0$ and so $\hat{g} \,.\, \hat{f} = 0$. Since $\hat{g}$ is never zero it follows that $\hat{f} = 0$ and so $f = 0$. □

317. For f in $L^2(\mathbb{R}, \lambda) \cap L^1(\mathbb{R}, \lambda)$, Tf is also in $L^2(\mathbb{R}, \lambda) \cap L^1(\mathbb{R}, \lambda)$ and so $(Tf)^\wedge$ is defined and $(Tf)^\wedge = \hat{g} \,.\, \hat{f}$. The Plancherel theorem states that the Fourier transform defined on $L^2(\mathbb{R}, \lambda) \cap L^1(\mathbb{R}, \lambda)$ is an isometry with respect to $\|\cdot\cdot\cdot\|_2$ and is extendible from the dense subset $L^2(\mathbb{R}, \lambda) \cap L^1(\mathbb{R}, \lambda)$ to all $L^2(\mathbb{R}, \lambda)$. Thus if the closure $\overline{T(B(0, 1))}$ of $T(B(0, 1))$ is compact in $L^2(\mathbb{R}, \lambda)$ then the image of $\overline{T(B(0, 1))}$ under the Fourier transform is also compact.

But the image is $A = \overline{\{\hat{g} \,.\, \hat{f} \colon f \in B(0, 1)\}}$. Since $g \neq 0$, $\hat{g} \neq 0$ and on some interval $[a, b]$, $|\hat{g}|^2 \geqq p^2 > 0$. It may be assumed for simplicity that $[a, b] = [0, 1]$. If $\hat{f}_n^2 = 2^{n+1} \chi_{[2^{-(n+1)}, 2^{-n}]}$, then $\{\hat{f}_n\}_{n=1}^\infty$ is an orthonormal set (hence contained in $B(0, 1)$) and so $\|\hat{f}_n - \hat{f}_m\|_2^2 \geqq 2$. Consequently $\|\hat{g}(\hat{f}_n - \hat{f}_m)\|_2 \geqq 2^{1/2} p > 0$ and neither A nor $\overline{T(B(0, 1))}$ is compact. □

318. If $F \in (L^1([1, \infty), \mu))^*$ and if δ_x is the map $[1, \infty) \ni y \mapsto \delta_{xy}$ then $F(\delta_x) = a_x$ and $|a_x| \leqq \|F\| \,.\, \|\delta_x\|_1 = \|F\| \,.\, x$. If $f \in L^1([1, \infty), \mu)$ then $F(f) = \sum_x f(x) \,.\, a_x$.

Conversely if $|a_x| \leqq M \,.\, x$ for some finite M then $F \colon f \mapsto \sum_x f(x) \,.\, a_x$ is in $(L^1([1, \infty), \mu))^*$ and $\|F\| \leqq M$.
(Note that $(L^1([1, \infty), \mu))^* \supsetneqq L^\infty([1, \infty), \mu)$.) □

319. It may be assumed that $\|f\|_\infty \neq 0$. Since

$$|\sigma_N(f)(x) - f(x)| \leqq \int_{-\pi}^{\pi} F_N(x - y)|f(y) - f(x)|\,dy$$

$$= \int_{x-\pi}^{x+\pi} F_N(z)|f(x - z) - f(x)|\,dz$$

$$= \int_{-\pi}^{\pi} F_N(z)|f(x - z) - f(x)|\,dz,$$

if $a>0$ let b be positive and such that for all z in $(-b, b)$ and all x, $|f(x-z)-f(x)|<a/3$. Then for some N_0, if $N \geqq N_0$, $|F_N|<a/6\|f\|_\infty$ in $[-\pi, \pi]\backslash(-b, b)$. Thus $\|\sigma_N(f)-f\|_\infty \leqq 2a/3<a$ if $N \geqq N_0$ and the result follows. □

320. If $f \in L^1(\mathbb{T}, \lambda)$ and $a>0$ there is in $C(\mathbb{T}, \mathbb{C})$ a g such that $\|f-g\|_1<a/2$. Then $\|\sigma_N(f)-f\|_1 \leqq \|\sigma_N(f)-\sigma_N(g)\|_1+\|\sigma_N(g)-g\|_1$. However,

$$\begin{aligned}\|\sigma_N(f)-\sigma_N(g)\|_1 &\leqq \int_{\mathbb{T}^2} F_N(x-y)|f(y)-g(y)|\,dy\,dx \\ &\leqq \int_{\mathbb{T}} |f(y)-g(y)|\left(\int_{\mathbb{T}} F_N(x-y)\,dx\right) dy \\ &= \|f-g\|_1\end{aligned}$$

and $\|\sigma_N(g)-g\|_1 \leqq 2\pi\|\sigma_N(g)-g\|_\infty$. For large N the last number is small and the result follows. □

321. If $p \geqq 1$ and $f \in L^1(\mathbb{R}, \lambda)$ then $T_f: L^p(\mathbb{R}, \lambda) \ni g \mapsto f * g$ maps $L^p(\mathbb{R}, \lambda)$ into $L^p(\mathbb{R}, \lambda)$ and $\|T_f(g)\|_p \leqq \|f\|_1 \,.\, \|g\|_p$. Thus if g is

$$y \mapsto \begin{cases} |\sin 1/y| \,.\, 1/|y|^{1/2}, & \text{if } y \neq 0 \\ 0, & \text{if } y=0 \end{cases}$$

and if $g \in L^p(\mathbb{R}, \lambda)$, then $f * g \in L^p(\mathbb{R}, \lambda)$.

However, $\|g\|_p^p=\int_{\mathbb{R}} (|\sin z|^p/|z|^p)|z|^{3p/2-2}\,dz=\int_{-1}^1+\int_{\mathbb{R}\backslash[-1,1]}$. The first integral is finite iff $p>2/3$ and the second integral is finite iff $p<2$. Thus if $2/3<p<2$, $f * g \in L^p(\mathbb{R}, \lambda)$ for all f in $L^1(\mathbb{R}, \lambda)$. □

12. $L^2(X, \mu)$ or $\mathfrak{H}$ (Hilbert Space)

322. If $z \in \mathbb{C}$ and $|z| < \frac{1}{2}$ then

$$\left|\sum_{n=k}^{m} \frac{a_n}{n-z}\right| \leqq \left(\sum_{n=k}^{m} |a_n|^2\right)^{1/2} \left(\sum_{n=k}^{m} (n-1)^{-2}\right)^{1/2} \leqq \left(\sum_{n=k}^{m} |a_n|^2\right)^{1/2} \frac{\pi^2}{6}.$$

The last member approaches zero as $k, m \to \infty$. Thus $f: z \mapsto \sum_{n=1}^{\infty} a_n/(n-z) \in H(B(0, \frac{1}{2})^0)$. Since $f = 0$ on $(-\frac{1}{2}, \frac{1}{2})$, $f = 0$ in $B(0, \frac{1}{2})^0$. On the other hand $f(z) = \sum_{n=1}^{\infty} a_n (\sum_{k=0}^{\infty} z^k/n^{k+1}) = \sum_{k=0}^{\infty} (\sum_{n=1}^{\infty} a_n/n^{k+1}) z^k$ and so for all $k, \sum_{n=1}^{\infty} a_n/n^{k+1} = 0$. In particular $a_1 = -\sum_{n=2}^{\infty} a_n/n^{k+1}$ and since the right member approaches zero as $k \to \infty$, $a_1 = 0$. By induction it follows that $a_n = 0$ for all n. □

323. It suffices to show: if K is a compact subset of $l^2(\mathbb{N})$ and if $a > 0$ there is in $\mathbb{N}$ an N_a such that if $\{a_n\}_{n=1}^{\infty} \in K$ then $\sum_{n \geqq N_a} |a_n|^2 < a^2$. Indeed, if the statement is false there is in K a sequence $\{\{k_{mn}\}_{n=1}^{\infty}\}_{m=1}^{\infty}$ and there is a positive a such that for all m, $\sum_{n \geqq m} |k_{mn}|^2 \geqq a^2$. Since K is compact, passage to subsequences as needed permits the assumption that there is in K an element $\{k_n\}_{n=1}^{\infty}$ such that $\|\{k_{mn}\}_{n=1}^{\infty} - \{k_n\}_{n=1}^{\infty}\|_2 \to 0$ as $m \to \infty$ and such that $\lim_{m \to \infty} k_{mn} = k_n$, n in $\mathbb{N}$. For some n_0, $\sum_{n \geqq n_0} |k_n|^2 < a^2/2^4$. If $\|\{k_{mn}\}_{n=1}^{\infty} - \{k_n\}_{n=1}^{\infty}\|_2^2 < a^2/2^6$ for m in $[m_0, \infty)$, then for m in $[m_0 + n_0, \infty), a < (\sum_{n \geqq n_0} |k_{mn}|^2)^{1/2} \leqq (\sum_{n \geqq n_0} |k_{mn} - k_n|^2)^{1/2} + (\sum_{n \geqq n_0} |k_n|^2)^{1/2} < \|\{k_{mn}\}_{n=1}^{\infty} - \{k_n\}_{n=1}^{\infty}\|_2 + a/2^2 < a/2^3 + a/2^2 < a$, a contradiction. □

324. Since all functions considered are in $L^2(\mathbb{R}^2, \lambda)$ it follows that all integrals given or introduced below exist. Green's theorem implies that for all positive R, $\int_{B(0,R)} (f(x, y)\Delta g(x, y) - \Delta f(x, y) g(x, y))\, dx\, dy = \oint_{\partial B(0,R)} (f(x, y)\, \partial g(x, y)/\partial n - g(x, y)\, \partial f(x, y)/\partial n)\, ds$ (line integral) $= \oint_{\partial B(0,R)} (f(x, y)\, \partial g(x, y)/\partial x - g(x, y)\, \partial f(x, y)/\partial x)\, dy + (g(x, y)\, \partial f(x, y)/\partial y - f(x, y)\, \partial g(x, y)/\partial y)\, dx$. A typical estimate for the integrals in these equations is:

$$\left|\int\!\!\int_{\partial B(0,R)} f(x,y)\,\partial g(x,y)/\partial x\,dy\right|$$

$$\leq \int_0^{2\pi} |f(R\cos\theta, R\sin\theta)|\,.\,|\partial g(R\cos\theta, R\sin\theta)/\partial x|\;|R\sin\theta|\,d\theta = A(R).$$

Since $\int_{B(0,R)} |f(r\cos\theta, r\sin\theta)\,.\,\partial g\,(r\cos\theta, r\sin\theta)/\partial x\,|r\,dr\,d\theta = \int_0^R A(r)r\,dr \leq \|f\|_2\,.\,\|\partial g/\partial x\|_2$ it follows that there is a sequence $\{R_n\}_{n=1}^{\infty}$ such that $R_n\to\infty$ and $A(R_n)\to 0$ as $n\to\infty$. In a similar manner the other three integrals may be estimated. Thus $|\int_{B(0,R_n)} (f(x,y)\Delta g(x,y)-g(x,y)\Delta f(x,y))\,dx\,dy|$ is dominated by the sum of four quantities each of which approaches zero as $n\to\infty$ and the result follows. □

325. If $\mathfrak{H}=L^2(I,\lambda)$ and γ is $t\mapsto\chi_{[0,t]}$ the required behavior is provided. □

326. For all x in $[0,\infty)$,

$$|f_n(x)-f_n(1)|\leq\left|\int_1^x f_n'(t)\,dt\right|\leq\left(\int_1^x |f_n'(t)|^2\right)^{1/2}\left|\int_1^x dt\right|^{1/2}\leq M|x-1|^{1/2}.$$

If $0\leq x\leq 1$, $|f_n(x)|\leq|f_n(1)|+M|x-1|^{1/2}\leq 1+M$ and if $1<x$, $|f_n(x)|\leq x^{-1}<1$, i.e., for all n and all x, $|f_n(x)|\leq 1+M$.

Similarly it follows that if $0\leq a, b<\infty$ then $|f_n(b)-f_n(a)|\leq M|b-a|^{1/2}$ and so with respect to a uniform Lipschitz constant (M) all f_n are in Lip($\frac{1}{2}$) on $[0,\infty)$.

Hence for each k in $\mathbb{N}$, $\{f_n\}_{n=1}^{\infty}$ is a uniformly bounded equicontinuous sequence on $[0,k]$ and, *via* the Arzelà–Ascoli theorem, there is a sequence $\{\{f_{kn}\}_{n=1}^{\infty}\}_{k=1}^{\infty}$ of sequences such that $\{f_{k+1,n}\}_{n=1}^{\infty}\subset\{f_{kn}\}_{n=1}^{\infty}\subset\{f_n\}_{n=1}^{\infty}$ and such that $\{f_{kn}\}_{n=1}^{\infty}$ converges uniformly on $[0,k]$. It will be shown that $\{f_{nn}\}_{n=1}^{\infty}=\{g_n\}_{n=1}^{\infty}$ converges uniformly on $[0,\infty)$.

Indeed, if $1>b>0$ and if $k>3/b$ then for some N, if $m,n\geq N$ and $0\leq x\leq k$ then $|g_m(x)-g_n(x)|<b/3$. If $k<x$, $|g_m(x)-g_n(x)|<2b/3$ and so $|g_m(x)-g_n(x)|<b$ for all x if $m,n\geq N$, i.e., $\|g_m-g_n\|_\infty\to 0$ as $m,n\to\infty$.

Again, if $b>0$ choose p so that $4/p<b/2$ and choose N so that if $m,n\geq N$ then $\|g_m-g_n\|_\infty<(b/2p)^{1/2}$. Then

$$\|g_m-g_n\|_2^2=\int_{[0,p)}+\int_{[p,\infty)}|g_m(x)-g_n(x)|^2\,dx\leq\frac{b}{2p}p+\frac{4}{p}<b$$

and it follows that $\|g_m-g_n\|_2\to 0$ as $m,n\to\infty$.

In sum, the assertions i), ii), and iii) are all true. □

327. Since $\int_{-y}^{y}|f(x,y)|\,dx\leq(\int_{-y}^{y}|f(x,y)|^2\,dx)^{1/2}(2y)^{1/2}$ it follows that $\infty>\|f\|_2^2=\int_0^1(\int_{-y}^{y}|f(x,y)|^2\,dx)\,dy\geq\int_0^1(1/2y)(\int_{-y}^{y}|f(x,y)|\,dx)^2\,dy$. If $\liminf_{y=0}\int_{-y}^{y}|f(x,y)|\,dx=a>0$ then for some y arbitrarily near zero $(1/2y)(\int_{-y}^{y}|f(x,y)|\,dx)^2\geq a^2/4y$ and thus $\|f\|_2^2=\infty$, a contradiction. □

328. If $E_n=\{x:|f_n(x)|\geq a>0\}$ then $\lambda(E_n)\to 0$ as $n\to\infty$ and $\int_I|f_n(x)|\,dx=\int_{I\setminus E_n}+\int_{E_n}|f_n(x)|\,dx\leq a+\|f_n\|_2(\lambda(E_n))^{1/2}\to a$ as $n\to\infty$. Since a is arbitrary the result follows. □

329. If $f_h = \Delta_h f - g$ then $\|f_{1/n}\|_2 \to 0$ as $n\to\infty$ and *via* subsequences as needed, it may be assumed that $f_{1/n}\to 0$ a.e. as $n\to\infty$. Since $\int_0^x |g(t)-\Delta_h f(t)|\,dt \leqq (\int_{\mathbb{R}} |g(t)-\Delta_h f(t)|^2\,dt)^{1/2}|x|^{1/2}$ it follows that for a and x off a null set, $\int_a^x g(t)\,dt = \lim_{n\to\infty}\int_a^x (n^{-1}(f(t+1/n)-f(t)))\,dt = \lim_{n\to\infty} n^{-1}\int_x^{x+n^{-1}} f(t)\,dt - \lim_{n\to\infty} n^{-1}\int_a^{a+n^{-1}} f(t)\,dt = f(x)-f(a)$ as required. □

330. i) Since $\|Tf\|_2^2 = \int_I |\int_0^x f(t)\,dt|^2\,dx \leqq \int_I x\|f\|_2^2\,dx = \frac{1}{2}\|f\|_2^2$, it follows that $\|T\|\leqq 2^{-1/2}$.

ii) Since

$$(Tf, g) = \int_I \left(\int_0^x f(t)\,dt\right)\overline{g(x)}\,dx = \int_I \left(\int_t^1 \overline{g(x)}\,dx\right) f(t)\,dt$$

$$= \int_I \left[\int_I \overline{g(x)}\,dx - \int_0^t \overline{g(x)}\,dx\right] f(t)\,dt,$$

if P is $h\mapsto \int_I h(x)\,dx$ then $T^* = -T+P$. Furthermore, $P^2h = Ph$ and $P(L^2(I,\lambda)) = \mathbb{C}$.

iii) If $\{f_n\}_{n=1}^\infty \subset B(0,1)$ (the unit ball of $\mathfrak{H}$) then $|Tf_n(x)|\leqq x^{1/2}\leqq 1$ and $|Tf_n(x)-Tf_n(y)|\leqq |x-y|^{1/2}$. Thus $\{Tf_n\}_{n=1}^\infty$ is a uniformly bounded equicontinuous sequence; the Arzelà–Ascoli theorem implies, again *via* a subsequence as needed, that the sequence is uniformly convergent and since $\lambda(I)=1$ the result follows. □

331. i) The fact that $\lambda(I)=1$ implies that convergence in the norm-induced topology follows from uniform convergence and so S is closed in $C(I,\mathbb{C})$.

ii) The identity map id from S regarded as a (closed) subspace of $L^2(I,\lambda)$ to S regarded as a closed subset of $C(I,\mathbb{C})$ is a closed map and $id(S)$ is closed (see i)). Hence the closed graph theorem implies id is continuous and so $\|f\|_\infty \leqq M\|f\|_2$ for some M and all f in S. Because $\lambda(I)=1$, $\|f\|_2\leqq\|f\|_\infty$.

iii) The map $L_y : S\ni f\mapsto f(y)$ is a continuous linear functional and so $|L_y(f)|\leqq K_y\|f\|_\infty \leqq K_y M\|f\|_2$. The Hahn–Banach theorem implies there is in $L^2(I,\lambda)$ a $\bar{k}_y$ such that for all f in S, $L_y(f) = (f,\bar{k}_y) = \int_I k_y(x)f(x)\,dx$. □

332. The hypothesis implies $\|f_{(h)}-f_{(-h)}\|_2^2 \leqq 4C|h|^{1+a} = C_1|h|^{1+a}$. If the Fourier series for f is $\frac{1}{2}a_0 + \sum_{n=1}^\infty (a_n\cos nx + b_n \sin nx)$ then the Fourier series for $f_{(h)}-f_{(-h)}$ is $2\sum_{n=1}^\infty (-a_n\sin nx + b_n\cos nx)\sin nh$ and $\|f_{(h)}-f_{(-h)}\|_2^2 = 4\pi\sum_{n=1}^\infty (|a_n|^2+|b_n|^2)\sin^2 nh \leqq C_1|h|^{1+a}$. If $|a_n|^2+|b_n|^2 = r_n^2$ and if $h=\pi/2N$ it follows that for all N, $\sum_{n=1}^\infty r_n^2 \sin^2 \pi n/2N \leqq C_2 N^{-(1+a)}$. If $N=2^k$ it follows that $\sum_{n=2^{k-1}+1}^{2^k} r_n^2 \leqq 2C_2 2^{-k(1+a)}$ and so, *via* the Schwarz inequality, $\sum_{n=2^{k-1}+1}^{2^k} r_n \leqq C_2^{1/2} 2^{1/2-(1/2k)(1+a)+(1/2)(k-1)}$, whence $\sum_{n=2}^\infty r_n < \infty$ and the result follows. □

333. For all n in $\mathbb{Z}\backslash\{0\}$, $x\mapsto x\,e^{2n\pi i x}\in A$. If $g\in A^\perp$ then $\int_I (x\,e^{2n\pi i x})g(x)\,dx = 0$ if $n\neq 0$. Hence $xg(x) = c = $ constant a.e. If $c\neq 0$ then $g(x) = c/x$ a.e. and then $g\notin L^2(I,\lambda)$. Thus $A^\perp = \{0\}$ and since A is a linear set, A is dense in $L^2(I,\lambda)$. □

334. By hypothesis $\int_I (f(t)-t)t^n\,dt = 0$ for n in $\mathbb{N}$. Hence, according to the Stone–Weirstrass theorem, for all continuous functions g,

$$\int_I (f(t)-t)g(t)\,dt = 0$$

and so $f(t) = t$ a.e. □

335. Since $\|f_n - f_{n+k}\|_2 \leqq 2^{-n} + \cdots + 2^{-(n+k-1)} < 2^{-(n-1)}$ it follows that $\{f_n\}_{n=1}^\infty$ is a Cauchy sequence. If $f_n \to f$ in $L^2(I, \lambda)$ as $n \to \infty$ then $\|f_n - f\|_2 < 2^{-(n-1)}$. Since $\sum_n \|f_n - f\|_2^2 < \sum_n 2^{-2(n-1)} < \infty$ it follows that $\sum_n |f_n(x) - f(x)|^2 < \infty$ a.e. and so $|f_n(x) - f(x)| \to 0$ a.e. as $n \to \infty$. □

336. (See Problem 164.) Only the case in which X is not σ-finite requires discussion.

If X is not σ-finite, then since each $f_n \in L^2(X, \mu)$, $\{x : f_n(x) \neq 0,\ n \text{ in } \mathbb{N}\}$, denoted C, is σ-finite and off C all f_n are zero. The proof above now applies with X replaced by C. □

337. If $\sup_n$ (variation of f_n) $= M < \infty$ the hypothesis implies that for all n and all x $|f_n(0)| - M \leqq |f_n(x)| \leqq |f_n(0)| + M$. If $\{|f_n(0)|\}_{n=1}^\infty$ is unbounded then for some large n, $\|f_n\|_2 > 1$ in contradiction of the normality of the f_n. Hence the Helly selection principle (see below) implies that it may be assumed that there is a measurable function f such that $f_n \to f$ as $n \to \infty$ and for some M_1, $|f_n| \leqq M_1$ for all n. Hence, (see Problem 336) $f = 0$ a.e., $f_n^2 \to 0$ a.e. as $n \to \infty$ and the bounded convergence implies $\|f_n\|_2 \to 0$ as $n \to \infty$, again in contradiction of the normality of the f_n. □

Helly Selection Principle. If $\{g_n\}_{n=1}^\infty$ is a uniformly bounded sequence of monotone increasing functions (defined on $\mathbb{R}$) there is a subsequence $\{g_{n_k}\}_{k=1}^\infty$ and a monotone function g such that $g = \lim_{k\to\infty} g_{n_k}$. Proof: The diagonal process used to prove the Arzelà–Ascoli theorem defines a subsequence converging on $\mathbb{Q}$. If $\bar{G}$ resp. $\underline{G}$ are lim sup and lim inf of this subsequence then each is monotone increasing and the intersection of their sets of points of continuity is dense. At each of these points $\bar{G} = \underline{G}$. On the (at most) countable set of points of discontinuity of $\bar{G}$ or of $\underline{G}$, a second application of the diagonal process to the subsequence already constructed provides a subsequence that converges at all the points of discontinuity and the limit function exists on all $\mathbb{R}$.

(Since the functions in question are all of bounded variation, each is the difference of two monotone increasing components. The Helly selection principle applied to these is the effective device in the solution of Problem 337.)

338. Since $\sum_{n=1}^\infty n^{-2} < \infty$ there is in $L^2(X, \mu)$ an f such that if $S_N = \sum_{n=1}^N n^{-1} f_n$ then $\|S_N - f\|_2 \to 0$ as $N \to \infty$. Furthermore $\|S_{(N+1)^2} - S_{N^2}\|_2^2 = \sum_{N^2+1}^{(N+1)^2} n^{-2} \leqq \int_{N^2}^{(N+1)^2} x^{-2}\,dx \leqq CN^{-6}$. Thus if $g_N = \sum_1^N |S_{(n+1)^2} - S_{n^2}|$ then the

Minkowski inequality implies that $\|g_{N_1}-g_{N_2}\|_2 \leqq C_1(N_2^{-2}-N_1^{-2})$ which approaches zero as $N_1, N_2 \to \infty$. Hence there is in $L^2(X, \mu)$ a g such that $g_N \to g$ as $N \to \infty$. Since $g_N \leqq g_{N+1}$ it follows that $g_N \uparrow g$ a.e. and $\lim_{N\to\infty} S_{N^2} = f$ exists. If $N^2 < p < (N+1)^2$ then $|S_p - S_{N^2}| \leqq M^2(\sum_{N^2+1}^{p} k^{-2}) \leqq M^2(p-N^2)(N^2+1)^{-2} \leqq M^2(2N+1)(N^2+1)^2$ which approaches zero as $N \to \infty$. Thus $|f-S_p| \leqq |f-S_{N^2}| + |S_{N^2}-S_p|$ and both terms in the right member approach zero as $p \to \infty$. □

339. i) The Schwarz inequality shows $\|Tf\|_2 \leqq \|K\|_2 \cdot \|f\|_2$.

ii) Let $\{g_n\}_{n=1}^{\infty}$ be a complete orthonormal sequence in $L^2(I, \lambda)$. Then $\{(x, y) \mapsto g_m(x)g_n(y)\}_{m,n=1}^{\infty}$ is a complete orthonormal sequence in $L^2(I^2, \lambda)$. If $K(x, y) = \sum_{m,n} a_{mn} g_m(x) g_n(y)$ (convergence in $L^2(I^2, \lambda)$) and if T_{MN} is $L^2(I, \lambda) \ni \mapsto \int_I (\sum_{m,n=1}^{M,N} a_{mn} g_m(x) g_n(y)) f(y)\, dy$ then $\|(T_{MN}-T)f\|_2^2 \leqq (\sum_{m \geqq M \text{ or } n \geqq N} |a_{mn}|^2) \cdot \|f\|_2^2$ from which the result follows. □

340. Let E_a be $\{x: f(x) \geqq a > 0\}$. If $\lambda(E_a) = b > 0$, then for n in $\mathbb{N}$, $\int_I f(T^n x) f(x)\, dx = 0 = \int_{I \setminus E_a} + \int_{E_a} f(T^n x) f(x)\, dx$, and since $f \geqq 0$ it follows that $f(T^n x) = 0$ a.e. in E_a. In other words, $\lambda((T^n E_a) \cap E_a) = 0$, n in $\mathbb{N}$, and so $\lambda((T^n(E_a) \cap T^m(E_a)) = 0$ for m in $\mathbb{N} \cup \{0\}$ and n in $\mathbb{N}$. Since $\lambda(T^n(E_a)) = \lambda(E_a)$ there emerges the inequality $\lambda(I) \geqq nb$ for all n in $\mathbb{N}$ and the contradiction shows $b = 0$ as required. □

341. Let $\mathcal{M}$ be $\{f$: there is in $\mathfrak{H}$ a g_f such that $\|(N+1)^{-1}(\sum_{n=0}^{N} U^n(f)) - g_f\| \to 0$ as $N \to \infty\}$. It will be shown that the subspace $\mathcal{M}$ is norm-closed.

If $\{f_n\}_{n=1}^{\infty} \subset \mathcal{M}$ and $\|f_m - f\| \to 0$ as $m \to \infty$, then

$$\left\|(N+1)^{-1}\sum_0^N U^n(f) - (M+1)^{-1}\sum_0^M U^n(f)\right\|$$

$$\leqq \left\|(N+1)^{-1}\sum_0^N U^n(f-f_m)\right\| + \left\|(M+1)^{-1}\sum_0^M U^n(f-f_m)\right\|$$

$$+\left\|(N+1)^{-1}\sum_0^N U^n(f_m) - (M+1)^{-1}\sum_0^M U^n(f_m)\right\|, \qquad m \text{ in } \mathbb{N}.$$

If $a > 0$ there is an m_0 such that if $m \geqq m_0$ then $\|f_m - f\| < a/3$. Since U is unitary (hence norm-preserving), if $m = m_0$,

$$\left\|(N+1)^{-1}\sum_0^N U^n(f) - (M+1)^{-1}\sum_0^M U^n(f)\right\| \leqq 2\frac{a}{3} + \left\|(N+1)^{-1}\sum_0^N U^n(f_{m_0}) - (M+1)^{-1}\sum_0^M U^n(f_{m_0})\right\|.$$

As $N, M \to \infty$ the last term approaches zero and the result follows.

Note that since $0 \in \mathcal{M}$ it follows that $\mathcal{M} \neq \emptyset$. The following steps lead to the desired result.

i) If $f = U(f)$ then $f \in \mathcal{M}$ since $(N+1)^{-1}\sum_0^N U^n(f) = f$.

ii) If $f \in \mathfrak{H}$ then $f - U(f) \in \mathcal{M}$ since

$$\left\|(N+1)^{-1} \sum_0^N U^n(f - U(f))\right\| = \frac{\|f - U^{N+1}(f)\|}{(N+1)} \leqq 2(N+1)^{-1}\|f\| \to 0$$

as $N \to \infty$.

iii) If $f \in \mathcal{M}$ then $U(f) \in \mathcal{M}$ since $\|(N+1)^{-1} \sum_0^N U^n(Uf)) - U(g_f)\| = \|(N+1)^{-1} \sum_0^N U^n(f) - g_f\|$.

iv) If $f \in \mathcal{M}$ then $U^{-1}(f) \in \mathcal{M}$ since

$$(N+1)^{-1} \sum_0^N U^n(U^{-1}(f)) = \frac{N}{N+1}\left[N^{-1} \sum_0^{N-1} U^{n-1}(f)\right] + \frac{U^{-1}(f)}{N+1} \to g_f$$

as $N \to \infty$. Thus $U(\mathcal{M}) = \mathcal{M}$.

v) If $f \in \mathcal{M}^\perp$ then $U(f) \in \mathcal{M}^\perp$ since if $h \in \mathcal{M} = U(\mathcal{M})$, $h = U(k)$ for some k in $\mathcal{M}$ and then $(U(f), h) = (U(f), U(k)) = (f, k) = 0$.

Hence if $f \in \mathcal{M}^\perp$ then $f - U(f) \in \mathcal{M} \cap \mathcal{M}^\perp$ whence $f = U(f)$, $f \in \mathcal{M} \cap \mathcal{M}^\perp$, $f = 0$. In sum, $\mathcal{M} = \mathfrak{H}$ as required. □

342. Let E be $B(0, \frac{1}{4})^0$. Then E is measurable. If μ is translation-invariant and $\mu(E) = 0$ then $\mu = 0$ since every set is covered by a countable union of translates of E. If $\mu(E) > 0$ let $\{f_n\}_{n=1}^\infty$ be an orthonormal sequence in $\mathfrak{H}$. Then $\{\frac{1}{2}f_n + E\}_{n=1}^\infty$ is a sequence of pairwise disjoint measurable subsets of $B(0, 1)^0$ and so $\mu(B(0, 1)^0) = \infty$, a contradiction. □

343. Since ρ^p is a translation-invariant measure the argument in Solution 342 shows $\rho^p(B(a, r)^0)$ is zero for all a in $\mathfrak{H}$ and all positive r or $\rho^p(B(a, r)^0)$ is infinite for all a in $\mathfrak{H}$ and all positive r.

If $p \in [0, \infty)$ choose n in $\mathbb{N}$ so that $p < n$. If $\rho^p(B(0, 1)) < \infty$ then (Problem 135) $\rho^n(B(0, 1)) < \infty$. Let $\{f_k\}_{k=1}^n$ be an orthonormal set in $\mathfrak{H}$ and let E_n be $\{\sum_{k=1}^n a_k f_k : a_k \text{ real}\}$. Then E_n and $\mathbb{R}^n$ are isomorphic. Let B_n be $B(0, \frac{1}{4}) \cap E_n$ and let a be positive. Choose a sequence $\{U_m\}_{m=1}^\infty$ of open sets such that $\operatorname{diam}(U_m) < \varepsilon$, $\bigcup_m U_m \supset B(0, \frac{1}{4})$, and $\sum_{m=1}^\infty (\operatorname{diam}(U_m))^n < \rho_\varepsilon^n(B(0, \frac{1}{4})) + a$. Then $\{V_m = U_m \cap E_n\}_m$ is a sequence of (relatively) open sets in E_n, $\operatorname{diam}(V_m) < \varepsilon$ and $\bigcup_m V_m \supset B_n$. Let $\tilde{\rho}^n$ denote n-dimensional Hausdorff measure derived in $\mathbb{R}^n$ (i.e., in E_n) from the open sets of $\mathbb{R}^n$ (i.e., from the (relatively) open sets of E_n). Then $\tilde{\rho}_\varepsilon^n(B_n) \leqq \sum_m (\operatorname{diam}(V_m))^n \leqq \sum_m (\operatorname{diam}(U_m))^n < \rho_\varepsilon^n(B(0, \frac{1}{4})) + a$ and so $\tilde{\rho}_\varepsilon^n(B_n) \leqq \rho^n(B(0, \frac{1}{4})) + a$, $\tilde{\rho}^n(B_n) \leqq \rho^n(B(0, \frac{1}{4})) + a$, $\tilde{\rho}^n(B_n) \leqq \rho^n(B(0, \frac{1}{4}))$. But (Problem 136) $0 < \lambda_n(B_n) \leqq \tilde{\rho}^n(B_n)$. Thus $\rho^n(B(0, \frac{1}{4})) > 0$ and so (Solution 342) $\rho^n(B(0, 1)) = \infty$. An argument based on homothety implies that if W is nonempty and open then $\rho^n(W) = \infty$. But then (Problem 135) $\rho^n(W) = \infty$ and the result follows. □

344. Let $\mathfrak{H}$ be $L^2(I, \lambda)$ and let $\|\cdot\cdot\cdot\|$ be $\|\cdot\cdot\cdot\|_2$ and $\|\cdot\cdot\cdot\|'$ be $\|\cdot\cdot\cdot\|_1$. Then $\{f_k : x \mapsto k^{-1/4}x^{-1/2+1/k}\}_{k=1}^{\infty} \subset \mathfrak{H}$, $\|f_k\|^2 = \|f_k\|_2^2 = \frac{1}{2}k^{1/2}$, and $\|f_k\|' = \|f_k\|_1 = 2k^{3/4}/(k+2)$. Thus $\|f_k\| \to \infty$ and $\|f_k\|' \to 0$ as $k \to \infty$.

Let f be $\sum_{k=1}^{\infty} k^{-11/8} f_k$. Then

$$\|f\| \leqq \sum_{k=1}^{\infty} \|k^{-11/8} f_k\| = 2^{-1/2} \sum_{k=1}^{\infty} k^{-11/8} k^{1/4} < \infty,$$

i.e., $f \in \mathfrak{H}$. On the other hand, $(f_k, f) = k^{3/4} \sum_{n=1}^{\infty} n^{-5/8}/(n+k) \geqq k^{3/4} \int_1^{\infty} dx/(x+k)^{13/8} = 8k^{3/4}/5(1+k)^{5/8} \to \infty$ as $k \to \infty$. Hence if $B'(0, r)^0 = \{f : f \in \mathfrak{H}, \|f\|' < r\}$ and $N(0) = \{g : (g, f) < 1\}$ then for large k, $f_k \in B'(0, r)^0 \backslash N(0)$ and so the topology induced by $\|\cdot\cdot\cdot\|'$ is not stronger than $\sigma(\mathfrak{H}, \mathfrak{H})$. □

345. Let $l^2(\mathbb{N})$ be $\mathfrak{H}$. Let S be $\mathfrak{H} \ni x = \{x_n\}_{n=1}^{\infty} \to S(x) = \{x_n/n\}_{n=1}^{\infty}$. Then $S(\mathfrak{H}) = D$ and $S(M^{\perp}) = A$. If A is not dense in $\mathfrak{H}$ then there is in $A^{\perp}$ a nonzero $y = \{y_n\}_{n=1}^{\infty}$ and so for all $z = \{z_n\}_{n=1}^{\infty}$ in $M^{\perp}$, $(S(z), y) = (z, S(y)) = 0$, i.e., $S(y) \in (M^{\perp})^{\perp} = M$. Since $M \cap D = \{0\}$, $S(y) = 0$ and so $y = 0$, a contradiction. □

346. The hypothesis and the uniform boundedness principle imply that if $x_n = \{x_{nm}\}_{m=1}^{\infty}$, n in $\mathbb{N}$, then for all m, $\lim_{n\to\infty} x_{nm} = 0$ and that for some M and all n, $\|x_n\| \leqq M$. It may be assumed that $M = 1$.

By induction there can be constructed three sequences $\{a_k\}_{k=1}^{\infty}$, $\{n_k\}_{k=1}^{\infty}$, and $\{m_k\}_{k=0}^{\infty}$ such that all a_k are positive, $n_1 = m_0 = 1$, $\{n_k\}_k \cup \{m_k\}_k \subset \mathbb{N}$, $n_k < n_{k+1}$, $m_k < m_{k+1}$, and satisfying:

$$\sum_{m>m_1} |x_{n_1 m}|^2 < 2^{-2} = a_1^2,\ a_2 < 2^{-3}/m_1,\ \sum_{m \leqq m_1} |x_{n_2 m}| < a_2;$$

$$\cdots\cdots\cdots$$

$$a_k < 2^{-(k+1)}/m_{k-1},\ \sum_{m \leqq m_{k-1}} |x_{n_k m}| < a_k,\ \sum_{k=1}^{p} \sum_{m>m_p} |x_{n_k m}|^2 < a_p^2;$$

etc.

If $y_K = K^{-1} \sum_{k=1}^{K} x_{n_k}$ and if $b_k = \sum_{p=k}^{\infty} a_p$ then $b_k < 2^{-k}/m_k$ and $|y_{Km}| \leqq K^{-1}(|x_{n_k m}| + b_k)$, $m_{k-1} < m \leqq m_k$, $1 \leqq k \leqq K$ and

$$\begin{aligned}\sum_{m=1}^{m_K} |y_{Km}|^2 &\leqq K^{-2} \sum_{k=1}^{K} \left(\sum_{m_{k-1}<m \leqq m_k} |x_{n_k m}|^2 + 2b_k \sum_{m_{k-1}<m \leqq m_k} |x_{n_k m}| + b_k^2 \right) \\ &\leqq \sum_{k=1}^{K} 1/K^2 + K^{-2} . 2 \sum_{k=1}^{K} b_k a_{k+1} + K^{-2} \sum_{k=1}^{K} b_k^2 \\ &< K^{-1} + 4 . K^{-2}.\end{aligned}$$

Furthermore,

$$\begin{aligned}\sum_{m>m_k} |y_{Km}|^2 &= \sum_{m>m_k} K^{-2} \left| \sum_{k=1}^{K} x_{n_k m} \right|^2 \\ &\leqq \sum_{m>m_k} K^{-2} \left(\sum_{k=1}^{K} |x_{n_k m}|^2 \right) . K \quad \text{(Schwarz inequality)} \\ &\leqq K^{-1} . a_K^2.\end{aligned}$$

In sum, $\|y_K\|_2^2 < 4/K^2 + 2/K$ and hence $y_K \to 0$ as $K \to \infty$. □

347. The hypothesis implies that for all f in S, $\|f\|_2^2 \leqq \frac{1}{2}\sum_n |a_n|^2 \leqq \frac{1}{2}\sum_n n^{-2} \leqq M < \infty$. If $\{f_k\}_{k=1}^\infty = \{x \mapsto \sum_n a_{kn} \sin 2n\pi x\}_{k=1}^\infty \subset S$ let $\{f_{k_i}\}_{i=1}^\infty$ be a subsequence such that $\{a_{k_i 1}\}_{i=1}^\infty$ is a Cauchy sequence, let $\{f_{k_{ij}}\}_{j=1}^\infty$ be a subsubsequence such that $\{a_{k_{ij}2}\}_{j=1}^\infty$ is a Cauchy sequence, etc. If $g_1 = f_{k_1}$, $g_2 = f_{k_{i_2}}$, etc., (diagonal process), $g_p(x) = \sum_n b_{pn} \sin 2n\pi x$, and if $p > r$ then $\|g_p - g_r\|_2^2 \leqq \sum_{n=1}^N |b_{pn} - b_{rn}|^2 + 2\sum_{n>N} n^{-2}$. The second term on the right is not more than $2/N$ and thus the entire right member approaches zero as $p, r \to \infty$. If $\|g_p - g\|_2 \to 0$ as $p \to \infty$ then the Fourier series for g must be $\sum_{n=1}^\infty b_n \sin 2n\pi x$ and $b_n = \lim_{p\to\infty} b_{pn}$. Thus

$$\sum_{n=1}^N n|b_n| = \lim_{p\to\infty} \sum_{n=1}^N n|b_{pn}| \leqq 1$$

whence $\sum_n n|b_n| \leqq 1$, i.e., $g \in S$ and so S is norm-compact. □

348. If $\int_0^x g_1(t)\,dt + c_1 = \int_0^x g_2(t)\,dt + c_2$ for all x then setting x to be zero shows $c_1 = c_2$ whence $\int_0^x (g_1(t) - g_2(t))\,dt = 0$ for all x and so $g_1 = g_2$. Hence T is well-defined.

If $\{f_n\}_{n=1}^\infty \subset S$, $f_n(x) = \int_0^x g_n(t)\,dt + c_n$, $\|f_n - f\|_2 \to 0$, and $\|g_n - g\|_2 \to 0$ as $n \to \infty$ then (see Solution 339) $\|f_n - c_n - (f_m - c_m)\|_2 \leqq \|g_n - g_m\|_2 \to 0$ as $n, m \to \infty$ and so, since $|c_n - c_m| = \|c_n - c_m\|_2 \leqq \|f_n - c_n - (f_m - c_m)\|_2 + \|f_n - f_m\|_2$, $c = \lim_{n\to\infty} c_n$ exists.

If F is $x \mapsto \int_0^x g(t)\,dt$ then $\|f - F - c\|_2 \leqq \|f - f_n\|_2 + \|F - f_n + c_n\|_2 + |c_n - c|$. The argument in the second paragraph shows $\|F - f_n + c_n\|_2 \to 0$ as $n \to \infty$ and so $f = F + c$, $Tf = g$ and the graph of T is closed. □

349. Let $\{f_n\}_{n=1}^N$ be an orthonormal set in M and let E be $\{x : \sum_{n=1}^N |f_n(x)|^2 \neq 0\}$. For x fixed in E let a_n be $\overline{f_n(x)}/(\sum_{n=1}^N |f_n(x)|^2)^{1/2}$. Then $|\sum_{n=1}^N a_n f_n(x)| = |\sum_{n=1}^N |f_n(x)|^2|^{1/2} \leqq C\|\sum_{n=1}^N a_n f_n\|_2$. But $\|\sum_{n=1}^N a_n f_n\|_2 = (\sum_{n=1}^N |a_n|^2)^{1/2} = 1$ and so for all x in E, $\sum_{n=1}^N |f_n(x)|^2 \leqq C^2$. Off E the inequality is *a priori* true and so $\int_I (\sum_{n=1}^N |f_n(x)|^2)\,dx = N \leqq C^2$. □

350. Since $\|f_n - f\|_2^2 = \|f_n\|_2^2 + \|f\|_2^2 - (f_n, f) - (f, f_n)$ and since the dominated convergence theorem implies that each of the last two terms in the right member converges to $\|f\|_2^2$ as $n \to \infty$ the result follows. □

351. i) If $f \in C([-1, 1], \mathbb{C})$ let Pf be $x \mapsto \frac{1}{2}(f(x) + f(-x))$. Then $\|Pf\|_2 \leqq \|f\|_2$ and since P is defined on a dense subset of $L^2([-1, 1], \lambda)$ it follows that P has a unique extension, again denoted P, to a continuous linear map defined on all $L^2([-1, 1], \lambda)$. Note that $P^2 = P$ and $P(M) = M$ whence $M = P(L^2([-1, 1], \lambda)$.

ii) If f_n is $x \mapsto \cos n\pi x$, $n = 0, 1, \ldots$, then $Pf_n = f_n$ and so $\{f_n\}_{n=0}^\infty$ is an orthonormal subset of M, If $f \in L^2([-1, 1], \lambda)$ let the Fourier series for f be $\frac{1}{2}a_0 + \sum_{n=1}^\infty (a_n \cos n\pi x + b_n \sin n\pi x)$. Then the Fourier series for Pf is $\frac{1}{2}a_0 + \sum_{n=1}^\infty a_n \cos n\pi x$ because the sine function is odd. Hence $\{f_n\}_{n=0}^\infty$ is an orthonormal basis for M. Similarly it follows that $\{x \mapsto \sin n\pi x\}_{n=1}^\infty$ is an orthonormal basis for $M^\perp$.

iii) The power series for f_n shows that the set of all polynomials in x^2, i.e. the set of all polynomials of the form $\sum_{k=0}^K a_k x^{2k}$, K in $\mathbb{N}$ is norm dense

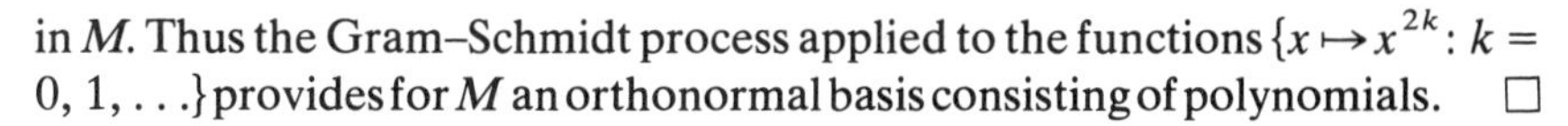

in M. Thus the Gram–Schmidt process applied to the functions $\{x \mapsto x^{2k}: k = 0, 1, \ldots\}$ provides for M an orthonormal basis consisting of polynomials. $\square$

352. Bessel's inequality shows that for all x in $\mathfrak{H}$, $\sum_n |(x, x_n)|^2 \leqq \|x\|^2$ and so for all x, $(x, x_n) \to 0$ as $n \to \infty$. $\square$

353. (See Problem 350.) The conclusion results from the following equation and convergence statement: $\|x_0 - x_n\|^2 = \|x_0\|^2 + \|x_n\|^2 - (x_n, x_0) - (x_0, x_n) \to 0$ as $n \to \infty$. $\square$

354. (See Problems 350 and 353.) The conclusion results from the following equation and convergence statement:

$$\|x_n - x_0\|^2 = \|x_n\|^2 + \|x_0\|^2 - (x_n, x_0) - (x_0, x_n) \to 0 \text{ as } n \to \infty. \quad \square$$

355. (See Solutions 350, 353, 354.) Let (x_n, y_n) be $1 + a_n$. Then $a_n \to 0$ as $n \to \infty$ and $\|x_n\| \cdot \|y_n\| \geqq 1 - |a_n|$. Since $\|x_n\|, \|y_n\| \leqq 1$, it follows that $\|x_n\|, \|y_n\| \to 1$ as $n \to \infty$. Then the calculations in the cited Solutions lead to the desired result. $\square$

356. For y fixed, $L_y: x \mapsto b(x, y)$ is a continuous linear functional since $|L_y(x)| \leqq C\|x\| \cdot \|y\|$. Thus there is an A such that $L_y(x) = (x, Ay)$. Since $y \mapsto Ay$ is linear and since $|(x, Ay)| \leqq C\|x\| \cdot \|y\|$ it follows that A is continuous and $\|A\| \leqq C$. $\square$

357. If u and v are two different points in $\mathfrak{H}$ then $N(u)$ resp. $N(v)$ given by $\{y: |(y - u, u - v)| < \frac{1}{2}\|u - v\|^2\}$ resp. $\{w: |((w - v, u - v)| < \frac{1}{2}\|u - v\|^2\}$ are weak neighborhoods of u resp. v. If $z \in N(u) \cap N(v)$ then $\|u - v\|^2 = |(u - z + z - v, u - v)| \leqq |(u - z, u - v)| + |(z - v, u - v)| < \|u - v\|^2$, a contradiction. $\square$

REMARK. A stronger result obtains. In the weak* topology $\mathfrak{H}$ (or the dual X^* of any Banach space X) is normal. The sequence of ideas is the following. i) The unit ball of $\mathfrak{H}$ is compact in the weak* topology. ii) In the weak* topology $\mathfrak{H}$ is the countable union of compact sets ($\mathfrak{H}$ is σ-compact). iii) In the weak* topology $\mathfrak{H}$ is Lindelöf. iv) In the weak* topology $\mathfrak{H}$ is regular. v) In Lindelöf spaces regularity and paracompactness are equivalent. vi) Every paracompact space is normal.

358. If $\mathfrak{H}$ is finite-dimensional its weak and strong (metric) topologies are the same. As a complete metric space $\mathfrak{H}$ is of the second category.

If $\mathfrak{H}$ is infinite-dimensional then every weakly open set is contained in no $B(0, n)$ since every weak neighborhood, e.g., $\{y: |(x, x_n) - (y, x_n)| < a, n = 1, 2, \ldots, N\}$ contains all elements $x + Rz$ if $R > 0$ and $z \in (\{x_n\}_{n=1}^N)^\perp$. It will be shown next that every $B(0, n)$ is weakly nowhere dense (and for this it will suffice to show that $B(0, 1)$ is weakly nowhere dense). It will follow that $\mathfrak{H}$, as the countable union of the $B(0, n)$, n in $\mathbb{N}$, is of the first (and hence not of the second) category.

Thus for x in $\mathfrak{H}$ and W a weakly open set containing x there is in W a y such that $\|y\| = 1 + b > 1$ and there is in $B(0, 1)$ a z such that $(y, z) > 1 + \frac{1}{2}b$. Then $U = \{u: |(u, z) - (y, z)| < \frac{1}{2}\}$ is a weak neighborhood of y and if $v \in B(0, 1)$ then $|(v, z)| \leqq 1$ and so $|(v, z)| - (y, z)| \geqq |(y, z)| - |(v, z)| > 1 + \frac{1}{2}b - 1 = \frac{1}{2}b$, i.e., $U \cap B(0, 1) = \varnothing$. Thus $W \cap U$ is a weak neighborhood of y, $W \cap U \subset W$ and $W \cap U \cap B(0, 1) = \varnothing$, i.e., $B(0, 1)$ is nowhere dense in the weak topology of $\mathfrak{H}$. □

359. i) If $W = \{x: |(x, y_p)| < a, p = 1, 2, \ldots, P\}$ is a weak neighborhood of zero then (see Problem 352) for some m, $|(x_m, y_p)| < a/3$, $p = 1, 2, \ldots, P$ and for some n, $n > m$, $|(x_n, y_p)| < a/3m$, $p = 1, 2, \ldots, P$ and so $|(x_m + mx_n, y_p)| < a/3 + a/3 < a$, i.e., $x_m + mx_n \in W$, i.e., $E \cap W \neq \varnothing$.

ii) If F is norm-bounded, say $F \subset B(0, R)$ then, since $\|x_m + mx_n\|^2 = \|x_m\|^2 + m^2\|x_n\|^2 = 1 + m^2$, it follows that if $x_m + mx_n \in F$ then $1 + m^2 \leqq R^2$ and hence for some M in $\mathbb{N}$, if $x_m + mx_n \in F$ then $m < M$.

On the other hand, if $a > 0$ and if $y = 2a \sum_{m=1}^{M-1} x_m$ then for all $x_m + mx_n$ in F, since $n > m$, $(x_m + mx_n, y) = 2a + 2am$ or $2a$ according as $m < n \leqq M - 1$ or $n > M - 1$. In either case $|(x_m + mx_n, y)| \geqq 2a > a$ and so $F \cap \{x: |(x, y)| < a\} = \varnothing$. Hence zero is not in the weak closure of F.

iii) The uniform boundedness principle implies that every weakly convergent sequence is norm-bounded. Thus ii) shows that zero is not in the weak closure of any weakly convergent subsequence of E and *a fortiori* zero is not the weak limit of any weakly convergent sequence in E. (A more direct proof is the following: The boundedness of a weakly convergent sequence implies that if $x_{m_k} + m_k x_{n_k} \to 0$ as $k \to \infty$ then for some M in $\mathbb{N}$, $m_k < M$ and so $x_{m_k} = 0$ for all k greater than some k_0, and a contradiction ensues.) □

360. If $x_0 \in M$ and $y \in M^\perp$ then $|(x_0, y)| = 0$ and $a = \|x_0 - x_0\| = 0$. If $x_0 \notin M$ the Hahn–Banach theorem shows that if $c > 0$ there is in $M^\perp \cap B(0, 1)$ a y such that $|(x_0, y)| > a - c$. Thus $b > a - c$ for all positive c, i.e., $b \geqq a$. On the other hand, if $y \in B(0, 1) \cap M^\perp$ and $x \in M$ then $\|x - x_0\| \geqq |(x - x_0, y)| = |(x_0, y)|$, i.e., $a \geqq b$, and so $a = b$. □

361. If $g = (\{f_{(t)}\}_{t \in \mathbb{T}})^\perp$, let h be $t \mapsto (g, f_{(t)})$. Then on the one hand $h = 0$ and on the other hand $\hat{h}(n) = \int_{\mathbb{T}^2} g(x)\overline{f(x + t)}\, e^{-int}\, dx\, dt/2\pi = 2\pi\hat{g}(-n)\bar{\hat{f}}(-n)$. Thus $\hat{g} = g = 0$ as required. □

NOTE. This is an "easy" case of the Wiener Tauberian theorem one form of which is the statement that the linear span of the set of translates of a function f is norm-dense in $L^1(\mathbb{R}, \lambda)$ iff $\hat{f}$ is never zero.

13. $L^p(X, \mu)$, $1 \leqq p \leqq \infty$

362. i) If $f, g \in E \cap C(I, \mathbb{C})$ then $\int_I |f(x) - g(x)|^p \, dx = 0$ and so $f = g$ a.e. Since f and g are continuous, $f = g$.

ii) If $a \in (0, 1)$ let C_a be a Cantor-like set contained in I and such that $\lambda(C_a) = a$. Then for some E, $\chi_{C_a} \in E$. If $f \in E \cap C(I, \mathbb{C})$ then $f = 0$ a.e. on $I \backslash C_a$, which is dense in I. Hence $f = 0$ since f is continuous. But $\|\chi_{C_a}\|_p > 0$ whereas $\|f\|_p = 0$. □

363. If i) obtains then for any step-function S, $\int_I f_n(x) S(x) \, dx \to \int_I f_0(x) S(x) \, dx$ as $n \to \infty$. Since the set of step-functions is dense in $L^q(I, \lambda)$, that ii) obtains is a corollary of the following general theorem: If X is a Banach space and $\{x_n\}_{n=0}^{\infty} \subset X$ then $x_n \to x_0$ weakly as $n \to \infty$ iff for some K, $\|x_n\| < K < \infty$ for all n and for all x^* in a norm-dense subset of X^*, $x^*(x_n) \to x^*(x_0)$ as $n \to \infty$. A sketch of the proof follows.

The uniform boundedness principle implies that if $x_n \to x_0$ weakly as $n \to \infty$ then for some K, $\|x_n\| < K < \infty$ for all n. Conversely if for some K, $\|x_n\| < K < \infty$ for all n and for all x^* in a norm-dense subset S of X^*, $x^*(x_n) \to x^*(x_0)$ as $n \to \infty$, then if $y^* \in X^*$ and $a > 0$ there is in S an x^* such that $\|x^* - y^*\| < a/3K$ and there is an n_0 such that if $n > n_0$ then $|x^*(x_n) - x^*(x_0)| < a/3$. Hence $|y^*(x_n) - y^*(x_0)| \leqq |y^*(x_n) - x^*(x_n)| + |x^*(x_n) - x^*(x_0)| + |x^*(x_0) - y^*(x_0)| < a$ if $n > n_0$.

If ii) obtains the theorem just cited and proved implies the result. □

364. If $f \in L^p(\mathbb{R}, \lambda) \cap C_{00}(\mathbb{R}, \mathbb{C})$ then $\|f - f_{(h)}\|_p \leqq \|f - f_{(h)}\|_\infty (\lambda(\text{supp}(f)))^{1/p}$ which approaches zero as $h \to 0$. If $g \in L^p(\mathbb{R}, \lambda)$ and $a > 0$ there is in $L^p(\mathbb{R}, \lambda) \cap C_{00}(\mathbb{R}, \mathbb{C})$ an f such that $\|g - f\|_p < a/3$ and there is a positive b such that if $|h| < b$ then $\|f - f_{(h)}\|_p < a/3$. Then $\|g - g_{(h)}\|_p \leqq \|g - f\|_p + \|f - f_{(h)}\|_p + \|f_{(h)} - g_{(h)}\|_p = 2\|f - g\|_p + \|f - f_{(h)}\|_p < a$.

In particular, $|\|g+g_{(h)}\|_p - 2\|g\|_p| \leqq \|g+g_{(h)}-2g\|_p = \|g_{(h)}-g\|_p < a$ and the result follows. □

365. According to problem 364 for each n in $\mathbb{N}$ there is an a_n such that $a_1 > a_2 > \cdots > a_n > \cdots > 0$ and such that if $|b_n| < a_n$ then $\|f_{(b_n)} - f\|_p < 2^{-n}$. Thus $\sum_n \|f_{(b_n)} - f\|_p^p < \infty$ if $|b_n| < a_n$; the monotone convergence theorem implies $\sum_n |f(x+b_n) - f(x)|^p < \infty$ a.e. from which the result follows. □

366. If $0 < a_1 < b_1 < 1$, $0 < a_2 < b_2 < 1$, and if p is arbitrary then $f \in L^p([a_1, b_1] \times [a_2, b_2], \lambda)$ and $\int_{[a_1,b_1]\times[a_2,b_2]} |f(x, y)|^p\, dx\, dy$ may be calculated *via* the variables change $u \mapsto 1 - xy$, $v \mapsto x$ $(x \mapsto v,\ y \mapsto (1-u)/v)$ for which the determinant of the Jacobian is v^{-1}. The corresponding integral is $\int_{a_1}^{b_1} (\int_{1-b_2 v}^{1-a_2 v} v^{-1} u^{-p}\, du)\, dv$. If $p = 1$ there emerges

$$\int_{a_1}^{b_1} v^{-1} \log((1-a_2 v)/(1-b_2 v))\, dv.$$

As $a_2 \to 0$ and $b_2 \to 1$ the integral converges to $\int_{a_1}^{b_1} -v^{-1} \log(1-v)\, dv = \int_{a_1}^{b_1} \sum_{n=1}^{\infty} (v^{n-1}/n)\, dv = \sum_n (b_1^n/n^2 - a_1^n/n^2) \to \pi^2/6$ as $b_1 \to 1$ and $a_1 \to 0$.

If $p > 1$, the integral is $\int_{a_1}^{b_1} (1-p)^{-1} v^{-1} [(1-a_2 v)^{1-p} - (1-b_2 v)^{1-p}]\, dv$. As $a_2 \to 0$ and $b_2 \to 1$ the integral approaches

$$\int_{a_1}^{b_1} (1-p)^{-1} v^{-1} (1-(1-v)^{1-p})\, dv$$

$$= (1-p)^{-1} \left[\left[\int_{a_1}^{b_1} v^{-1}\, dv - \int_{a_1}^{b_1} v^{-1} (1-v)^{1-p}\, dv \right] \right].$$

As $b_1 \to 1$ the second integral converges iff $-1 < 1-p$, i.e., iff $p < 2$. On the other hand, $v^{-1}(1-(1-v)^{1-p})$ is bounded as $v \to 0$ (e.g., *via* L'Hôpital's rule or *via* the binomial series for $(1-v)^{1-p}$) and so as $a_1 \to 0$,

$$\int_{a_1}^{1} (1-p)^{-1} v^{-1} (1-(1-v)^{1-p})\, dv$$

converges. In sum, $f \in L^p(I^2, \lambda)$ iff $1 \leqq p < 2$. □

367. The result is directly verifiable if A and B are measurable sets of finite measure, $f = \chi_A$, and $g = \chi_B$. If g is a simple function, e.g., $g = \sum_{k=1}^{K} b_k \chi_{B_k}$, it may be assumed that $0 \leqq b_1 < b_2 < \cdots < b_K$ and that the B_k are pairwise disjoint. If F is $t \mapsto \int_{E_t} f(x)\, dx$ and if $f = \chi_A$ then

$$F(t) = \begin{cases} 0, & \text{if } t > b_K \\ \lambda(A \cap B_K), & \text{if } b_K \geqq t > b_{K-1} \\ \sum_{k=p}^{K} \lambda(A \cap B_k), & \text{if } b_p \geqq t > b_{p-1}. \\ \sum_{k=1}^{K} \lambda(A \cap B_k), & \text{if } b_1 \geqq t \end{cases}$$

It follows that

$$\int_0^\infty F(t)\,dt = \sum_{p=1}^{K-1} \lambda\left(A \cap \left(\bigcup_{k=p}^{K} B_k\right)\right)(b_p - b_{p-1}) = \sum_{k=1}^{K} b_k \lambda(A \cap B_k)$$

(Abel summation) $= \int_{\mathbb{R}^n} f(x)g(x)\,dx$.

If $f = \chi_A$, $g \in L^q(\mathbb{R}^n, \lambda)$, and $g \geqq 0$ let $\{g_m\}_{m=1}^\infty$ be a sequence of nonnegative simple functions monotonely increasing everywhere to g. For each m in $\mathbb{N}$ let E_{mt} be $\{x: g_m(x) > t\}$. Then for each t, $E_{m+1,t} \supset E_{mt}$ and furthermore for all t, $|\lambda(E_{mt}) - \lambda(E_t)| \to 0$ as $m \to \infty$. If F_m is $t \mapsto \int_{E_{mt}} f(x)\,dx$, then $F_m(t) \uparrow F(t)$ as $m \to \infty$ and $\int_0^\infty F_m(t)\,dt = \int_{\mathbb{R}^n} f(x) g_m(x)\,dx \to \int_0^\infty F(t)\,dt$ as $m \to \infty$. Then $|\int_{\mathbb{R}^n} f(x)g_m(x)\,dx - \int_{\mathbb{R}^n} f(x)g(x)\,dx| \leqq \int_{\mathbb{R}^n} |f(x)| \,.\, |g_m(x) - g(x)|\,dx \to 0$ as $m \to \infty$ and so $\int_{\mathbb{R}^n} f(x)g(x)\,dx = \int_0^\infty F(t)\,dt$.

Finally if $f \in L^p(\mathbb{R}^n, \lambda)$ and $g \in L^q(\mathbb{R}^n, \lambda)(f, g \geqq 0)$ let $\{f_m\}_{m=1}^\infty$ be a sequence of simple nonnegative functions increasing monotonely to f everywhere. Then $\int_{\mathbb{R}^n} f_m(x)g(x)\,dx \uparrow \int_{\mathbb{R}^n} f(x)g(x)\,dx$ as $m \to \infty$. If G_m is $t \mapsto \int_{E_t} f_m(x)\,dx$ then $\int_0^\infty G_m(t)\,dt = \int_{\mathbb{R}^n} f_m(x)g(x)\,dx$. Furthermore, for all t, $G_m(t) \uparrow F(t)$ as $m \to \infty$ and hence $\int_0^\infty G_m(t)\,dt \uparrow \int_0^\infty F(t)\,dt$ as $m \to \infty$ and the result follows. □

368. i) Since $f * g = g * f$ it suffices to prove $S(f * g) = S(f) * g$ since then $S(f * g) = S(g * f) = S(g) * f$. By hypothesis $S(f) * g(x) = \int_{\mathbb{T}} S(f)(x - t)g(t)\,dt = \int_{\mathbb{T}} (S(f))_{(-t)}(x)g(t)\,dt = \int_{\mathbb{T}} (S(f_{(-t)}))(x)g(t)\,dt$. However for x fixed, $\int_{\mathbb{T}} (S(f_{(-t)}))(x)g(t)\,dt$ is the limit of sums of the form $\sum_k a_k (S(f_{(-t_k)}))(x)g(t_k) = (S(\sum_k a_k f_{(-t_k)} g(t_k)))(x)$ and such sums converge to $S(\int_{\mathbb{T}} f(x - t)g(t)\,dt)$, whence $S(f) * g = S(f * g)$ as required.

ii) By definition $S(-f)\hat{}(n) = (2\pi)^{-1} \int_0^{2\pi} S(f)(t)\,e^{-int}\,dt$. If g_n is $t \mapsto e^{int}$, then

$$S(f)\hat{}(n) = e^{-inx} S(f) * g_n(x) = (2\pi)^{-1} \int_{\mathbb{T}} S(f)\hat{}(n)\,dx$$

$$= (2\pi)^{-1} \int_{\mathbb{T}} e^{-inx} S(f) * g_n(x)\,dx$$

$$= (2\pi)^{-1} \int_{\mathbb{T}} e^{-inx} (f * S(g_n))(x)\,dx = \hat{f}(n) S(g_n)\hat{}(n).$$

If $a_n = S(g_n)\hat{}(n)$ the result follows. □

369. The map $T: \{P \mapsto \sum_{k=1}^K b_k t^k\} \mapsto \sum_{k=1}^K a_k b_k$ of polynomials into $\mathbb{C}$ is by hypothesis a bounded linear functional defined on a dense subset of $L^p(I, \lambda)$ and so T is extendible without increase of its norm to all $L^p(I, \lambda)$. Since $(L^p(I, \lambda))^* = L^q(I, \lambda)$ the result follows. □

370. Consider first the case for $L^p_{\mathbb{R}}(X, \mu) = \{f: f \in L^p(X, \mu),\ f(X) \subset \mathbb{R}\}$. It will be shown that $E_1 = \varnothing$. Since in all instances the norm of an extreme point f is one, there is a measurable set A of finite positive measure and a positive number a such that $|f| \geqq a$ on A. Furthermore, at least one of $A \cap \{x: f(x) \gtrless 0\} = A^{\pm}$ is a measurable set of positive measure. If, e.g.,

$\mu(A^+) > 0$ let A^+ be decomposed into disjoint measurable subsets of positive measure: $A^+ = A_1 \cup A_2$, $A_1 \cap A_2 = \emptyset$, $\mu(A_1) \,.\, \mu(A_2) > 0$. If $0 < b$, $c < a$, and $b\mu(A_1) = c\mu(A_2)$ and if

$$g_1 = \begin{cases} f + b \text{ on } A_1 \\ f - c \text{ on } A_2, \\ f \text{ elsewhere} \end{cases} \qquad g_2 = \begin{cases} f - b \text{ on } A_1 \\ f + c \text{ on } A_2 \\ f \text{ elsewhere} \end{cases},$$

then $g_1 \neq g_2$, $g_1 + g_2 = 2f$, and $\|g_1\|_1 = \|g_2\|_1 = 1$ and so f is not an extreme point of $B(0, 1)$.

If $p > 1$ it will be shown that $E_p = \{f : \|f\|_p = 1\}$. Indeed if $\|f\|_{pp} = 1$, if $\|g\|_p, \|h\|_p \leqq 1$, $0 < a < 1$, and if $f = ag + (1-a)h$, then $a\|g\|_p + (1-a)\|h\|_p \leqq 1$ and so according to the criterion for equality in the Minkowski inequality, there are constants A, B not both zero and such that $Ag + Bh = 0$ a.e., whence $(aB - A(1-a))h = -Af$ (if $A \neq 0$, f is a multiple of h) or

$$((1-a)A - Ba)g = -Bf$$

(if $B \neq 0$, f is a multiple of g). Whichever multiplier is used, e.g., if $f = k \,.\, h$, then $k = \pm 1$ which implies that $f = h$ or $a = 1$ and then $f = g$.

If the case $L^p(X, \mu)$ rather than $L^p_{\mathbb{R}}(X, \mu)$ is treated, again the two cases, $p = 1$ and $p > 1$, are treated separately.

If $f \in B(0, 1)$ in $L^1(X, \mu)$ then $|f| = (\operatorname{sgn} f)f \in B(0, 1)$. If f is an extreme point and if $|f|$ is not then $|f| = ag + (1-a)h$, $0 < a < 1$, $g, h \in B(0, 1)$, $g \neq h$. But then $\overline{(\operatorname{sgn} f)}|f| = f = a\overline{(\operatorname{sgn} f)}g + (1-a)\overline{(\operatorname{sgn} f)}h$, whence $\overline{(\operatorname{sgn} f)}g = \overline{(\operatorname{sgn} f)}h$ and so $g = h$ a.e. where $f \neq 0$. If $D = \{x : f(x) \neq 0\}$ then $\|f\|_1 = \||f|\|_1 = a \int_D g(x)\, d\mu(x) + (1-a) \int_D h(x)\, d\mu(x) = 1$ and so $\int_D g(x)\, d\mu(x) = \int_D h(x)\, d\mu(x) = 1$. Since $\|g\|_1, \|h\|_1 \leqq 1$ it follows that $g = h = 0$ a.e. off D and so $g = h = f$ a.e., a contradiction. Thus if f is an extreme point so is $|f|$. However the argument given for $L^1_{\mathbb{R}}(X, \mu)$ is applicable for $|f|$ and shows it cannot be an extreme point.

If $f \in B(0, 1)$ in $L^p(X, \mu)$, $p > 1$, the argument given for $L^p_{\mathbb{R}}(X, \mu)$ then shows that the constant k must be of the form $e^{i\theta}$. It follows that, e.g., $a = |a - 1 + e^{i\theta}|$ whence $a^2 = a^2 + 2 - 2a + 2a \cos\theta - 2\cos\theta$. If $\cos\theta \neq 1$ it follows that $a = 1$. If $\cos\theta = 1$ then $e^{i\theta} = 1$ and inexorably, in every circumstance, the conclusion follows. □

NOTE. If X contains atoms then E_1 can fail to be empty. For example, in $l^1_{\mathbb{R}}(\{0, 1\})$, $B(0, 1)$ contains four extreme points.

371. If $A = B$ then $A^\perp = B^\perp$. Conversely, if $A^\perp = B^\perp$ then $(A^\perp)^\perp = (B^\perp)^\perp$. Since $A \subset (A^\perp)^\perp$, if $f \in (A^\perp)^\perp \backslash A$ there is in $L^q(X, \mu)$ a g such that $\int_X \overline{g(x)} f(x)\, d\mu(x) = 1$ and $g \in A^\perp$, whence $\int_X \overline{f(x)} g(x)\, d\mu(x) = 0$, a contradiction. □

372. As shown in Solution 370 there is no extreme point in the unit ball $B(0, 1)$ of $L^1(X, \mu)$ if μ is nonatomic. If $(L^\infty(X, \mu))^* = L^1(X, \mu)$ then

$L^1(X, \mu)$ is reflexive and its unit ball is weakly compact. The Krein–Milman theorem implies that the unit ball is the closed convex hull of its extreme points and the contradiction implies the result.

Alternatively, for the case $(I, \mathsf{S}_\lambda, \lambda)$ let T be the map $C(I, \mathbb{C}) \ni f \mapsto f(0)$. Then $|Tf| \leqq \|f\|_\infty$ and *via* the Hahn–Banach theorem there is a norm-preserving extension, again denoted T, such that $|Tg| \leqq \|g\|_\infty$ for all g in $L^\infty(I, \lambda)$. Hence if $L^1(I, \lambda) = L^\infty(I, \lambda)$ it follows that there is in $L^1(I, \lambda)$ an h such that for all f in $C(I, \mathbb{C})$, $f(0) = \int_I f(x)\overline{h(x)}\, dx$. Consequently if $0 < a < b \leqq 1$, $\int_{[a,b]} h(x)\, dx = 0$ and so $h = 0$ a.e. But then if $f = 1$, there emerges the contradiction: $1 = 0$. □

373. Let F be the equivalence class to which f belongs, i.e., $F = \{g : g = f$ a.e.$\}$. If $\{f_k\}_{k=1}^K \subset F$ and if $\{\varphi_k\}_{k=1}^K$ is a partition of unity (subordinate to some open cover $\{N_k\}_{k=1}^K$ of $I\}$ then $\sum_k \varphi_k f_k \in F$. Indeed, if $E_k = \{x : f_k(x) = f(x)\}$ and $E = \bigcap_{k=1}^K E_k$ then $\lambda(E) = 1$ and on E, $f_k = f$, $k = 1, 2, \ldots, K$. Hence $\sum_k \varphi_k f_k = f \sum_k \varphi_k = f$ on E. (The same kind of argument shows that F is convex.)

If $\lim_{t \to x} g_x(t) = v_x$ and if h_x is

$$t \mapsto \begin{cases} v_x, & \text{if } t = x \\ g_x(t), & \text{if } t \neq x \end{cases}$$

then $h_x \in F$ and h_x is continuous at x. For each n in $\mathbb{N}$ and x in I let $N_n(x)$ be a neighborhood of x and such that $\operatorname{osc}_{N_n(x)}(h_x) < 1/n$. The compactness of I implies that for some n in $\mathbb{N}$, if $J_{nk} = (k/n - 1/n, k/n + 1/n)$ then $\operatorname{osc}_{J_{nk}}(h_{k/n}) < (1/n)$, $k = 0, 1, \ldots, n$. If $\{f_{nk}\}_{k=0}^n$ is a partition of unity subordinate to $\{J_{nk}\}_{k=0}^n$ let H_n be $\sum_{k=0}^n f_{nk} h_{k/n}$. It will be shown that: i) for each x in I there is a neighborhood $N(x)$ such that $\operatorname{osc}_{N(x)}(H_n) < 2/n$; ii) for all n and for all x and for every neighborhood $U(x)$ there is in $U(x)$ a y such that $H_m(y) = H_n(y)$ for all m.

Ad i) If $x \in I$ let $S(x)$ be $\{k : x \in J_{nk}\}$. Then $\operatorname{card}(S(x)) \leqq 2$ and there is a neighborhood $N(x)$ of x and such that $N(x) \cap \bigcup_{k' \notin S(x)} J_{nk'} = \varnothing$. Then if $y_1, y_2 \in N(x)$, $H_n(y_i) = \sum_{k \in S(x)} f_{nk}(y_i) h_{k/n}(y_i)$, $i = 1, 2$, and so

$$|H_n(y_1) - H_n(y_2)| \leqq \sum_{k \in S(x)} f_{nk}(y_1) |h_{k/n}(y_1) - h_{k/n}(y_2)|$$

$$+ \sum_{k \in S(x)} |f_{nk}(y_1) f_{nk}(y_2)| . |h_{k/n}(y_2)|.$$

If $A = \sup_{k \in S(x)} |h_{k/n}(y_2)|$ there is a positive b such that if $\operatorname{diam}(E) < b$ then $\sup_k \operatorname{osc}_E f_{nk} < 1/nA$ and if $\operatorname{diam}(N(x)) < b$ (which may be assumed), $|H_n(y_1) - H_n(y_2)| < 1/n + 1/n = 2/n$, i.e., $\operatorname{osc}_{N(x)} H_n < 2/n$.

Ad ii) If $G_n = \{x : H_n(x) = f\}$ then $\lambda(\bigcap_{n=1}^\infty G_n) = 1$ and so $G = \{x : H_n(x) = f,\ n \text{ in } \mathbb{N}\}$ is dense in I. Hence if $U(x)$ is given let y be in $G \cap U(x)$ and the result follows.

Next it will be shown that for all x, $\{H_n(x)\}_{n=1}^\infty$ is a Cauchy sequence. Indeed, if $a > 0$ and $x \in I$ then $|H_m(x) - H_n(x)| \leqq$

$|H_m(x) - H_m(y)| + |H_m(y) - H_n(y)| + |H_n(y) - H_n(x)|$. If $2/m$, $2/n < a/3$ and if $U(x)$ is such that $\operatorname{osc}_{U(x)}(H_m)$, $\operatorname{osc}_{U(x)}(H_n) < a/3$ there is in $U(x)$ a y such that $H_m(y) = H_n(y)$ whence $|H_m(x) - H_n(x)| < a$ if m, $n > 6/a$. Hence $\lim_{n \to \infty} H_n(x) = H(x)$ exists everywhere and uniformly, whence H is continuous and $H = f$ a.e., as required. □

14. Topological Vector Spaces

374. The uniform boundedness principle applied to the sequence $\{T_n(x)\}_{n=1}^{\infty}$, regarded as a subset of Y^{**}, implies $\sup_n \|T_n(x)\| < \infty$. A second application of the principle yields the result. □

375. It will be shown that the graphs $\{(x, T(x)): x \text{ in } E\}$ and $\{(x^*, S(x^*)): x^* \text{ in } E^*\}$ are closed in the norm-derived product topologies. Thus assume $\|x_n - x\| + \|x_n^* - x^*\| + \|T(x_n) - y\| + \|S(x_n^*) - y^*\| \to 0$ as $n \to \infty$. Then for all x^* in E^*, $x^*(T(x_n)) = S(x^*)(x_n) \to S(x^*)(x) = x^*(T(x))$ as $n \to \infty$. Thus $T(x) = y$. Similarly for all x in E, $S(x_n^*)(x) = x_n^*(T(x) \to x^*(T(x)) = S(x^*)(x)$ as $n \to \infty$ and so $S(x^*) = y^*$. The closed graph theorem implies the result. □

376. Granted the continuum hypothesis (there is no cardinal number strictly between card $(\mathbb{N})$ and card $(\mathbb{R})$), the following argument provides a valid counter-example. An infinite-dimensional Banach pace cannot have a countable Hamel basis (see Problem 408). Thus, since the cardinality of separable Banach space is card $(\mathbb{R})$, it follows that the cardinality of every Hamel basis of such a space is card $(\mathbb{R})$. Hence the separable Banach spaces $c_0(\mathbb{N})$ and $l^1(\mathbb{N})$ have Hamel bases $\{x_t\}_{t\in\mathbb{R}}$ and $\{y_t\}_{t\in\mathbb{R}}$ and the map $x_t \mapsto y_t$ extended linearly provides a not necessarily continuous isomorphism $T: c_0(\mathbb{N}) \to l^1(\mathbb{N})$. Since $T^{-1}(0) = \{0\}$, $\ker(T)$ is closed. However, since $(c_0(\mathbb{N}))^* = l^1(\mathbb{N})$ is separable and $(l^1(\mathbb{N}))^* = l^\infty(\mathbb{N})$ is not separable, neither T nor T^{-1} is continuous and so T is neither continuous nor open. □

(For a proof independent of the continuum hypothesis see Solution 489).

377. It may be assumed that $0 < a < 1$. If $b = (2a-1)/2(1-a)$ there is in Y a v such that $\|x - v\| < (1+b)d(x, Y)$. If $z = (x-v)/\|x-v\|$ then

$z \in \operatorname{span}(x, Y)$ and $\|z\| = 1$. Furthermore if $y \in Y$ then

$$\|z-y\| = \|x-v\|^{-1} . \|x-v-\|x-v\| . y\| > \frac{\|x-v-\|x-v\| . y\|}{(1+b)d(x, Y)}$$

$$\geqq \frac{d(x, Y)}{(1+b)d(x, Y)} = \frac{1}{1+b}$$

$$= 2(1-a) > 1-a$$

as required. □

378. If such a $\#$ exists there is in $X \times X^*$ a pair x, x^* such that $x^*(x) = 1$. Then $x^*(x^\#) = 1$, $(ix^*)(x) = i$, $(ix^*)(x^\#) = -i$ whereas $(ix^*)(x^\#) = i(x^*(x^\#)) = i$. The contradiction $i = -i$ shows $\#$ cannot exist. □

379. If $x^* \in M$ and $x \in M_\perp$ then $x^*(x) = 0$ whence $M \subset (M_\perp)^\perp$. Let x^* be in the weak* closure, denoted $\bar{M}^{w*}$, of M. Then for all x in $M_\perp$ and any positive a, $\{y^*: |(y^* - x^*)(x)| < a\} \cap (M_\perp)^\perp \neq \emptyset$. Thus there is in $(M_\perp)^\perp$ a y^* such that $|(y^* - x^*)(x)| = |x^*(x)| < a$. Thus $x^* \in (M_\perp)^\perp$ and so $\bar{M}^{w*} \subset (M_\perp)^\perp$. If $x^* \in (M_\perp)^\perp \backslash \bar{M}^{w*}$ then according to the Hahn–Banach theorem for locally convex vector spaces there is in $(\bar{M}^{w*})_\perp$ an x such that $x^*(x) = 1$, a contradiction. □

Remark. Note that $E^{**} = E$ when E^* is regarded as a (locally convex) topological vector space in the topology $\sigma(E^*, E)$. This is true whether or not E is reflexive. [25])

380. It may be assumed that $M \neq \emptyset$ and that $0 \in M$. If $\bar{M}^w$ denotes the weak closure of M and if $x \notin M$ there are positive numbers a and b and in X^* an x^* such that $x^*(B(x, a)) \subset (-\infty, -b)$ while $x^*(M) \subset [0, \infty)$. Thus $\{y: |x^*(y-x)| < \frac{1}{2}b\} \cap M = \emptyset$ and so $x \notin \bar{M}^w$, i.e., $\bar{M}^w = M$.

By definition, the norm-closure, denoted $\bar{M}$, is a subset of $\bar{M}^w$. Since $\bar{M}$ is convex and norm-closed it is, according to the previous paragraph, weakly closed, whence $\bar{M} = \bar{M}^w$. □

381. Since $M \subset (M^\perp)_\perp$ if $x \in (M^\perp)_\perp \backslash M$ there is, according to the Hahn–Banach theorem, in $M^\perp$ an x^* such that $x^*(x) = 1$. Hence $x \notin (M^\perp)_\perp$, a contradiction. □

382. Let E be a nonreflexive Banach space, e.g., $E = c_0(\mathbb{N})$. Then E, regarded as a proper norm-closed subspace of E^{**}, is such that $E_\perp = \{0\}$, $(E_\perp)^\perp = E^{**} \supsetneqq E$. □

383. Since F is finite-dimensional it has a finite basis $\{x_n\}_{n=1}^N$. Because F is finite-dimensional it is norm-closed and so is a Banach space. Hence (see problem 398) the coefficient maps are continuous linear functionals, which may be extended without increase of norm to elements $\{x_n^*\}_{n=1}^N$ of E. Then $P: E \ni x \mapsto \sum_{n=1}^N x_n^*(x)x_n$ is such that $P^2 = P$ and $P(E) = F$. Furthermore, if $M = \sup_n \|x_n^*\|$ then $\|P\| \leqq M(\sum_n \|x_n\|)$. □

384. If f is continuous then $\ker(f)=f^{-1}(0)$ is closed. Conversely if $\ker(f)=M$ is closed it is closed subspace of E. If $M=E$ then $f=0$ and f is continuous. If $M\neq E$ there is in $E\backslash M$ an x such that $f(x)=1$. For all z in E, $z-f(z)x\in M$ and z is uniquely representable as $f(z)x+y$, y in M. The Hahn–Banach theorem implies there is in $M^\perp$ an x^* such that $x^*(x)=1$. Then for all z, $x^*(z)=f(z)$ and so $f=x^*$ and f is continuous.

If $\ker(f)$ is not closed then f is not continuous and $f\neq 0$. If the closure M of $\ker(f)$ is not E there is in $E\backslash M$ an x such that $f(x)=1$ and there is in $M^\perp$ an x^* such that $x^*(x)=1$. The argument of the preceding paragraph shows $x^*=f$ and a contradiction results. □

385. Let $\{x_{\gamma_n}\}_{\gamma_n\in\Gamma}$ be a countable, infinite, proper subset of $\{x_\gamma\}_{\gamma\in\Gamma}$ and let y be $\sum_{n=1}^{\infty}2^{-n}x_{\gamma_n}/\|x_{\gamma_n}\|$. Then y also has a finite representation $\sum_{\gamma\in\delta}a_\gamma x_\gamma$. Let γ_{n_0} be in $\{\gamma_n\}_{n=1}^{\infty}\backslash\delta$. If all coefficients functionals are continuous then there emerges the contradiction $2^{-n_0}=0$. □

386. Let $\{y_n\}_{n=1}^{N}$ be a basis for F/M. If $x_n/M=y_n$ the x_n are linearly independent and if N is their span then N is a closed subspace of F. Let G be $E\oplus N$ normed according to $(u,v)\mapsto\|(u,v)\|=\|u\|+\|v\|$. Then G is a Banach space and if T is $G\ni(u,v)\mapsto K(u)+v$ then $T\in\operatorname{Hom}(G,F)$. If $z\in F$ and $z/M=\sum_n a_ny_n$ then $(z-\sum_n a_nx_n)/M=0$, i.e., there is in M a w such that $z=w+\sum_n a_nx_n$. In other words $T(G)=F$ and so T is open.

If $\{w_n\}_{n=1}^{\infty}\subset M$ and $w_n\to w_0$ as $n\to\infty$ then since T is open there is in G a sequence $\{(u_n,v_n)\}_{n=0}^{\infty}$ such that $(u_n,v_n)\to(u_0,v_0)$ as $n\to\infty$ and $T((u_n,v_n))=w_n=K(u_n)+v_n$, $n=0,1,\ldots$. Then for n in $\mathbb{N}$, $v_n=0$ and so $v_0=0$ whence $w_0=K(u_0)\in M$ and so M is closed.

Let $\{x_\gamma\}_{\gamma\in\Gamma}$ be a Hamel basis for an infinite-dimensional Banach space F and let γ_0 be such that the coefficient functional $x\mapsto a_{\gamma_0}$ is not continuous. Thus its kernel M is not closed, $\{x_\gamma\}_{\gamma\neq\gamma_0}$ is a Hamel basis for M, and the dimension of F/M is one but M is not closed. □

387. Let $\{x_n^*\}_{n=1}^{\infty}$ be a norm-dense subset of E^*. For each n there is in E an x_n such that $\|x_n\|\leqq 1$ and $|x_n^*(x_n)|>\|x_n^*\|-1/n$. It will be shown that the closure M of span $(\{x_n\}_{n=1}^{\infty})$ is E, from which the result will follow. Indeed, if $x\in E\backslash M$ there is in $M^\perp$ an x^* such that $x^*(x)=1$. Since $\{x_n^*\}_{n=1}^{\infty}$ is dense in E^* there is a subsequence $\{x_{n_k}^*\}_{k=1}^{\infty}$ such that $x_{n_k}^*\to x^*$ as $k\to\infty$. But then $\|x^*-x_{n_k}^*\|\cdot\|x_{n_k}\|\geqq|x^*(x_{n_k})-x_{n_k}^*(x_{n_k})|=|x_{n_k}^*(x_{n_k})|>\|x_{n_k}^*\|-1/n_k$. Thus $\|x_{n_k}^*\|\to 0$ as $k\to\infty$ whereas $x^*\neq 0$ and the contradiction implies the result.

Although $l^1(\mathbb{N})$ is separable, $(l^1(\mathbb{N}))^*=l^\infty(\mathbb{N})$ is not. □

388. Let $\mathcal{T}$ be the norm-induced topology of E^*. Since $\sigma(E^*,E)\subset\mathcal{T}$ it follows that $\sigma\mathsf{A}(\sigma(E^*,E))\subset\sigma\mathsf{A}(\mathcal{T})$. On the other hand every ball $B(a^*,r)$ in E^* is weak*-compact (Alaoglu's theorem). By hypothesis E^* is norm-separable whence it follows that every norm-open set is the countable union of sets in $\sigma\mathsf{A}(\sigma(E^*,E))$, i.e., $\sigma\mathsf{A}(\sigma(E^*,E))=\sigma\mathsf{A}(\mathcal{T})$. □

389. Two observations are in order first. i) If M is a proper subspace of a topological vector space E then M contains no nonempty open subset. Indeed, if U is a nonempty subset of M and $x \in U$ then $-x+U$ is a neighborhood of the origin. Hence if $y \in E$ there is a nonzero t such that $ty \in -x+U \subset M$ and so $y \in M$, i.e., $M=E$, a contradiction. ii) If M is a closed proper subspace of a topological vector space E then M is nowhere dense. Indeed, if $x \in E \backslash M$ then since M is closed there is an open set U such that $x \in U$ and $U \cap M = \varnothing$. If $x \in M$, U is open, and $x \in U$ then, according to i) $U \not\subset M$ and so there is in $U \backslash M$ a y. Since M is closed, $U \backslash M$ is open and there is an open set W such that $y \in W \subset U \backslash M$ whence M is nowhere dense.

If $\{x_n\}_{n=1}^{\infty}$ is dense in E then $\{B(x_n, r_m)^0: n \text{ in } \mathbb{N}, r_m \text{ in } \mathbb{Q}, r_m > 0\}$ is a countable set $\{A_k\}_{k=1}^{\infty}$ of open sets. Let y_1 be an arbitrary nonzero element of E and let Y_1 be span (y_1). There is in $A_1 \backslash Y_1$ a y_2. If $y_1, y_2, \ldots, y_n$ have been defined so that they are linearly independent, $Y_m = \text{span}(\{y_k\}_{k=1}^{m})$, $m=1,2,\ldots,n$, and $y_k \in A_{k-1} \backslash Y_{k-1}$, $k=2,3,\ldots,n$, there is in $A_n \backslash Y_n$ a y_{n+1}. The sequence $\{y_n\}_{n=1}^{\infty}$ contains no finite linearly independent subset. Since every open set is the union of some of the A_k, it follows that $\{y_n\}_{n=1}^{\infty}$ is dense. □

390. The following will be shown. i) If T_N is $X^* \ni x^* \mapsto \{x^*(x_1), x^*(x_2), \ldots, x^*(x_N), 0, \ldots\} \in l^1(\mathbb{N})$ then for all N, $T_N \in \text{Hom}(X^*, l^1(\mathbb{N}))$. ii) If T is $X^* \ni x^* \mapsto \{x^*(x_n)\}_{n=1}^{\infty} \in l^1(\mathbb{N})$ then for all x^* in X^*, $\|(T_N - T)(x^*)\| \to 0$ as $N \to \infty$. iii) The map T is in $\text{Hom}(X^*, l^1(\mathbb{N}))$.

If i)–iii) are granted, if $a>0$, $M = \sup_N \|T_N\|$, and $\{a_n\}_{n=1}^{\infty} \in c_0(\mathbb{N})$ let N be such that $|a_n| < a/2M$ if $n > N$. Then for all x^* in $B(0,1)$ (the unit ball of X^*) $|x^*(\sum_{n=p}^{q} a_n x_n)| < (a/2M) \sum_{n=p}^{q} |x^*(x_n)| = (a/2M)\|(T_p - T_q)(x^*)\| < a$ if $p, q > N$. Thus $\|\sum_{n=p}^{q} a_n x_n\| = \sup\{|x^*(\sum_{n=p}^{q} a_n x_n)|: \|x^*\| \leqq 1\} \leqq a$ and so $\sum_n a_n x_n$ exists.

The proofs of i)–iii) follow.

Ad i) Since $\|T_N(x^*)\| \leqq (\sum_{n=1}^{N} \|x_n\|)\|x^*\|$, $T_N \in \text{Hom}(X^*, l^1(\mathbb{N}))$.

Ad ii) Since $\|(T_N - T)(x^*)\| = \sum_{n=N+1}^{\infty} |x^*(x_n)|$ the result follows from the hypothesis.

Ad iii) Since $T(x^*) = \lim_{N\to\infty} T_N(x^*)$ for all x^* in X^*, $\|T(x^*)\| = \lim_{N\to\infty} \|T_N(x^*)\|$. The uniform boundedness principle implies that $\|T_N\| \leqq M < \infty$ for some M and all N and so $\|T\| \leqq M$. □

391. i) If $x \in B(0,1)$ choose in $\{x_n\}_{n=1}^{\infty}$ an x_{n_1} such that $\|x_{n_1} - x\| < \frac{1}{2}$. Then choose in $\{x_n\}_{n=1}^{\infty}$ an x_{n_2} such that $n_2 > n_1$ and $\|x_{n_2} - 2(-x_{n_1} + x)\| < \frac{1}{2}$ (since $\|-x_{n_1} + x\| < \frac{1}{2}$ such an x_{n_2} exists). Thus $\|x - x_{n_1} - \frac{1}{2}x_{n_2}\| < (\frac{1}{2})^2$. By induction there can be developed a sequence $\{x_{n_k}\}_{k=1}^{\infty}$ such that $n_k < n_{k+1}$ and $\|x - \sum_{k=1}^{K} x_{n_k}/2^{k-1}\| < (\frac{1}{2})^K$. Hence $T \in \text{Sur}(l^1(\mathbb{N}), X)$.

ii) Note that the argument in i) can be given with $\frac{1}{2}$ replaced by any r in $(0,1)$. It follows that if $a>0$ and $x \in X$ there is in $l^1(\mathbb{N})$ some $\{a_n\}_{n=1}^{\infty}$ such that $\sum_n |a_n| < \|x\| + a$ and $\sum_n a_n x_n = x$. Hence $\inf\{\|\{a_n\}_n\|_1 : \{a_n\}_n \in T^{-1}(x)\} \leqq \|x\|$. In other words if $\{a_n\}_n \in l^1(\mathbb{N})$ then the quotient norm of

$\{a_n\}_n/\ker(T)$ does not exceed $\|T(\{a_n\}_n\|$. Since $\|x\| \leqq \sum_n |a_n| \cdot \|x_n\| \leqq \|\{a_n\}_n\|_1$ the converse is also true and the result follows. □

392. i) If $x \in C$ there are in $A, x_1, x_2, \ldots, x_Q$ and in $I, t_1, t_2, \ldots, t_Q$ such that $\sum_q t_q = 1$ and $x = \sum_q t_q x_q$. If $b > 0$ there are in A finitely many points $a_1, a_2, \ldots, a_P$ such that $\bigcup_p B(a_p, b/2) \supset A$. For each q there is in $\{a_p\}_{p=1}^P$ an α_q and there is in $B(0, b/2)$ a β_q such that $x_q = \alpha_q + \beta_q$. Hence $x = \sum_q t_q \alpha_q + \sum_q t_q \beta_q = \alpha + \beta$, α in the convex hull C_1 of $\{a_p\}_{p=1}^P$ and β in $B(0, b/2)$. As the convex hull of a finite set of points C_1 is compact and hence there is in C_1 a finite set $\{c_r\}_{r=1}^R$ such that $\bigcup_r B(c_r, b/2) \supset C_1$. Thus $C \subset \bigcup_r B(c_r, b)$, C is totally bounded and so its closure K is compact.

ii) If $M = \max_{x \in A} |f(x)|$, $N = \max_{x \in K} |f(x)|$ then there is in A an x_0 such that $M = |f(x_0)|$ and there is in K a y_0 such that $N = |f(y_0)|$. Since $\sup_{x \in C} |f(x)| = M$, if $N > M$ then $y_0 \in K \backslash C$ and

$$\{z : |f(z) - f(y_0)| < \tfrac{1}{2}(N - M)\} \cap C = \emptyset.$$

Hence y_0 is not in the weak closure of C, i.e., (see Problem 380) the norm-closure K of C. Thus $N = M$ and the result follows.

iii) The map $C(A, \mathbb{C}) \ni g \mapsto g(x)$ is a positive continuous linear functional and hence by the Riesz representation theorem there is a positive Borel measure μ_x such that $g(x) = \int_A g(y)\, d\mu_x(y)$. Since $|g(x)| \leqq \|g\|_\infty$ it follows that $\|\mu_x\| \leqq 1$. However ii) implies $\|\mu_x\| = 1$.

iv) See iii). □

393. It suffices to prove that ii) implies i). Let B_y be $E \ni x \mapsto B(x, y) \in G$. Hence there is a constant K_y such that for all x, $\|B_y(c)\| \leqq K_y \|x\|$. The map $y \mapsto B_y$ is linear and the closed graph theorem implies the map is continuous. Hence $\|B_y\| \leqq C\|y\|$ for some constant C and finally $\|B(x, y)\| \leqq C\|x\| \cdot \|y\|$. □

394. Fix y^* in F^*, x, h in E and let g be $\mathbb{R} \ni t \mapsto y^*(f(x + th))$. Then by definition of differentiability there is an a depending on $x + t_1 h$, $t_2 - t_1$, and h, approaching zero as $|t_2 - t_1| \to 0$, and such that $|g(t_2) - g(t_1)| \leqq \|y^*\| \cdot a \cdot \|h\| \cdot |t_2 - t_1|$. Hence g' exists everywhere and $g' = 0$, i.e., $g = g(0) = y^*(f(x))$. If f is not constant there are in E an x and an h such that $f(x + h) \neq f(x)$. The Hahn–Banach theorem implies there is in E^* a y^* such that the corresponding g is not constant and the contradiction implies the result. □

395. Assume T is norm-continuous. If $x_n \to 0$ weakly as $n \to \infty$ and $T(x_n) \not\to 0$ weakly as $n \to \infty$ then, *via* a subsequence as needed, it may be assumed there is in E^* an x^* and there is a positive a such that $|x^*(T(x_n))| \geqq a$ for all n. But $x^*(T(x_n)) = T^*(x^*)(x_n) \to 0$ as $n \to \infty$ and a contradiction results.

Conversely, assume $T(x_n) \to 0$ weakly whenever $x_n \to 0$ weakly as $n \to \infty$. Let T_1 be $E^* \ni x^* \mapsto T_1(x^*) \in E^*$ defined by the formula: for all x, $T_1(x^*)(x) = x^*(T(x))$. If $\|x_n\| \to 0$ as $n \to \infty$ then $x_n \to 0$ weakly as $n \to \infty$ whence $T(x_n) \to 0$ weakly and $x^*(T(x_n)) \to 0$ as $n \to \infty$. Thus $x \mapsto x^*(T(x))$

is in E^* and so T_1 is well-defined. According to Problem 375 T and T_1 are norm continuous. □

396. If $f=0$ the result is clear. If $f\neq 0$ there are sequences $\{a_n\}_n$, $\{h_n\}_n$, and $\{y_n\}_n$ such that $a_n\downarrow 0$, $\{h_n\}_n\subset H$, $d(x,H)>\|x+h_n\|-a_n$, $\|y_n\|\leqq 1$, and $|f(y_n)|\geqq\|f\|-a_n>0$. Then $x-f(x)y_n/f(y_n)\in H$ and

$$d(x,H)\leqq\frac{|f(x)|}{|f(y_n)|}\leqq\frac{|f(x)|}{\|f\|-a_n}.$$

On the other hand $|f(x)|=|f(x+h_n)|\leqq\|f\|(d(x,H)+a_n)$ and so

$$\frac{|f(x)|}{\|f\|}-a_n\leqq d(x,H)\leqq\frac{|f(x)|}{\|f\|-a_n}$$

and the result follows. □

397. The open mapping theorem implies that T^{-1} is continuous. Thus if $\|T(x_n)\|\not\to\infty$ as $n\to\infty$ then, *via* subsequence as needed, it may be assumed that for some M and all n, $\|T(x_n)\|\leqq M<\infty$. Since T^{-1} is continuous, for some N and all n, $\|x_n\|\leqq N<\infty$, a contradiction. □

398. Let E_1 be $\{y=\{a_n\}_{n=1}^{\infty}$: a_n in $\mathbb{C}$, $\sum_n a_nx_n$ is norm-convergent$\}$. If $\|\cdots\|'$: $E_1\ni y\mapsto\|y\|'=\sup_N\|\sum_{n=1}^{N}a_nx_n\|$ is used as a norm for E_1 then with respect to this norm E_1 is a Banach space (see Solution 399.) Furthermore T:$E_1\ni y\mapsto\sum_n a_nx_n\in E$ is in Sur(E_1,E) and indeed T is bijective. Thus $T^{-1}\in\text{Hom}\,(E,E_1)$. In particular for all n, $\frac{1}{2}|a_n|\leqq\|T^{-1}(x)\|'/\|x_n\|$ and so each $x\mapsto a_n$, denoted x_n^*, is in E^*. □ Note that $x_n^*(x_m)=\delta_{nm}$.

399. If $E_2=\{y=\{a_\gamma\}_{\gamma\in\Gamma}$: a_γ in $\mathbb{C}$, the net $\sum_{\gamma\in\delta\in\Delta}a_\gamma x_\gamma$ converges weakly$\}$ then the uniform boundedness principle implies that $\|\cdots\|'$: $E_2\ni y\mapsto\sup_\delta\|\sum_\delta a_\gamma x_\gamma\|=\|y\|'$ is a norm with respect to which E_2 is a Banach space. (If $\{y_n\}_n$ is a Cauchy sequence in E_2 it follows that for all γ, $\{a_{n\gamma}\}_n$ is a Cauchy sequence with limit, say a_γ. If $b>0$ there is an n_0 such that if $p,q>n_0$ then $\|y_p-y_q\|'<b$. If $\delta\in\Delta$, $\|\sum_{\gamma\in\delta}(a_{p\gamma}-a_{q\gamma}\|\leqq 2b$ if $p,q>n_0$. As $p\to\infty$ there emerges the inequality $\|\sum_{\gamma\in\delta}(a_\gamma-a_{q\gamma})x_\gamma\|\leqq 2b$ and so if $x^*\in E^*$, $|x^*(\sum_{\gamma\in\delta}a_\gamma x_\gamma)|\leqq 2b\|x^*\|+|x^*(\sum_{\gamma\in\delta}a_{q\gamma}x_\gamma)|$. For some δ_0 the last term is less than $b\|x^*\|$ if $\delta\cap\delta_0=\varnothing$. Hence $y=\{a_\gamma\}_\gamma\in E_2$ and $\|y-y_q\|'\to 0$ as $q\to\infty$.)

Again T: $E_2\ni y\mapsto\sum_\gamma a_\gamma x_\gamma$ is a bijective map of E_2 onto E and T is continuous with respect to the norm-induced topologies in E_2 and E. The argument in Solution 398 shows x_γ^*: $x\mapsto a_\gamma$ are all in E^*. Note that $x_\gamma^*(x_{\gamma'})=\delta_{\gamma\gamma'}$.

For each δ let T_δ be $E\ni x\mapsto\sum_{\gamma\in\delta}x_\gamma^*(x)x_\gamma$. Then $T_\delta\in\text{End}\,(E)$ and the uniform boundedness principle implies there is an M such that for all δ, $\|T_\delta\|\leqq M<\infty$. Furthermore $\{T_\delta(E)\}_{\delta\in\Delta}=S$ is a linear (hence convex) weakly dense subset of E. According to problem 380, S is norm-dense in E. Thus if $x\in E$ and $a>0$ there is a δ_0 such that $\|x-\sum_{\gamma\in\delta}x_\gamma^*(x)x_\gamma\|<a$ if $\delta>\delta_0$ and the result follows. □

400. Since for all x, $S_N(x) \to x$ as $N \to \infty$ the uniform boundedness principle implies there is an M such that $\|S_N\| \leqq M < \infty$ for all N. Since $P_N = S_N - S_{N-1}$ it follows that $\|P_N\| \leqq 2M$. The biorthogonality relations imply $S_N^2 = S_N$ and $P_N^2 = P_N$. □

401. The Hahn–Banach theorem implies there are x and x^* such that $x^*(x) = 1$. Hence the set of biorthogonal systems is nonempty. If the set is partially ordered by inclusion, Zorn's lemma is applicable and yields a maximal biorthogonal system. □

402. If $\{x_\gamma, x_\gamma^*\}_\gamma$ is not maximal there is an x^* in $(\{x_\gamma\}_\gamma)^\perp \backslash \{0\}$. Since the closure of span $(\{x_\gamma\}_\gamma)$ is E there is a contradiction. □

403. If span $(\{x_\gamma\}_\gamma)$ is not dense its closure is a closed proper subspace F or E. If $x \in E \backslash F$ the Hahn–Banach theorem implies there is in $F^\perp$ an x^* such that $x^*(x) = 1$. The result follows now from problem 402. □

404. By hypothesis, if $x \in E$ and $a > 0$ there is a finite sequence $\{b_n\}_{n=1}^N$ such that $\|x - \sum_{n=1}^N b_n x_n\| a$. Hence if $N_1 > N$,

$$\|x - S_{N_1}(x)\| \leqq \left\|x - \sum_{n=1}^N b_n x_n\right\| + \left\|\sum_{n=1}^N b_n x_n - S_{N_1}\left(\sum_{n=1}^N b_n x_n\right)\right\|$$

$$+ \left\|S_{N_1}\left(\sum_{n=1}^N b_n x_n - x\right)\right\| < a + 0 + aM$$

and the result follows. □

405. If $\{x_n\}_n$ is a basis for E and $\|S_N\| \leqq S$ (see Problem 400) then if $m \leqq n$, $\|\sum_{i=1}^m a_i x_i\| = \|S_m(\sum_{i=1}^n a_i x_i)\| \leqq S \|\sum_{i=1}^n a_i x_i\|$, i.e., S may serve for M.

Conversely, given the hypothesis, let F_N be span $(\{x_n\}_{n=1}^N)$, N in $\mathbb{N}$. For each N there is (see Problem 383) a projection Q_N (called P in Problem 383) of E on F_N. These Q_N can be defined so that $Q_N Q_{N+K} = Q_N$, i.e., so that the Hahn–Banach extensions of the coefficient functionals in the various F_N are coherent.

If $\|x\| \leqq 1$, $1 > a > 0$, $N \in \mathbb{N}$ there is in $\mathbb{N}$ an N_1 such that $N_1 > N$ and there is a finite sequence $\{a_i\}_{i=1}^{N_1}$ such that $\|x - \sum_{i=1}^{N_1} a_i x_i\| < a$. Then $Q_N(\sum_{i=1}^{N_1} a_i x_i = \sum_{i=1}^N a_i x_i$ and so $\|Q_N(\sum_{i=1}^{N_1} a_i x_i)\| \leqq M \|\sum_{i=1}^{N_1} a_i x_i\|$. Furthermore,

$$\|Q_N(x)\| \leqq \left\|Q_N\left(x - \sum_{i=1}^{N_1} a_i x_i\right)\right\| + \left\|Q_N\left(\sum_{i=1}^{N_1} a_i x_i\right)\right\| \leqq \left\|Q_N\right\| . a + M\left\|\sum_{i=1}^{N_1} a_i x_i\right\|$$

$$\leqq \|Q_N\| . a + M(a + \|x\|).$$

Hence $\|Q_N\| \leqq M(1+a)/(1-a)$, i.e., for all N, $\|Q_N\| \leqq M$.

Furthermore if $x \in E$ and $a > 0$ there is an N such that if $N_1 > N$ then there is in F_{N_1} a y such that $\|x - y\| < a/(M+1)$. Hence $\|Q_{N_1}(x) - x\| \leqq \|Q_{N_1}(x) - y\| + \|y - x\| = \|Q_{N_1}(x - y)\| + \|y - x\| < (M+1)a/(M+1) = a$, i.e., $Q_N(x) \to x$ as $N \to \infty$. If $P_N = Q_N - Q_{N-1}$ then $P_N(x) = a_N x_N$ and $|a_N| \leqq$

$2M\|x\|$. Hence the maps $x_N^*: x \to a_N$ are in E^* and $\{x_n, x_n^*\}_{n=1}^\infty$ is a biorthogonal system. The result follows. □

406. Let $f_n - g_n$ be h_n. If $f \in \mathfrak{H}$ and $\|f\| \leqq 1$ then $f = \sum_{n=1}^\infty A_{0n} f_n = \sum_n A_{0n}(g_n + h_n)$. Then

$$\sum_n |A_{0n}| \cdot \|h_n\| \leqq \left(\sum_n |A_{0n}|^2\right)^{1/2} \left(\sum_n \|h_n\|^2\right)^{1/2} \leqq 1 \cdot \left(\sum_n c^{2n}\right)^{1/2}$$
$$= \left(\frac{c^2}{1-c^2}\right)^{1/2} = r < 1.$$

Thus $z_1 = \sum_n A_{0n} h_n$ exists and $\|z_1\| \leqq r$. If $z_1 = \sum_n A_{1n} f_n = \sum_n A_{1n}(g_n + h_n)$ the preceding argument shows $z_2 = \sum_n A_{in} h_n$ exists and $\|z_2\| \leqq r^2$. In general $f = \sum_{n=1}^\infty (\sum_{p=0} A_{pn}) g_n + z_{P+1}$, $\|z_{P+1}\| \leqq r^{P+1}$ and $\sum_{n=1}^\infty (\sum_{p=0}^\infty |A_{pn}|^2) = \sum_{p,q=0}^\infty (\sum_{n=1}^\infty |A_{pn}| \cdot |A_{qn}|) \leqq \sum_{p,q} (r^{p+q}) = \sum_{s=0}^\infty (s+1) r^s < \infty$. Hence $a_n = \sum_p A_{pn}$ exists and $\sum_n |a_n|^2 < \infty$. Finally, if $a_{Pn} = \sum_{p=0}^P A_{pn}$ then $\sum_n |a_{Pn}|^2 < \infty$ and $\sum_n |a_n - a_{Pn}|^2 \to 0$ as $P \to \infty$. Thus

$$\left\| f - \sum_{n=1}^N a_n g_n \right\| \leqq \|z_{P+1}\| + \left\| \sum_{n=N+1}^\infty a_{Pn} g_n \right\| + \left\| \sum_{n=1}^N (a_{Pn} - a_n) g_n \right\|.$$

The last term is not more than

$$\left\| \sum_{n=1}^N (a_{Pn} - a_n) f_n \right\| + \left\| \sum_{n=1}^N (a_{Pn} - a_n) h_n \right\|$$
$$\leqq \left(\sum_{n=1}^N |a_{Pn} - a_n|^2 \right)^{1/2} + \sum_{n=1}^N |a_{Pn} - a_n| c^n$$
$$\leqq \left(\sum_{n=1}^\infty |a_{Pn} - a_n|^2 \right)^{1/2} \left(1 + \left(\sum_{n=1}^\infty c^{2n} \right)^{1/2} \right).$$

The second term is not more than $(\sum_{n=N+1}^\infty |a_{Pn}|^2)^{1/2} (1 + (\sum_{n=N+1}^\infty c^{2n})^{1/2}$. In conclusion, for large N, $\|f - \sum_{n=1}^N a_n g_n\|$ is small.

If $\sum_n b_n g_n = 0$ then since $\|g_n\|$ is near $\|f_n\| = 1$ and $|b_n| \cdot \|g_n\| \leftarrow 0$ as $n \to \infty$ it follows that $b_n \to 0$ as $n \to \infty$. Thus $\sum_n |b_n| \cdot \|h_n\| < \infty$ and

$$\sum_n b_n (f_n - h_n) + \sum_n b_n h_n = \sum_n b_n f_n = \sum_n b_n h_n = h$$

exists. Hence $\|h\|^2 = \sum_n |b_n|^2 \leqq (\sum_n |b_n| \cdot \|h_n\|)^2 \leqq (\sum_n |b_n|^2) \cdot r^2$ and so there results the contradiction $1 \leqq r^2 < 1$ unless $\sum_n |b_n|^2 = 0$. The proof is complete. □

407. If $a_n = 2^{-n} / \|x_n^*\|$, if $x_n - y_n = z_n$, and if $\|z_n\| < a_n$, then $\sum_n \|z_n\| \cdot \|x_n^*\| = r < 1$ and $\sum_n |x_n^*(x)| \cdot \|z_n\| < r\|x\|$, i.e., $y = \sum_n x_n^*(x) y_n$ exists.

If $S = \text{span}(\{y_n\}_n)$ then it will be shown that S is dense in E. Indeed, if $v_0 \in E$, then

$$v_0 = \sum_n x_n^*(v_0) x_n = \sum_n x_n^*(v_0)(y_n + z_n) = \sum_n x_n^*(v_0) y_n + \sum_n x_n^*(v_0) z_n$$
$$= u_1 + v_1, \ \|v_1\| \leqq r\|v_0\|.$$

Iteration of the argument as in Solution 406 leads to the equation $v_0 = \sum_n (\sum_{p=0}^{P} x_n^*(v_p))y_n + v_{P+1}$, $\|v_{P+1}\| \leqq r^{P+1}\|v_0\|$ and so S is dense in E.
in E.

It will be shown that Problem 405 is applicable. Indeed if $m \leqq n$,

$$\left\|\sum_{j=1}^{m} a_j y_j\right\| \leqq \left\|\sum_{j=1}^{m} a_j x_j\right\| + \sum_{j=1}^{m} |a_j| \,.\, \|x_j - y_j\|.$$

However if $|a_j| = e^{i\theta_j} a_j$ then

$$\sum_{j=1}^{m} |a_j| \,.\, \|x_j - y_j\| = \left(\sum_{j=1}^{m} e^{i\theta_j} x_j^* \,.\, \|x_j - y_j\|\right)\left(\sum_{j=1}^{n} a_j x_j\right) \leqq r\left\|\sum_{j=1}^{n} a_j x_j\right\|.$$

Thus $\|\sum_{j=1}^{m} a_j y_j\| \leqq (1+r)\|\sum_{j=1}^{n} a_j x_j\| \leqq (1+r)(\|\sum_{j=1}^{n} a_j y_j\| + \sum_{j=1}^{n} |a_j| \,.\, \|x_j - y_j\|)$. If $A = \|\sum_{j=1}^{n} a_j y_j\|$ and $B = \|\sum_{j=1}^{n} a_j x_j\|$, the preceding argument shows that $B \leqq A + rB \leqq A + r(A + rB) \leqq \cdots \leqq A/(1-r)$. Thus $\|\sum_{j=1}^{m} a_j y_j\| \leqq (1+r)B \leqq (1+r)A/(1-r) \times (1+r)(1-r)^{-1} \|\sum_{j=1}^{n} a_j y_j\|$ whence Problem 405 applies to yield the result. □

NOTE. Only the inequality $\sum_n \|x_n^*\| \,.\, \|z_n\| = r < 1$ and the basis property of $\{x_n\}_n$ are used in the proof.

408. Only the case card (Γ) = card $(\mathbb{N})$ needs discussion. The following generalization of the Gram–Schmidt process may be applied to $\{x_n\}_{n=1}^{\infty}$.

Let y_1 be $x_1/\|x_1\|$ and let y_1^* be such that $y_1^*(y_1) = 1$. Having chosen y_k and y_k^*, $k = 1, 2, \ldots, n$ so that $y_k^*(y_{k'}) = \delta_{kk'}$, $\|y_k\| = 1$, span $(\{y_k\}_{k=1}^{n}) =$ span $(\{x_k\}_{k=1}^{n})$, let y_{n+1} be $(x_{n+1} - \sum_{k=1}^{n} y_k^*(x_{n+1})y_k)/\|x_{n+1} - \sum_{k=1}^{n} y_k^*(x_{n+1})y_k\|$ and let y_{n+1}^* be chosen in $(\{y_k\}_{k=1}^{n})^{\perp}$ ($\neq \{0\}$ because E is infinite-dimensional) and so that $y_{n+1}^*(y_{n+1}) = 1$. Proceed by induction.

Then $\{y_n\}_{n=1}^{\infty}$ is also a Hamel basis for E. If $y = \sum_{n=1}^{\infty} 2^{-n} y_n$ then $y = \sum_{n=1}^{N} b_n y_n$ and so if $M > N$, $y_M^*(y)$ is both 0 and 2^{-M}, a contradiction. □

409. Let X be $\mathbb{C}[x]$ endowed with the norm it inherits from $L^2(I, \lambda)$. Then $\{f_n : x \mapsto x^n\}_{n=0}^{\infty}$ is a Hamel basis for X and X is infinite-dimensional. □

410. Let G_1 be the (additively written) group of order two: $G_1 = \{0, 1\}$; let G be the countably infinite product of G_1 with itself: $G = \{g = \{e_n\}_{n=1}^{\infty} :$ e_n in $G_1\}$. Normalized Haar measure μ_1 on G_1 assigns measure one-half to each of $\{0\}$ and $\{1\}$. Product (Haar) measure μ on G permits a measure-preserving surjection $T : G \ni g \mapsto \sum_{n=1}^{\infty} e_n 2^{-n} \in I$ to be defined; T is also continuous with respect to the product topology on G and the usual topology on I; moreover T is bijective on the complement of a null set in G.

i) For g in G let $\tilde{S}_M(g)$ be $x \mapsto \sum_{m=1}^{M} e_m x_m^*(x) x_n$. Then $G \ni g \mapsto \tilde{S}_M(g) \in$ End (E) and $G \ni g \mapsto \tilde{S}_M(g)(x) \in E$ (x fixed) are continuous, hence Bochner-measurable, vector-valued maps. Since

$$C_1(x) = \left\{g : \lim_{M_1, M_2 \to \infty} \|\tilde{S}_{M_1}(g)(x) - \tilde{S}_{M_2}(g)(x)\| = 0\right\}$$

it follows that $C_1(x)$ is measurable for each x. Since $C(x)=T(C_1(x))$, $C(x)$ is also measurable.

ii) Since E has a countable basis E is separable. Let $\{z_m\}_{m=1}^{\infty}$ be dense in E. Then $B_0=\{g: \limsup_{M=\infty}\|S_M(g)\|<\infty\}$ is measurable as are $B_m=\{g: \lim_{M\to\infty} S_m(g)(z_m) \text{ exists}\}$, m in $\mathbb{N}$. Hence $B=\bigcap_{m=0}^{\infty} B_m$ is measurable and $C_1=\bigcap_{x\in E} C_1(x)\subset B$. If $g\in B$, $x\in E$, $S=\sup_M\|S_M(g)\|$, and $a>0$, choose z_m such that $\|z_m-x\|<a/2S$. Then $\|(S_{M_1}(g)-S_{M_2}(g))(x)\|\leqq \|(S_{M_1}(g)-S_{M_2}(g))(z_m)\|+a/2$. Since $g\in B$, $\|(S_{M_1}(g)-S_{M_2}(g))(z_m)\|\to 0$ as $M_1, M_2\to\infty$ and so $g\in C_1$. Hence $B=C_1$ and both C_1 and $C=T(C_1)$ are measurable.

Furthermore C_1 is a subgroup of G. Indeed, for x in E and g in C_1 let $x(g)$ be $\sum_n e_n x_n^*(x)x_n$. Then if $h\in C_1$, $x(g+h)=x(g)(h)$ and so $g+h\in C_1$. Since C_1 contains all $\{e_n\}_n$ such that the elements of the sequence are ultimately constant (0 or 1) it follows that C_1 is (nonempty and) dense in G. Hence C is dense in I.

iii) If κ is a subset of $\mathbb{N}$, a set in G is a κ-cylinder iff membership in the set is determined completely by conditions specifying the coordinates having indices in κ. Thus, e.g., the basic neighborhoods for the (product) topology of G are κ-cylinders for finite subsets κ. (If $\kappa=\{n+1, n+2, \ldots\}$ then a κ-cylinder is sometimes called a J_n-cylinder [13].) The σ-algebra of (Haar) measurable sets in G is generated by the set of all κ-cylinders, κ finite. If E is a measurable set that is a J_n-cylinder for every n and if F is a κ-cylinder for some finite κ then $\mu(E)=\mu(E)\mu(F)$. Since the sets of the type F generate the σ-algebra of measurable sets it follows that in particular, $\mu(E)=(\mu(E))^2$, i.e., $\mu(E)=0$ or 1. Since C_1 is a J_n-cylinder for every n it follows that $\mu(C_1)=0$ or 1. Hence $\lambda(C)=0$ or 1.

iv) If $\mu(C_1)=1$ and if $C_1\neq G$, then there is at least one coset $g_0+C_1\neq C_1$. But then $\mu(g_0+C_1)=1$ and $1=\mu(G)\geqq\mu(C_1)+\mu(g_0+C_1)=2$, a contradiction. Thus $C_1=G$ and $C=I$. □ [4]

411. The basic properties of "sup" insure that $L(af)=|a|L(f)$ and $L(f+g)\leqq L(f)+L(g)$. If $L(f)=0$ then f is constant and since $f(0)=0$ it follows that $f=0$. Thus L is a norm.

If $\{f_n\}_n$ is a Cauchy sequence (L) then for all s, $|f_n(s)-f_m(s)|=|f_n(s)-f_n(0)-(f_m(s)-f_m(0))|\leqq L(f_n-f_m)|s|^a$. Hence $\lim_{n\to\infty} f_n(s)=f(s)$ exists. Furthermore if $s\neq t$,

$$|f(s)-f(t)|/|s-t|^a=\lim_{n\to\infty}|f_n(s)-f_n(t)|/|s-t|^a\leqq\lim_{n\to\infty} L(f_n)<\infty$$

and so $f\in E_a$. Finally,

$$\begin{aligned}|f(s)-f(t)-(f_n(s)-f_n(t))|/|s-t|^a\\ &=\lim_{m\to\infty}|f_m(s)-f_m(t)-(f_n(s)-f_n(t))|/|s-t|^a\\ &\leqq\lim_{m\to\infty} L(f_m-f_n)\end{aligned}$$

and so $L(f-f_n)\to 0$ as $n\to\infty$. The result follows. □

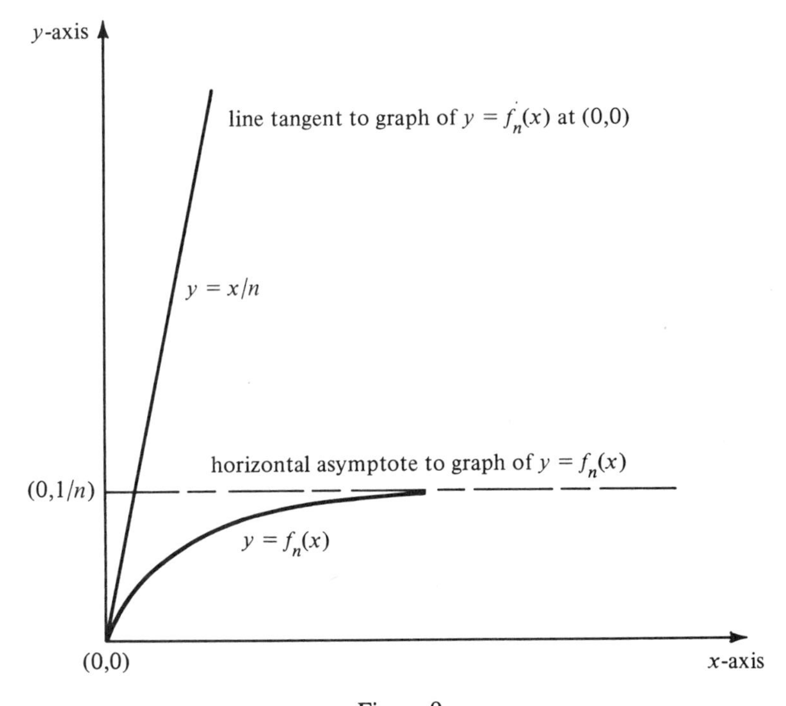

Figure 9

412. The argument of Solution 411 may be repeated to show that L is a norm and so N is also a norm.

If f_n is the map having the graph in Figure 9, then $L(f_n) = n$, $\|f_n\|_\infty = n^{-1}$ and so $N(f_n) = n + 1/n \to \infty$ while $\|f_n\|_\infty \to 0$ as $n \to \infty$ and the result follows. $\square$

413. By definition there is an M such that for all x, t, $|f_{(t)}(x)| = |f(t+x)| \leqq M < \infty$. Hence $\{f_{(t)}\}_{t\in\mathbb{R}}$ is a uniformly bounded set.

Since K is compact there is an n such that $K \subset [-n, n]$. Then on $[(-n+1), n+1]$ f is uniformly continuous, i.e., if $a > 0$ there is in $(0, 1)$ a b such that $|f(x+t) - f(y+t)| = |f_{(t)}(x) - f_{(t)}(y)| < a$ if $|x-y| < b$ and $t \in K$. Hence the set $\{f_{(t)}\}_{t\in K}$ is equicontinuous and the result follows. $\square$

414. i) If $1 - a/n = 1/q - r$ then $r > 0$ and $qn - qa = n - nqr = n - d$, $n = 1, 2$. Then

$$\int_{B(0,b)} |k(x-y)|^q \, dx \leqq c^q \int_{B(0,b)} |x-y|^{qa-qn} \, dx$$

$$\leqq c^q \int_{B(y,1)^0} + c^q \int_{B(0,b)\setminus B(y,1)^0} |x-y|^{qa-qn} \, dx.$$

If $n=1$ the first term in the last expression is

$$\int_{B(y,1)^0} |x-y|^{-1+d}\,dx = \int_{-1}^{1} |u|^{-1+d}\,du = 2/d < \infty.$$

If $n=2$ the corresponding calculation leads to

$$\int_{B(y,1)^0} |x-y|^{-2+d}\,dx = \int_0^{2\pi}\int_0^1 r^{-1+d}\,dr\,d\theta = 2\pi/d < \infty.$$

Thus in both cases $\int_{B(y,1)^0} < \infty$. Since $x \mapsto |x-y|^{-n+d}$ is continuous and bounded in $B(0,b)\backslash B(y,1)^0$ it follows that for all b in $(0,\infty)$, $k \in L^q(B(0,b))$.

ii) The argument in i) shows that for all f in $C_{00}(\mathbb{R}^n, \mathbb{R})$, $|k * f(x)| \leqq \|f\|_\infty$ and thus $k * f \in L^q(B(0,b), \lambda)$. Furthermore $C_{00}(\mathbb{R}^n, \mathbb{R})$ is a dense subset of $L^1_{\mathbb{R}}(B(0,b), \lambda)$ and the result follows. □

415. Since K is the intersection of $B(0,1)$ of $L^1(I,\lambda)$ with a finite-dimensional subspace it follows that K is compact. □

416. If f is $x \mapsto (2\pi)^{-1/2} e^{-x^2/2}$ then $\hat{f}$ is $t \mapsto (2\pi)^{-1/2} e^{-t^2/2}$. Thus $(L^1(\mathbb{R},\lambda))\hat{}$ is a subalgebra of $C_0(\mathbb{R}, \mathbb{C})$ (Riemann–Lebesgue) and contains a real separating function. The Stone–Weierstrass theorem implies the result. □

417. i) In P_n^2 define the map $p, q \mapsto (p,q) = L(pq)$. The map is a true inner product because $(p,p) > 0$ if $p \neq 0$. Apply the Gram-Schmidt process to $\{g_k\}_{k=0}^n = \{x \mapsto x^k; 0 \leqq k \leqq n\}$ to produce for P_n an orthonormal basis $\{f_k\}_{k=0}^n$ and let p_{n+1} be $x \mapsto g_{n+1} - \sum_{q=0}^n L(g_{n+1}f_k)f_k$. Then for all q in P_n, $L(p_{n+1}q) = 0$.

ii) Let p_{n+1} have k distinct real zeros, $0 \leqq k \leqq n+1$. If $k=0$, then $n+1$ is even, say $2m$ and there are real numbers $\{r_j\}_{j=1}^m, \{s_j\}_{j=1}^m$ such that $p_{n+1}(x) = \prod_{j=1}^m ((x-r_j)^2 + s_j^2)$, which is a sum of squares of polynomials and hence $L(p_{n+1}) > 0$ whereas $L(p_{n+1} \,.\, 1) = 0$ as found in i). Thus $1 \leqq k$. If the real distinct zeros of p_{n+1} are $\{z_j\}_{j=1}^k$, then $n+1-k$ is even, again say $2m$, and *mutatis mutandis*, $p_{n+1}(x) = \prod_{j=1}^k (x-z_j) \prod_{i=1}^m ((x-r_i)^2 + s_i^2)$. If $m=0$ there is nothing to be proved. If $m>0$ then $Q(x) = p_{n+1}(x)\prod_{j=1}^k (x-z_j) = \prod_{j=1}^k (x-z_j)^2 \prod_{i=1}^m ((x-r_i)^2 + s_i^2)$ is a sum of squares of polynomials; $L(Q) > 0$ whereas again, as found in i), $L(Q) = 0$. Thus $m=0$ and the result follows.

iii) If π_{n+1} is a polynomial of degree $n+1$ and $L(\pi_{n+1} \cdot q) = 0$ for all q in P_n, it may be assumed that $\pi_{n+1}(x) = x^{n+1} + \cdots$ and so $\deg(\pi_{n+1} - p_{n+1}) \leqq n$. Thus $L((\pi_{n+1} - p_{n+1})^2) = L(\pi_{n+1}(\pi_{n+1} - p_{n+1})) - L(p_{n+1}(\pi_{n+1} - p_{n+1})) = 0 - 0 = 0$, whence $\pi_{n+1} = p_{n+1}$. □

418. Define T_x in $\mathfrak{H}^*(=\mathfrak{H})$ by $T_x\colon y \mapsto (f(x) - f(x_0), y)$. For y fixed and g denoting $x \mapsto \partial(f(x), y)/\partial x$, there is a map $x, y \mapsto a(x,y)$ such that $a(x,y) \to 0$ as $x \to x_0$ and $T_x(y) = g(x_0, y)(x - x_0) + a(x,y)(x-x_0)$. If $x \neq x_0$ then $|T_x(y)| \,.\, \|x - x_0\|^{-1/2} \leqq \|x - x_0\|^{1/2} |g(x_0,y) + a(x,y)| \to 0$ as $x \to x_0$. The uniform boundedness principle implies that for x near x_0 there is an M such that $\|T_x\| \,.\, \|x-x_0\|^{-1/2} \leqq M < \infty$. Hence $|f(x)-f(x_0)| \leqq M\|x-x_0\|^{1/2}$ and the result follows. □

419. If f_n is

$$x \mapsto \begin{cases} n, & 0 \leqq x \leqq n^{-1} \\ 0, & n^{-1} < x \leqq 1 \end{cases}$$

then $\|f_n\|_{1/2} = n^{-1} \to 0$ as $n \to \infty$. Hence if $g \in (L^{1/2}(I, \lambda))^*$ then $g(f_n) \to 0$ as $n \to \infty$. But $|g(\chi_{[a,b]})| \leqq \|g\| \,.\, (b-a)^2$. Hence if $g(\chi_{[a,b]}) = \mu([a, b])$ then μ is a measure, $\mu \ll \lambda$ and so there is a function h_g such that for all f in $L^{1/2}(I, \lambda)$, $g(f) = \int_I f(x)\overline{h_g(x)}\, dx$. In particular, $g(f_n) = n \int_0^{1/n} \overline{h_g(x)}\, dx$ and $g(f_{n(-t)}) = n \int_t^{t+1/n} \overline{h_g(x)}\, dx \to 0$ as $n \to \infty$ for all t in I. Hence $h_g = 0$ a.e. and $g = 0$. □

420. A helpful notation for X is $l^4(\{1, 2, \ldots, n\}, \nu, L^2(\mathbb{R}, \mu))$, i.e., X is the set of all $L^2(\mathbb{R}, \mu)$-valued functions on $\{1, 2, \ldots, n\}$ and the norm for X is the analog of that for $l^4(\{1, 2, \ldots, n\}, \nu)$, ν being counting measure. The Minkowski and Hölder inequalities and the completeness of $L^2(\mathbb{R}, \mu)$ insure that X is a Banach space.

If $g \in l^{4/3}(\{1, 2, \ldots, n\}, \nu, L^2(\mathbb{R}, \mu))$ then $f \mapsto \sum_{k=1}^n (f_k, g_k) \in X^*$. Conversely if $T \in X^*$, then $L^2(\mathbb{R}, \mu) \ni f \mapsto T((0, \ldots, 0, \underset{\substack{\uparrow \\ k\text{th component}}}{f}, 0, \ldots, 0))$ defines a g_k in $L^2(\mathbb{R}, \mu)$ and $T((f_1, f_2, \ldots, f_n)) = \sum_{k=1}^n (f_k, g_k)$. The Minkowski and Hölder inequalities imply $\|T\| = (\sum_{k=1}^n \|g_k\|^{4/3})^{3/4}$. □

421. i) The open mapping theorem implies that T is open and hence there is a positive r such that $T(B(0, r)) \supset B(0, 1)$. Hence there is in $B(0, r)$ a sequence $\{x_n\}_{n=1}^\infty$ such that $T(x_n) = e_n = \{\delta_{nm}\}_{m=1}^\infty$. If $T(x) = \{a_n\}_{n=1}^\infty$ then $\sum_n |a_n| \cdot \|x_n\| < \infty$, $\sum_n a_n x_n$ exists, and $x - \sum_n a_n x_n \in \ker(T)$. Furthermore $\|\sum_n a_n x_n\| \leqq \|\{a_n\}_n\| \,.\, r \leqq \|T\| \,.\, \|x\| \,.\, r$ and so the map $Q: x \mapsto \sum_n a_n x_n$ is continuous. Since $Q(x_n) = x_n$ it follows that $Q^2 = Q$, i.e., that Q is a continuous projection. Then $id - Q = P$ is also a continuous projection. Furthermore $P(X) = \ker(T)$.

ii) Note that $P(X) \cap Q(X) = \{0\}$ and so every x may be expressed uniquely as $u + v$, $u = P(x) \in \ker(T)$, $v = Q(x) \in Q(X)$. Furthermore $Q(X)$ is closed since if $\{y_n\}_{n=1}^\infty \subset Q(X)$ and $y_n \to y$ as $n \to \infty$ then $Q(y_n) = y_n \to Q(y)$ as $n \to \infty$ and so $y = Q(y) \in Q(X)$. Hence $Q(X)$ is a Banach space and if the product topology is used for the direct sum, $\ker(T) \oplus Q(X) = Z$, Z too is a Banach space. The map $Z \ni x, y \mapsto x + y \in X$ is in $\mathrm{Sur}(Z, X)$ and hence is also open. □

422. If $K = 1$ it may be assumed that $y_1 \notin M$ and the Hahn–Banach theorem implies there is in $M^\perp$ a y^* such that $y^*(y_1) = 1$. Then any y in span $(y_1, M) = Y$ is of the form $ay_1 + m$, m in M. If $\{z_n\}_{n=1}^\infty \subset Y$ and $z_n \to z$ as $n \to \infty$ then $z_n = a_n y_1 + m_n$, m_n in M, and $y^*(z_n) = a_n \to y^*(z) = a$, whence $a_n y_1 \to a y_1$ and $m_n = z_n - a_n y_1 \to z - a y_1$ as $n \to \infty$. It then follows that $z = a y_1 + m$, m in M. The result follows by induction. □

15. Miscellaneous Problems

423. If $\langle a, b\rangle$ designates any one of $[a, b)$, $(a, b]$, $(a, b]$, or (a, b) then $\mu_g(\langle a, b\rangle) \leqq \mu_f(\langle a, b\rangle) \,.\, (f(b+0)+f(a-0))$ and so $\mu_g \ll \mu_f$.

If $\mu_f([a, b]) \neq 0$ and if both $f'(a)$ and $g'(a)$ exist then $\mu_g([a, b])/\mu_f([a, b]) = (f(b)+f(a)) \to 2f(a)$ as $b \to a$ and so $d\mu_g/d\mu_f = 2f$. □

424. If $f \in \mathrm{Lip}(1)$ on $[a, b]$ then $f \in AC([a, b])$ and for all x in $[a, b]$, $f(x) = \int_a^x f'(t)\, dt + k$. Since $f' = 0$ a.e., f is constant whereas f is assumed not to be constant. □

425. In the construction of *the* Cantor set C in I the intervals deleted may be enumerated, e.g., they are $\{J_n\}_{n=1}^{\infty}$. In each of $\bar{J}_n$ there may be constructed a Cantor-like null set and the intervals deleted in the process may be denoted $\{J_{nm}\}_{m=1}^{\infty}$. Thus for every finite sequence of integers $n_1, n_2, \ldots, n_k$ there is defined an interval $J_{n_1 n_2 \cdots n_k}$ and on each member of the countable set $\{J_{n_1 n_2 \cdots n_k}\}_{n_i, k \text{ in } \mathbb{N}}$ there may be constructed a Cantor-like null set. Let $f_{n_1 n_2 \cdots n_k}$ be the Cantor-like function on $\bar{J}_{n_1 n_2 \cdots n_k}$ and extend $f_{n_1 n_2 \cdots n_k}$ to a function continuous on I and constant off $\bar{J}_{n_1 n_2 \cdots n_k}$. Enumerate the functions so constructed as $\{g_n\}_{n=1}^{\infty}$ and let g be $\sum_n 2^{-n} g_n$. Then g is monotone increasing, $g' = 0$ a.e. and g is constant on no nondegenerate interval. The function g may be described as the sum of intercalated Cantor-like functions. □

426. i) If $0 < b - a < n^{-1}$ then

$$\begin{aligned} |f_n(b) - f_n(a)| &= n\left|(\lambda([0, n^{-1}] \cap (E - b)) - \lambda([0, n^{-1}] \cap (E - a)))\right| \\ &= n\left|(\lambda([b, b+n^{-1}] \cap E) - \lambda([a, a+n^{-1}] \cap E))\right| \\ &\leqq n \,.\, 2(b-a). \end{aligned}$$

Hence if $c>0$, $a_k<b_k<a_{k+1}$ and $\sum_k (b_k-a_k)<c/3n^2$ then

$$\sum_k |f_n(b_k)-f_n(a_k)| \leqq \frac{2c}{3b}$$

and so each f_n is absolutely continuous.

ii) Since $f_n(x)=\lambda([0,n^{-1}]\cap(E-x))/n^{-1}$, the metric density theorem implies that $f_n \to \chi_E$ a.e. as $n\to\infty$.

iii) Since $\lambda([0,n^{-1}]\cap(E-x)) \leqq n^{-1}$ it follows that $0\leqq f_n \leqq 1$.

iv) Since $E\subset I$ and $f_n=0$ off $[-n^{-1},1]$ it follows that

$$\int_{\mathbb{R}} |f_n(x)-\chi_E(x)|\,dx = \int_{-1}^{1} |f_n(x)-\chi_E(x)|\,dx$$

and the bounded convergence theorem implies $\|f_n-\chi_E\|_1\to 0$ as $n\to\infty$. □

427. Since each f is absolutely continuous, if $[p,q]\subset I$ the Schwarz inequality implies $|f(q)-f(p)|\leqq (q-p)^{1/2}(\int_I |f'(x)|^2\,dx)^{1/2}\leqq (q-p)^{1/2}$. Thus F is an equicontinuous set (if $c>0$ and $|q-p|<c^2$ then for all f in F, $|f(q)-f(p)|<c$). Furthermore $|f(x)-f(0)|\leqq |x|^{1/2}$ and so $|f(0)|-1\leqq |f(0)|-|x|^{1/2}\leqq |f(x)|\leqq |f(0)|+|x|^{1/2}\leqq |f(0)|+1$. Since $\{\|f\|_2\}_{f \text{ in } F}$ is a bounded set the preceding inequalities imply that $\{f(0)\}_{f\text{ in }F}$ is also a bounded set and so F is a uniformly bounded set. The Arzelà–Ascoli theorem implies that the closure of F is compact in $C(I,\mathbb{C})$. □

428. There is a sequence $\{f_{nm}\}_{n,m=1}^{\infty}$ such that each f_{nm} is even and continuous, $0\leqq f_{nm}\leqq 1$,

$$f_{nm}(x)=\begin{cases}1, & \text{if } |x-a|<1/m \text{ or } |x+a|<1/m\\ 0, & \text{if } |x-a|>2/m \text{ or } |x+a|>2/m'\end{cases}$$

and $f_{nm}\downarrow \chi_{[-a-1/m,-a+1/m]}+\chi_{[a-1/m,a+1/m]}$ as $n\to\infty$. Then

$$0=\lim_{n\to\infty}\int_{-1}^{1} f_{nm}(x)g(x)\,dx = \int_{-a-1/m}^{-a+1/m} + \int_{a-1/m}^{a+1/m} g(x)\,dx.$$

Since g is continuous at a and $-a$, for large m,

$$\left|m\left(\int_{-a-1/m}^{-a+1/m} + \int_{a-1/m}^{a+1/m} g(x)\,dx-(g(-a)+g(a))\right|=|g(-a)+g(a)|$$

is small and the result follows. □

429. If $[p,q]\subset[c,d]$ and $r=f^{-1}(p)$, $s=f^{-1}(q)$ then $\mu([p,q])=\int_{f^{-1}([p,q])} f'(x)\,dx=f(s)-f(r)=q-p=\lambda([p,q])$. In other words the countably additive measure μ coincides with λ on all intervals and the result follows. □

430. Since E is a null set there is a sequence $\{U_n\}_{n=1}^{\infty}$ of open sets such that $\bigcap_n U_n = G_\delta \supset E$ and $\lambda(G_\delta)=0$. It may be assumed that $\sum_n \lambda(U_n)=$

$M < \infty$ and that $U_n \supset U_{n+1}$. Let g be

$$x \mapsto \begin{cases} \sum_n \chi_{U_n}, & \text{if } x \notin G_\delta \\ 0, & \text{otherwise} \end{cases}.$$

Then g is finite and nonnegative and if f is $x \mapsto \int_0^x g(t)\,dt$ then $f(x) \leqq \sum_n \lambda(U_n \cap [0, x]) \leqq M < \infty$. Furthermore f is monotone increasing, absolutely continuous, and $f' = g$ a.e. If D is the null set off of which $f' = g$ then $D \cup G_\delta$ is a null set and, *a fortiori*, $f' = g$ off $D \cup G_\delta$. Hence $F = I \backslash (D \cup G_\delta)$ is dense in I and if $x \in E$ there is in F a sequence $\{x_n\}_{n=1}^\infty$ such that $x_n \to x$ as $n \to \infty$. It may be assumed that $x_n \in U_n \backslash \{0\}$, whence $f'(x_n) = g(x_n) \geqq n$, and even that $x_n \uparrow x$, whence if $c_n > 0$ there is in (x_n, x) a y_n such that $(f(y_n) - f(x_n))/(y_n - x_n) > n - \lambda(U_n)$ and $(x - x_n)/(y_n - x_n) < 1 + c_n$. Thus

$$n - \lambda(U_n) < \frac{f(y_n) - f(x_n)}{y_n - x_n} \leqq \frac{f(x) - f(x_n)}{x - x_n} \cdot \frac{x - x_n}{y_n - x_n}$$
$$< \frac{f(x) - f(x_n)}{x - x_n} + c_n \frac{f(x) - f(x_n)}{x - x_n}$$

and so $(f(x) - f(x_n))/(x - x_n) > n - \lambda(U_n) - c_n(f(x) - f(x_n))/(x - x_n)$. In particular, if $c_n = n^{-1}(x - x_n)/(f(x) - f(x_n))$ then $(f(x) - f(x_n))/(x - x_n) \geqq n - \lambda(U_n) - n^{-1} \to \infty$ as $n \to \infty$. Similarly if $x_n \downarrow x$, $(f(x) - f(x_n))/(x - x_n) \to \infty$ as $n \to \infty$ and the result follows. □

431. If f is strictly increasing and if $x > y$ then $(f(x) - f(y))/(x - y) > 0$ and so $f'(y) \geqq 0$ for all y. If $D = \{x : f'(x) = 0\}$ is not totally disconnected then D contains a nondegenerate closed interval $[a, b]$. If $a < c < d < b$ then according to Rolle's theorem $f(d) - f(c) = 0$ and then f is not strictly increasing.

Conversely, if $f' \geqq 0$ then Rolle's theorem implies f is monotone increasing. If $x < y$ and $f(x) = f(y)$ then for all z in (x, y) $f(x) \leqq f(z) \leqq f(y)$ and so $(x, y) \subset D$ and D is not totally disconnected. Hence if D is totally disconnected then f is strictly increasing. □

432. The map

$$f : x \mapsto \begin{cases} 0, & \text{if } 0 \leqq x < 1 \\ 1, & \text{if } x = 1 \end{cases}$$

has the stated properties. □

433. The map

$$f : x \mapsto \begin{cases} (x - 1)\sin(1/(x - 1)), & \text{if } 0 \leqq x < 1 \\ 0, & \text{if } x = 1 \end{cases}$$

has the stated properties. □

434. If f is g_C, *the* Cantor function for *the* Cantor set in I, then f is monotone and thus of bounded variation. Furthermore $f(I)=I$ and $\lambda(f(I\backslash C))=0$ whence $\lambda(f(C))=1$. □

435. If f is absolutely continuous then i) f is continuous, and ii) f is of bounded variation. Furthermore if E is a null set and $a>0$ there is a sequence $\{J_n\}_{n=1}^{\infty}$ of disjoint open intervals such that $\bigcup_n J_n \supset E$ and $\sum_n \lambda(J_n)<a$. Hence if $b>0$ and if $J_n=(a_n, b_n)$ then a can be chosen so that $\sum_n |f(b_n)-f(a_n)|<b$. For each n there are in $[a_n, b_n]$ points c_n, d_n such that $f(c_n)=\min\{f(x): x\in[a_n, b_n]\}$ and $f(d_n)=\max\{f(x): x\in[a_n, b_n]\}$. Then $\sum_n |d_n-c_n|<a$ and $\sum_n |f(d_n)-f(c_n)|=\sum_n \lambda(f(J_n))<b$ and so $\lambda(f(E))\leqq \sum_n \lambda(f(J_n))<b$, i.e., $\lambda(f(E))=0$ and iii) follows.

Conversely, if i), ii), and iii) hold then the interval function $\mu:[a,b]\mapsto f(b)-f(a)$ may be extended to a signed Borel measure such that $|\mu|([a,b])=T_f([a,b])$. The hypotheses imply $\mu \ll \lambda$ and so there is a measurable function g such that $\mu([a,b])=\int_a^b g(t)\,dt=f(b)-f(a)$. Hence $f'=g$ a.e., $\int_0^x f'(t)\,dt=f(x)-f(0)$ and so f is absolutely continuous. □

436. If i) is dropped then $\chi_{[0,1/2]}$ satisfies ii) and iii) and is not (absolutely) continuous. If ii) is dropped then

$$f:x\mapsto\begin{cases} x\sin(1/x), & \text{if } x\neq 0\\ 0, & \text{if } x=0\end{cases}$$

is absolutely continuous on $[a,1]$ if $a\in(0,1)$ and so i) and iii) hold, ii) does not and f is not absolutely continuous. If iii) is dropped, *the* Cantor function g_C for *the* Cantor set C satisfies i) and ii) but $\lambda(g_C(C))=1$ and g_C is not absolutely continuous. □

437. If A is Lebesgue measurable there is a Borel set B and a null set N such that $B\cap N=\varnothing$ and $A=B\cup N$. Then $f(A)=f(B)\cup f(N)$. From Problem 435 it follows that $f(N)$ is a null set. Since $f(B)$ is analytic (see Problems 508–517) and analytic sets are Lebesgue measurable it follows that $f(A)$ is Lebesgue measurable.

An alternative proof using less sophisticated tools is the following.

There are sequences $\{K_n\}_{n=1}^{\infty}$ resp. $\{V_n\}_{n=1}^{\infty}$ of compact resp. open sets such that $\cdots K_n\subset K_{n+1}\subset\cdots\subset A\subset\cdots\subset V_{n+1}\subset V_N\cdots$ and such that if $F_\sigma=\bigcup_n K_n$ and $G_\delta=\bigcap_n V_n$ then $\lambda(F_\sigma)=\lambda(A)=\lambda(G_\delta)$. The sets $f(K_n)$ are compact and thus are Borel sets. Each V_n is the countable union of pairwise disjoint open intervals: $V_n=\bigcup_n J_{nm}$ and $f(V_n)=\bigcup_n f(J_{nm})$. Furthermore, $f(J_{nm})$ is a connected set, hence some sort of interval and hence a Borel set. Consequently $f(V_n)$ is a Borel set. Since $f(K_n)\subset f(A)\subset f(V_n)$ it suffices to show that $\lambda(f(V_n)\backslash f(K_n))\to 0$ as $n\to\infty$. But $V_n\backslash K_n$ is an open set W_n and so is the countable union of pairwise disjoint open intervals: $W_n=\bigcup_n I_{nm}$. Since $\lambda(W_n)\to 0$ as $n\to\infty$ the argument for iii) in Solution 435 shows $\lambda(f(W_n))\to 0$ as $n\to\infty$. Finally, $f(W_n)\supset f(V_n)\backslash f(K_n)$ and the result follows.

□

438. By hypothesis there is a partition P_0 such that $T_{fP_0} > W + \frac{1}{2}(V - W)$. Let P_0 consist of $[0, x_1), [x_1, x_2), \ldots, [x_N, 1]$ and let c be positive and such that if $|x - x_n| < c$ then $|f(x) - f(x_n)| < (V - W)/8N$, $n = 1, 2, \ldots, N$. If $|P| < \min(c, |P_0|)$ and if P_1 is the partition arising from the use of all partition points of P and P_0, then $T_{fP_1} > W + \frac{1}{2}(V - W)$ and $T_{fP} > W + \frac{1}{2}(V - W) - N \,.\, 2 \,.\, (V - W)/8N = W + \frac{1}{4}(V - W)$.

If $f = \chi_{[1/2, 1]}$ and $P_n = \{[k/2^n, (k+1)/2^n),\ 0 \leqq k \leqq 2^n - 2, \text{ and } [1 - 1/2^n, 1]\}$, $|P_n| = 2^{-n}$, $T_{fP_n} = 0$. □

439. Since f is of bounded variation all its (at most countably many) discontinuities are jumps. Since $f = g'$ Darboux's theorem implies that f enjoys the intermediate value problem, *viz.*, if $a < b$ and if γ is between $f(a)$ and $f(b)$ there is in (a, b) a c such that $f(c) = \gamma$. Thus f can have no jump discontinuities and so is continuous. □

440. Let J_{kn} be $[k, 2^{-n}, (k+1)2^{-n}]$, n in $\mathbb{N}$, $k = 0, 1, \ldots, 2^n - 1$. If $m_{kn} = \inf\{f(x)\colon x \text{ in } J_{kn}\}$, $M_{kn} = \sup\{f(x)\colon x \text{ in } J_{kn}\}$, and if g_{kn} is

$$y \mapsto \begin{cases} 1, & \text{if } m_{kn} \leqq y \leqq M_{kn} \\ 0, & \text{otherwise} \end{cases}$$

then $G_n = \sum_k g_{kn}$ counts the number of J_{kn} on which the equation $f(x) = y$ has at least one solution. Each G_n is measurable and if $m = \inf\{f(x)\colon x \text{ in } I\}$ and $M = \sup\{f(x)\colon x \text{ in } I\}$ then $\int_m^M G_n(y)\,dy = \sum_k \int_m^M g_{kn}(y)\,dy$. However $\int_m^M g_{kn}(y)\,dy = M_{kn} - m_{kn}$ is the oscillation $\operatorname{osc}(f, J_{kn})$ of f on J_{kn} whence $\int_m^M G_n(y)\,dy = \sum_k \operatorname{osc}(f, J_{kn})$. If P_n is the partition corresponding to the endpoints of the intervals J_{kn} then, since f is continuous, for large n, $\sum_k \operatorname{osc}(f, J_{kn})$ is near T_{fP_n}. On the other hand, $G_n \leqq G_{n+1}$ and so if $G = \lim_{n\to\infty} G_n$ then $\int_m^M G(y)\,dy = \lim_{n\to\infty} T_{fP_n} = T_f$ (see Problem 438).

Since $G_n(y) \leqq M(y)$ it follows that $G(y) \leqq M(y)$. If $p \in \mathbb{N}$ and $p \leqq M(y)$ then for some n, $G_n(y) \geqq p$ and since M is $\mathbb{N}$-valued it follows that $G = M$ as required. □

441. If E and F are measurable sets, $f = \chi_E$, and $g = \chi_F$ then

$$A_y = \begin{cases} \mathbb{R}, & \text{if } y \leqq 0 \\ E, & \text{if } 0 < y \leqq 1 \\ \varnothing, & \text{if } 1 < y \end{cases}$$

and $\int_{\mathbb{R}} f(x)g(x)\,dx = \lambda(E \cap F)$ whereas

$$h(y) = \int_{A_y} g(x)\,dx = \begin{cases} \lambda(F), & \text{if } y \leqq 0 \\ \lambda(E \cap F), & \text{if } 0 < y \leqq 1. \\ 0, & \text{if } 1 < y \end{cases}$$

Thus $\int_0^\infty h(y)\,dy = \int_0^1 h(y)\,dy = \lambda(E \cap F) = \int_{\mathbb{R}} f(x)g(x)\,dx$.

If $\{E_i\}_{i=1}^m$ is a set of pairwise disjoint measurable sets and similarly if $\{F_j\}_{j=1}^n$ is a set of pairwise disjoint measurable sets and if $0 < a_1 < a_2 < \cdots <$

a_m and $0<b_1<b_2<\cdots<b_n$ let f be $\sum_i a_i\chi_{E_i}$ and g be $\sum_j b_j\chi_{F_j}$. Then $fg = \sum_{i,j} a_ib_j\chi_{E_i\cap F_j}$ and $\int_{\mathbb{R}} f(x)g(x)\,dx = \sum_{i,j} a_ib_j\lambda(E_i\cap F_j)$. On the other hand

$$A_y = \begin{cases} \mathbb{R}, & \text{if } y \leqq 0 \\ E_1\cup\cdots\cup E_m, & \text{if } 0<y\leqq a_1 \\ E_1\cup\cdots\cup E_{m-1}, & \text{if } a_1<y\leqq a_2 \\ \quad\cdots & \\ E_1, & \text{if } a_{m-1}<y\leqq a_m \\ \varnothing, & \text{if } a_m<y \end{cases}$$

and

$$h(y) = \begin{cases} \sum_{j=1}^{n} b_j\lambda(F_j), & \text{if } y\leqq 0 \\ \sum_{i,j=1}^{m,n} b_j\lambda(E_i\cap F_j), & \text{if } 0<y\leqq a_1 \\ \sum_{i,j=1}^{m-1.n} b_j\lambda(E_i\cap F_j), & \text{if } a_1<y\leqq a_2. \\ \quad\cdots & \\ 0, & \text{if } a_m<y \end{cases}$$

Abel summation (integration by parts with respect to counting measure) leads to the equation $\int_0^\infty h(y)\,dy = \int_{\mathbb{R}} f(x)g(x)\,dx$.

If f, g are arbitrary nonnegative measurable functions and if $E_{km} = \{x : k.2^{-m} \leqq f(x) < (k+1).2^{-m}\}$ and $F_{ln} = \{x : l.2^{-n} \leqq g(x) < (l+1)2^{-n}\}$ then $f_m = \sum_{k=-m.2^m}^{m.2m} k.2^{-m}\chi_{E_{km}} \uparrow f$ as $m\to\infty$ and $g_n = \sum_{l=-n.2^n}^{n.2^n} l.2^{-n}\chi_{F_{ln}} \uparrow g$ as $n\to\infty$. If $A_{ym} = \{x : f_m(x) \geqq y\}$ and h_m is $y \mapsto \int_{A_{ym}} g_m(x)\,dx$ then

$$\int_{\mathbb{R}} f_m(x)g_m(x)\,dx = \int_0^\infty h_m(y)\,dy$$

and

$$\int_{\mathbb{R}} f(x)g(x)\,dx = \lim_{m\to\infty}\int_{\mathbb{R}} f_m(x)g_m(x)\,dx = \lim_{m\to\infty}\int_0^\infty h_m(y)\,dy = \int_0^\infty h(y)\,dy.$$

□

442. If $f\in \mathrm{Lip}(1)$ on I and f is real-valued then $f\in AC(I)$ and hence there is a g such that for all x, y in I, $f(x)-f(y) = \int_x^y g(t)\,dt$. If $g\notin L^\infty(I)$ there are Lebesgue measurable sets E_n such that $\lambda(E_n)>0$ and $g\geqq n$ on E_n or $g\leqq -n$ on E_n, say $g\geqq n$ on E_n. Then $n\lambda(E_n) \leqq \int_{E_n} g(x)\,dx$. If $a>0$ there is an open set U_n containing E_n, $\lambda(U_n)<\lambda(E_n)(1+a)$, and $\int_{U_n} g(x) \geqq n\lambda(E_n)(1-a)$. Since U_n is the countable union of disjoint open intervals, if K is the Lipschitz constant for f ($|f(x)-f(y)|\leqq K|x-y|$) then $\int_{U_n} g(x)\,dx \leqq K\lambda(U_n)$ and so $n\lambda(E_n)(1-a) \leqq K\lambda(U_n) \leqq K\lambda(E_n)(1+a)$ or $n(1-a)<K(1+a)$, whence $n\leqq K$, a contradiction. If $f = u+iv$, u, v real-valued, the argument applied to u and v yields a contradiction.

Conversely, if $g \in L^\infty(I)$ and $f(x) = \int_0^x g(t)\,dt$ then $\|g\|_\infty$ serves as a Lipschitz constant for f. □

443. A complex measure is necessarily bounded whereas if $f \in C(I, \mathbb{R}) \backslash BV(I, \mathbb{R})$ the putative measure μ is not bounded. □

444. It may be assumed that $f \geqq 0$. If E_{km} is the set defined in the first sentence of the last paragraph of Solution 441, and f_m is the corresponding function then $f = \lim_{m\to\infty} f_m = f_1 + \sum_{n=2}^{\infty} (f_n - f_{n-1})$ and the result follows. □

445. It may be assumed that $f \geqq 0$. If E is a measurable set and $\lambda(E) < \infty$ there are compact sets K_n and open sets U_n such that $\cdots K_n \subset K_{n+1} \subset \cdots \subset E \subset \cdots \subset U_{n+1} \subset U_n \subset \cdots$ and such that $\lambda(K_n) \uparrow \lambda(E)$, $\lambda(U_n) \downarrow \lambda(E)$. There are in $C_{00}(\mathbb{R}, \mathbb{C})$ functions g_n such that $0 \leqq g_n \leqq 1$, $g_n = 1$ on K_n and $g_n = 0$ off U_n. Thus $g_n \uparrow \chi_E$ a.e. as $n \to \infty$ and $\chi_E - (g_1 + \sum_{n\geqq 2} (g_n - g_{n-1})) = 0$ a.e. Combined with Solution 444 the argument just given yields the result. □

446. If u_γ, v_γ are the real and imaginary parts of f_γ, then $|\int_E u_\gamma(x)\,d\mu(x)|$, $|\int_E v_\gamma(x)\,d\mu(x)| \leqq |\int_E f_\gamma(x)\,d\mu(x)|$ and so the discussion may be confined to that of $\mathbb{R}$-valued functions. It will be shown that $\{f_\gamma^+\}_\gamma$ is a uniformly integrable set.

Indeed, if there is a positive a and if there is a sequence $\{E_n\}_{n=1}^\infty$ of measurable sets such that $\mu(E_n) < 1/n$ and there are f_{γ_n} such that for all n $|\int_{E_n} f_{\gamma_n}^+(x)\,d\mu(x)| \geqq a$ let F_n be $\{x : f_{\gamma_n}^+(x) \neq 0\}$ and let G_n be $F_n \cap E_n$. Then $\mu(G_n) < 1/n$ and $|\int_{G_n} f_{\gamma_n}(x)\,d\mu(x)| = |\int_{G_n} f_{\gamma_n}(x)\,d\mu(x)| = |\int_{E_n} f_{\gamma_n}^+(x)\,d\mu(x)| \geqq a$, a contradiction. Thus $\{f_\gamma^+\}_\gamma$ and similarly $\{f_\gamma^-\}_\gamma$ are uniformly integrable whence $\{|f_\gamma|\}_\gamma$ is also uniformly integrable. □

447. The Schwarz inequality shows that $L^2(I, \mathbb{C}) \subset L^1(I, \mathbb{C})$ and that if $f \in L^2(I, \mathbb{C})$ then $|\int_a^b f(x)\,dx|^2 \leqq |b-a| \int_a^b |f(x)|^2\,dx$. If g is $x \mapsto \int_0^x |f(t)|^2\,dt$ then $|\int_a^b f(x)\,dx|^2 \leqq (g(b) - g(a))|b-a|$.

Conversely if $f \in L^1(I, \mathbb{C})$, g is such that $|\int_a^b f(x)\,dx|^2 \leqq (g(b)-g(a))|b-a|$, and g is monotone increasing it follows that if $a < b$ then

$$\left((b-a)^{-1}\int_a^b f(x)\,dx\right) \cdot \left((b-a)^{-1}\int_a^b \overline{f(y)}\,dy\right) \leqq (g(b)-g(a))/(b-a).$$

For all a off a null set the limits on both sides exist as $b \to a$, i.e., $|f(a)|^2 \leqq g'(a)$. Furthermore $\int_0^1 |f(a)|^2\,da \leqq \int_0^1 g'(a)\,da \leqq g(1) - g(0) < \infty$ and so $f \in L^2(I, \mathbb{C})$. □

448. The Schwarz inequality implies that $\int_0^1 |\int_0^x g(t)\,dt|^2\,dx \leqq \int_0^1 (\int_0^x 1^2\,dt)(\int_0^x |g(t)|^2\,dt)\,dx \leqq \int_0^1 x\|g\|_2^2\,dx = \frac{1}{2}\|g\|_2^2$ and the result follows. □

449. By definition $\int_I f(x)\,dx = 3^{-1}(1 + 2\,.\,(2/3) + 3\,.\,(2/3)^2 + \cdots)$. If $|x| < 1$, $\sum_{n=0}^\infty x^n = (1-x)^{-1}$ and $((1-x)^{-1})' = (1-x)^{-2} = \sum_{n=1}^\infty nx^{n-1}$. Thus

$$\left(1 - \frac{2}{3}\right)^{-2} = \sum_{n=1}^{\infty} n\left(\frac{2}{3}\right)^{n-1} = 9$$

and $\int_I f(x)\,dx = 3$. □

450. If B is a fixed Borel set let μ_B be $\mathsf{S}_\beta(\mathbb{R}) \ni A \mapsto \mu(A \times B)$. Then $\mu_B \ll \nu$ and thus the *LRN* theorem implies there is in $L^1(I, \nu)$ an f_B such that for all Borel sets A, $\mu_B(A) = \int_A f_B(x)\, d\nu(x)$. For x fixed the map $\mathsf{S}_\beta(\mathbb{R}) \ni B \mapsto f_B(x)$ has the required properties. □

451. Let $\{r_n\}_{n=1}^{\infty}$ be an enumeration of $\mathbb{Q} \cap I$, let E_m be

$$\bigcup_n (r_n - 2^{-(n+m)}, r_n + 2^{-(n+m)}),$$

and let E be $\bigcap_{m=1}^{\infty} E_m$. Then $\lambda(E_m) \leqq 2^{-(n-1)}$ and $\lambda(E) = 0$. Furthermore E is a G_δ and $I \backslash E$ is an F_σ. Since $\text{cont}(f)$ is a G_δ for any function f and since $\lambda(I \backslash \text{cont}(f)) = 0$ if f is Riemann integrable, it follows that $\text{cont}(f)$ is a dense G_δ if f is Riemann integrable. The Baire category theorem implies $\text{cont}(f)$ is of the second category. However $I \backslash E_m$ is nowhere dense and $I \backslash E = \bigcup_{m=1}^{\infty} (I \backslash E_m)$ is of the first category. Thus $\text{cont}(f) \not\subset I \backslash E$, i.e., $\text{cont}(f) \cap E \neq \varnothing$ if f is Riemann integrable. □

452. If $\|p_n - f\|_\infty \to 0$ as $n \to \infty$ then $\{p_n\}_n$ is a Cauchy sequence and so there is an n_0 such that $\|p_n - p_m\|_\infty < 1$ if $n, m \geqq n_0$. Hence the polynomials $p_n - p_m$ are bounded and must be constant if $n, m \geqq n_0$, i.e., there are constants a_m such that $p_m = p_{n_0} + a_m$. Hence $\lim_{m\to\infty} a_m = a$ exists and $f = p_{n_0} + a$, a polynomial. □

453. If f_n is $x \mapsto \sum_{k=1}^{n} k^{-2} |x - r_k|^{-1/2}$ then $\{f_n\}_n$ is an increasing sequence of nonnegative functions and $\int_I f_n(x)\, dx = \sum_{k=1}^{n} 2 \,.\, k^{-2}(r_k^{1/2} + (1 - r_k)^{1/2})$. The monotone convergence theorem shows that $\int_I f(x)\, dx \leqq 8 \sum_k k^{-2} < \infty$ and so $f < \infty$ a.e. □

454. If $t, s \in A$ and if $u = t - s$ then $\lim_{n\to\infty} e^{ic_n u}$ exists. Since $\lambda(A) > 0$ there is in $(0, 1)$ an a such that $A - A \supset (-a, a)$ and so for all u in $(-a, a)$, $g(u) = \lim_{n\to\infty} e^{ic_n u}$ exists.

If $\{c_n\}_n$ is unbounded it may be assumed that $|c_n| \to \infty$ as $n \to \infty$. Then if $[p, q] \subset (-a, a)$, $\int_p^q e^{ic_n u}\, du = (e^{ic_n q} - e^{ic_n p})/ic_n$ approaches both zero (since $|c_n| \to \infty$) and $\int_p^q g(u)\, du$ (by virtue of the bounded convergence theorem) as $n \to \infty$. Thus $g = 0$ a.e. whereas $|e^{ic_n u}| = 1$ and so $|g(u)| = 1$ for all u. Hence the sequence $\{c_n\}_n$ is bounded, say $|c_n| \leqq M < \infty$.

If $\lim_{n\to\infty} c_n$ does not exist it may be assumed that for some b in $(0, \pi/M)$ and for all pairs n, m of different indices, $|c_n - c_m| > b$. If $\min(\frac{1}{2}a, 5\pi/12M) < |u| < \min(a, 5\pi/6M)$ then $0 < d = \min(ba/4, b5\pi/24M) < |c_n - c_m| \,.\, |u|/2 < 5\pi/6$ and so $0 < \sin d < \sin \frac{1}{2}(c_n - c_m)u < \frac{1}{2}$ whereas $|e^{ic_n u} - e^{ic_m u}| = |2i\, e^{i(1/2)(c_n + c_m)u} \sin \frac{1}{2}(c_n - c_m)u| \geqq 2 \sin d$, a contradiction. □

455. Hölder's inequality implies that

$$\left(\int_a^b |f(x)|\, dx\right)^p \leqq \left(\int_a^b 1^q\, dx\right)^{p/q} \left(\int_a^b |f(x)|^p\, dx\right) = (b-a)^{p-1} \int_a^b |f(x)|^p\, dx. \square$$

456. For all a off a null set, $\lim_{b\to a, b\neq a}(b-a)^{-1}\int_a^b f(x)\,dx = f(a)$ and $\lim_{b\to a, b\neq a}(b-a)^{-1}\int_a^b |f(x)|^p\,dx = |f(a)|^p$. Thus

$$\left|(b-a)^{-1}\int_a^b |f(x)|\,dx\right|^p \leqq c|b-a|^{p-1}\int_a^b |f(x)|^p\,dx/|b-a|^p$$

$$= c\left|(b-a)^{-1}\int_a^b |f(x)|^p\,dx\right|$$

and so $|f(a)|^p \leqq c|f(a)|^p$. Since $0<c<1$, $f=0$ a.e. □

457. (The following solution was found by Harvey Diamond and Gregory Gellès.) If C is *the* Cantor set then $C-C=[-1, 1]$ (see [9]). If $x\in\mathbb{R}$ there is in $\mathbb{Q}\backslash\{0\}$ an r such that $rx\in(-1, 1)$ and there are in C points c_1, c_2 such that $c_1-c_2=rx$. Zorn's lemma implies that among all subsets of C there is at least one that is linearly independent over $\mathbb{Q}$ and properly contained in no other linearly independent subset of C. Thus if H is such a maximal linearly independent subset of C then H is a Hamel basis for $\mathbb{R}$ over $\mathbb{Q}$. Furthermore $\lambda(H)=0$.

Let S be $\bigcup_{r\in\mathbb{Q}} rH$. Then $\lambda(S)=0$ and $S=-S$. Let S_1 be S and S_{n+1} *be* $S_n+S_n=S_n-S_n$. Note that if $x\in S_n$ and $x=\sum_{h\in H} a_h\,.\,h$ $(a_h\in\mathbb{Q})$ then the number of nonzero coefficients among the a_h is not more than 2^{n-1}. Since $\bigcup_n S_n=\mathbb{R}$ if each S_n is measurable then for some n_0, $\lambda(S_{n_0})>0$. Hence $S_{n_0}-S_{n_0}=S_{n_0}+S_{n_0}$ contains a nonempty open set U containing $\{0\}$. If $(-2^{-M}, 2^{-M})\subset U$, if $\{r_k\}_k\subset\mathbb{Q}$, if $2^{n_0}<\text{card}(\{r_k\}_k)=K<\text{card}(N)$, and if $K\,.\,\sup_k|r_k|<2^{-M}$ then whenever $\{h_k\}$ is a K-element set in H,

$$\sum_{k=1}^{K} r_k h_k \in U\backslash S_{n_0+1},$$

a contradiction.

Thus some S_n is nonmeasurable and if $n_1=\inf\{n: S_n \text{ is nonmeasurable}\}$ then $n_1>1$, S_{n_1-1} is measurable and may serve for the E required. □

458. If $\text{card}(D)\leqq\text{card}(\mathbb{N})$ then $\text{card}(A\backslash D)>\text{card}(\mathbb{N})$ and thus Problem 19 implies there is in $A\backslash D$ an x of the type described in Problem 19 for A. i.e., $x\in D$ and $x\in A\backslash D$, a contradiction.

Choose z_0 in D. Then if $U(z_0)$ is a neighborhood of z_0, $\text{card}(U(z_0)\cap D)>\text{card}(\mathbb{N})$. If $U(z_0)=(z_0-a, z_0+a)$, $0<a<1$, and if $\text{card}((z_0, z_0+a)\cap D)>\text{card}(\mathbb{N})$ let y_0 be z_0. Otherwise $\text{card}(z_0-a, z_0)\cap D)>\text{card}(\mathbb{N})$. In that case choose z_1 in $(z_0-a, z_0)\cap D$. If for all n in $\mathbb{N}$, $\text{card}((z_1, z_1+n^{-1})\cap D)\leqq\text{card}(\mathbb{N})$, a contradiction results. Thus (z_1, z_0+a) is a neighborhood of z_0 and $\text{card}((z_1, z_0+a)\cap D)>\text{card}(\mathbb{N})$ and if $y_0=z_1$ the result follows. □

459. Let m be a strict (local) maximum value of f and let A_m be $\{x: x \text{ is a strict local maximum}\}\cap f^{-1}(m)$. By virtue of Problem 19 if card $(A_m)>\text{card}(\mathbb{N})$ there is in A_m an x of the type described in Problem 19. But then x cannot be a strict local maximum since in every neighborhood $U(x)$ of x there are other elements of A_m.

Let M be the set of strict (local) maximum values of f. If $\text{card}(M) > \text{card}(\mathbb{N})$ the arguments in Solutions 19 and 458 imply there is in M a y_0 such that for every positive a, $\text{card}((y_0, y_0+a)\cap M) > \text{card}(\mathbb{N})$. Thus $\text{card}(f^{-1}((y_0, y_0+a)\cap M)) > \text{card}(\mathbb{N})$ and there is in

$$T = f^{-1}((y_0, y_0+a)\cap M)$$

an x_0 such that for every neighborhood $U(x_0)$ of x_0, $\text{card}(U(x_0)\cap T) > \text{card}(\mathbb{N})$. Since $\text{card}(A_{y_0}) \leqq \text{card}(\mathbb{N})$ there is in $U(x_0)\cap T$ an x_1 such that $f(x_1) \neq y_0$ whence $f(x_1) > y_0$ and thus y_0 is not a strict maximum value. Hence $\text{card}(M) \leqq \text{card}(\mathbb{N})$. If $M = \{z_p\}_{p=1}^{\infty}$ and $A_{z_p} = \{w_{pq}\}_{q=1}^{\infty}$ then $f^{-1}(M) = \{w_{pq}\}_{p,q=1}^{\infty}$ and so $\text{card}(f^{-1}(M)) \leqq \text{card}(\mathbb{N})$. □

460. The set A is a subgroup of the additive group $\mathbb{R}$. The image $A/\mathbb{Z}$ in $\mathbb{R}/\mathbb{Z}$ is a finite or countable subgroup of $\mathbb{R}/\mathbb{Z}$ which may be regarded as $[0, 1)$ under the group operation of "addition modulo one". If $A/\mathbb{Z}$ is finite and if $b = \min(A/\mathbb{Z}\setminus\{0\})$ then $1/b = k \in \mathbb{N}$ and for all m, n in $\mathbb{Z}$ there is in $\mathbb{Z}$ a p and in $\mathbb{N}\cap[0, k-1]$ a q such that $m + na = p + q/k$ and so, in particular $a \in \mathbb{Q}$, a contradiction. Hence $A/\mathbb{Z}$ is (countably) infinite and hence dense in $\mathbb{R}/\mathbb{Z}$ and thus A is dense in $\mathbb{R}$. □

461. If $a < b$ there is a sequence of pairwise disjoint "dyadic" intervals $\{[a_n, b_n)\}_{n=1}^{\infty}$ such that for some p_n, m_n in $\mathbb{N}$, $a_n = 2^{-m_n}\cdot p_n$, $b_n = 2^{-m_n}(p_n+1)$ and $(a, b) = \bigcup_n [a_n, b_n)$. Furthermore there is an n_0 such that if $a_* = \inf_{n\leqq n_0} a_n$ and $b^* = \sup_{n\leqq n_0} b_n$ then $|f(a)-f(a_*)|$, $|f(b)-f(b^*)| < b-a$. Thus

$$|f(b)-f(a)| \leqq |f(b)-f(b^*)| + \sum_{n=1}^{n_0} |f(b_n)-f(a_n)| + |f(a_*)-f(a)|$$

$$< 2(b-a) + \sum_{n=1}^{n_0} 2^{-m_n}\cdot M < (M+2)(b-a).$$

Hence $f \in \text{Lip}(1)$ on $\mathbb{R}$ with Lipschitz constant $M+2$. In particular $f \in AC(\mathbb{R})$; for all x,

$$f(x+2^{-n}) - f(x) = \int_x^{x+2^{-n}} f'(t)\,dt$$

and $\Delta_n(x) = \int_x^{x+2^{-n}} f'(t)\,dt/2^{-n} \to 0$ as $n \to \infty$, whence $f' = 0$ a.e. From Problem 424 it follows that f is constant. □

462. The function $\chi_{\mathbb{Q}}$ is nowhere continuous and yet $\chi_{\mathbb{Q}} = 0$ a.e. The function $\chi_{[0,\infty)}$ is continuous a.e., yet for no continuous function f is it true that $\chi_{[0,\infty)} = f$ a.e. □

463. Since $f \in C^1([0, \pi], \mathbb{C})$ and $f(0) = 0$ there is in $C^1([-\pi, \pi], \mathbb{C})$ an F such that $F(x) = f(x)$ on $[0, \pi]$ and $F(x) = -f(-x)$ on $[-\pi, 0]$ and if $b_n = \pi^{-1/2}\int_{-\pi}^{\pi} F(x)\sin nx\,dx$, n in $\mathbb{N}$, then the Fourier series for $F(x)$ is

$\sum_n b_n\pi^{-1/2}\sin nx$. Furthermore, integration by parts shows

$$b_n = \pi^{-1/2}\Big(\int_{-\pi}^{0} f'(-x)\cos nx\,dx + \int_0^{\pi} f'(x)\cos nx\,dx\Big)/n.$$

Hence $|b_n| \leqq 2^{1/2}\|f'\|_2/n$ (Schwarz) and $\sum_n |b_n|^2 = \|F\|_2^2 = 2\int_0^\pi |f(x)|^2\,dx = 2\|f\|_2^2 \leqq 2\sum_n n^{-2}\|f'\|_2^2$ as required. □

464. Taylor's formula shows

$$f(x) = f(0) + f'(0)x + \cdots + f^{(n)}(0)x^n/n! + x^{n+1}\int_0^1 (1-t)^n f^{(n+1)}(tx)\,dt/n!.$$

If $R_n(x)$ denotes the last term above then since $f^{(n+2)} \geqq 0$, $x \mapsto f^{(n+1)}(tx)$ is monotone increasing for each t and so if $0 \leqq x < c$

$$\begin{aligned} 0 \leqq R_n(x) &\leqq \frac{x^{n+1}\int_0^1 (1-t)^n f^{(n+1)}(tc)\,dt}{n!} \\ &= \frac{x^{n+1}(f(0) - f'(0)c - \cdots - f^{(n)}(0)c^n/n!)}{c^{n+1}} \\ &\leqq f(0)x^{n+1}/c^{n+1}. \end{aligned}$$

Thus $R_n(x) \to 0$ for all x as $n \to \infty$ and so f is (real) analytic. □

465. There is in $(0, 1)$ an a such that if $0 < x < a$ then $|x^{-1}f(x)| \leqq 2(|f'(0)| + 1) = K$. Hence $|x^{-3/2}f(x)| \leqq Kx^{-1/2}$ if $0 < x < a$. In $[a, 1]$, $|f(x)| \leqq \|f\|_\infty$, $x^{-3/2} \leqq a^{-3/2}$ and so if g is $x \mapsto x^{-3/2}f(x)$ then $\int_I |g(x)|\,dx \leqq K\int_0^a x^{-1/2}\,dx + a^{-3/2}(1-a)\|f\|_\infty < \infty$ as required. □

466. It may be assumed that Tf is $\mathbb{R}$-valued. By hypothesis, $|T_A f| \leqq \|T_A\| \,.\, \|f\|$ and also $|T_A f| \leqq \|Tf\|_1$ and the uniform boundedness principle implies there is a positive K such that for all A, $\|T_A\| \leqq K$. But if $A^\pm = \{x : (Tf)(x) \gtrless 0\}$ then

$$\|Tf\|_1 = \int_X (Tf)^+(x)\,dx + \int_X (Tf)^-(x)\,dx = |T_{A^+}f| + |T_{A^-}f| \leqq 2K\|f\|,$$

i.e., T is continuous and the result follows. □

467. If $\{B_p\}_{p=1}^P \cup \{C_q\}_{q=1}^Q$ is a subset of $\{B_\gamma\}_\gamma$ and if $A = (\bigcap_p (X\backslash B_p)) \cap (\bigcap_q C_q) = (X\backslash \bigcup_p B_p) \cap (\bigcap_q C_q) = \bigcap_q C_q \backslash \bigcup_p (B_p \cap (\bigcap_q C_q))$, then

$$\begin{aligned} \mu(A) = \mu\Big(\bigcap_q C_q\Big) - \sum_p \mu\Big(B_p \cap \Big(\bigcup_q C_q\Big)\Big) \\ + \sum_{p \neq p', p, p'=1}^{P} \mu\Big(B_p \cap B_{p'} \cap \Big(\bigcap_q C_q\Big)\Big) - \cdots \pm \mu\Big(\Big(\bigcap_p B_p\Big) \cap \Big(\bigcap_q C_q\Big)\Big) \end{aligned}$$

$$=\prod_q \mu(C_q)\,.\prod_p (1-\mu(B_p))$$

$$=\prod_p \mu(X\backslash B_p)\,.\prod_q \mu(C_q)$$

as required. □

468. For any Borel set E and any set B in S, $\chi_B^{-1}(E)$ is $\emptyset$, B, $X\backslash B$, or X according as $\{0,1\}\subset\mathbb{R}\backslash E$, $\{1\}\subset E$ and $\{0\}\not\subset E$, $\{1\}\not\subset E$ and $\{0\}\subset E$, or $\{0,1\}\subset E$. Hence if $\{E_p\}_{p=1}^P$ is a set of Borel sets and if $\{F_p\}_{p=1}^P\subset\{B_\gamma\}_\gamma$ then $\bigcap_p \chi_{F_p}^{-1}(E_p)$ is a set of the form of A in Solution 467 and the result there provides the result required here. □

469. i) If σ is a finite subset of Γ_1 and if for γ in σ, A_γ is a Borel set then $T_{\Gamma_1}^{-1}((\prod_{\gamma\in\sigma}A_\gamma)\times(\prod_{\gamma'\notin\sigma}Y_{\gamma'}))=\bigcap_{\gamma\in\sigma}f_\gamma^{-1}(A_\gamma)\in\mathsf{S}$. Since sets of the form $(\prod_{\gamma\in\sigma}A_\gamma)\times(\prod_{\gamma'\notin\sigma}Y_{\gamma'})$ generate $\mathsf{S}_{\beta\Gamma_1}$ the measurability of T_{Γ_1} follows. The proof for S_{Γ_1} is similar.

ii) By definition $\{f_\gamma\}_\gamma$ is an independent set of functions iff (in the notation of i) $\mu(\bigcap_{\gamma\in\sigma}f_\gamma^{-1}(A_\gamma))=\prod_{\gamma\in\sigma}\mu(f_\gamma^{-1}(A_\gamma))$. Since $\bigcap_{\gamma\in\sigma}f_\gamma^{-1}(A_\gamma)=T_\sigma^{-1}(\prod_{\gamma\in\sigma}A_\gamma)$ and $\prod_{\gamma\in\sigma}\mu(f_\gamma^{-1}(A_\gamma))=\mu_\sigma(S_\sigma^{-1}(\prod_{\gamma\in\sigma}A_\gamma))$ the result follows.

iii) The hypothesis may be reformulated as: there is in $L^1(X,\mu)$ an f such that $\|\sum_{n=1}^N f_{\gamma_n}-f\|_1\to 0$ as $N\to\infty$. The latter statement may be expressed by: $\int_{X_{\Gamma_N}}|\sum_{n=1}^N \tilde f_{\gamma_n}(\tilde x_{\Gamma_N})-\tilde f(\tilde x_{\Gamma_N})|\,d\mu_{\Gamma_N}(\tilde x_{\Gamma_N})\to 0$ as $N\to\infty$. The independence hypothesis implies that μ_{Γ_N}, product measure, is to be used. Since

$$\tilde f-\sum_{n=1}^N \tilde f_{\gamma_n}=\sum_{n>N}\tilde f_{\gamma_n},$$

which is independent of the coordinates (variables) $\tilde x_{\gamma_1}, \tilde x_{\gamma_2},\ldots,\tilde x_{\gamma_N}$ and since all spaces involved have total measure equal to one, the result follows. □

470. If $\{A_p\}_{p=1}^P$ are Borel sets in $\mathbb{R}$ and if $h_p=g_p(f_{p1},f_{p2},\ldots,f_{pQ_p})$ then $h_p^{-1}(A_p)=C_p$ is a subset of X and $x\in C_p$ iff $(f_{p1}(x),f_{p2}(x),\ldots,f_{pQ_p}(x))\in g_p^{-1}(A)$, a Borel set in $\mathbb{R}^{Q_p}$. According to Problem 469 the product formula is valid for cylinders (see Problem 410) and by extension for all Borel sets and the result follows. □

471. For any Borel set A, $f^{-1}(A)$ is $\emptyset$ or X according as $f(X)\notin A$ or $f(X)\in A$. Hence if B is a Borel set $f^{-1}(A)\cap g^{-1}(B)=\emptyset$ or $g^{-1}(B)$ and the result follows. □

472. It may be assumed that f and g are $\mathbb{R}$-valued. Let F_{pn} be $\{x: 2^{-n}\,.\,p\leqq f(x)<2^{-n}(p+1)\}$ and let G_{qm} be $\{y: 2^{-m}\,.\,q\leqq g(y)<2^{-m}(q+1)\}$. Then for all p, n, q, m, $\chi_{F_{pn}}$ and $\chi_{G_{qm}}$ are independent. If $f_n=\sum_{p=-n.2^n}^{n.2^n}2^{-n}\,.\,p\chi_{F_{pn}}$ and $g_m=\sum_{q=-m.2^m}^{m.2^m}2^{-m}\,.\,q\chi_{G_{qm}}$ then $f=\lim_{n\to\infty}f_n$ and $g=\lim_{m\to\infty}g_m$ and $\int_X f(x)\,d\mu(x)=\lim_{n\to\infty}\int_X f_n(x)\,d\mu(x)$,

$$\int_X g(x)\,d\mu(x)=\lim_{m\to\infty}\int_X g_m(x)d\mu(x).$$

The hypothesis of independence shows

$$\int_X f_n(x)g_m(x)\,d\mu(x) = \int_X f_n(x)\,d\mu(x)\,.\int_X g_m(x)\,d\mu(x)$$

and the result follows. □

473. It suffices to show that if $\emptyset \subsetneqq A, B \subseteqq \mathbb{N}\backslash\{1\}$ then χ_A and χ_B are not independent. Since $\nu(\chi_A^{-1}(1)) = \sum_{n \text{ in } A} 2^{-n!}$, $\nu(\chi_B^{-1}(1)) = \sum_{m \text{ in } B} 2^{-m!}$ and $\nu(\chi_{A\cap B}^{-1}(1)) = \sum_{p \text{ in } A\cap B} 2^{-p!}$, and since $(\sum_{n \text{ in } A} 2^{-n!})(\sum_{m \text{ in } B} 2^{-m!})$ is a real number having a binary representation with nonzero entries only in those places with indices $n!+m!$, n in A and m in B, it suffices to prove $n!+m!$ is never $p!$ if $m, n \geqq 2$.

However, if $p! = n!+m!$ and $m \leqq n$, then $p = n+k$, $k>0$, $(n+k)! = (n+k)\cdots(n+1)n! = n!+m!$ or $n!((n+k)\cdots(n+1)-1) = m!$ and so $2 \leqq (n+k)\cdots(n+1)-1 \leqq m!/n! \leqq 1$, a contradiction. □

474. For any Borel subset S of I let F_S be $x \mapsto x\chi_S(x)$. It will be shown that if $0 \leqq a < b \leqq 1$ and if g and $F_{[a,b]}$ are independent then g is constant a.e. Indeed, otherwise there is a Borel set A such that $0 < \lambda(g^{-1}(A)) < 1$. If $a \leqq x-c < x+c \leqq b$ then $[x-c, x+c] = F_{[a,b]}^{-1}([x-c, x+c])$ and since g and F are independent, $\lambda(g^{-1}(A) \cap [x-c, x+c])/2c = \lambda(g^{-1}(A))$. The metric density theorem implies that the right member of the last equation approaches 0 or 1 for all x off a null set. The left member is neither 0 nor 1 and the contradiction implies the assertion.

For f as given in the problem and g independent of f let k be $y \mapsto f^{-1}(y)\,.\,\chi_{f([a,b])}$ and let h be $k \circ f$. Then $h = F_{[a,b]}$. Since h is Borel measurable, if A is a Borel set. $h^{-1}(A) = f^{-1}(k^{-1}(A))$ and since f and g are independent it follows that h and g are independent. The conclusion of the previous paragraph implies g is constant a.e. □

475. Let $\mathcal{P}$ be the set of all possible products of different f_γ's: $\mathcal{P} = \{f_{\gamma_1} f_{\gamma_2} \cdots f_{\gamma_n} : n \text{ in } \mathbb{N},\ f_{\gamma_i} \neq f_{\gamma_j} \text{ if } i \neq j\}$. Then $|\int_X f_{\gamma_1}(x) \cdots f_{\gamma_n}(x)\,d\mu(x)| = \prod_{k=1}^n |\int_X f_{\gamma_k}(x)\,d\mu(x)| < \infty$. If no two among $f_{\gamma_1}, \ldots, f_{\gamma_n}, f_\gamma$ are the same then

$$\begin{aligned}
&\int_X f_{\gamma_1}(x)\cdots f_{\gamma_n}(x) f_\gamma(x)\,d\mu(x) \\
&\quad = \prod_{k=1}^n \int_X f_{\gamma_k}(x)\,d\mu(x)\,.\int_X f_\gamma(x)\,d\mu(x) \\
&\quad = \int_X f_{\gamma_1}(x) f_{\gamma_2}(x)\,d\mu(x) \prod_{k=3}^n \int_X f_{\gamma_k}(x)\,d\mu(x)\,.\int_X f_\gamma(x)\,d\mu(x) \\
&\quad = 0
\end{aligned}$$

and so $\dim(\{f_\gamma\}_\gamma)^\perp \geqq 1$ unless some product in $\mathcal{P}$ and with more than one factor is an f_γ, e.g., $f_{\gamma_1} f_{\gamma_2} \cdots f_{\gamma_n} = f_\gamma$. But then $\int_X f_{\gamma_1}(x) \cdots f_{\gamma_n}(x) f_\gamma(x)\, d\mu(x) = \int_X f_\gamma(x)^2\, d\mu(x) = 0$ rather than 1.

If Γ is infinite then so is $\mathscr{P}$ and so $\dim(\{f_\gamma\}_\gamma)^\perp = \infty$. □

REMARK. The assumption about orthonormality is superfluous. The functions $g_\gamma = f_\gamma - E(f_\gamma)$ are pairwise orthogonal and independent. Normalized they constitute an orthonormal and independent set and thus cannot span $L^2(X, \mu)$, whence neither can the f_γ.

476. In Solution 158 it is shown that if $\sum_n \mu(A_n) < \infty$ then $\mu(\limsup_{n=\infty}(A_n)) = 0$ (even if the A_n are not independent). Assume that it has been shown that whenever $\sum_n \mu(A_n) = \infty$, $\mu(\limsup_{n=\infty} A_n) = 1$. According to problem 158 if $\mu(\limsup_{n=\infty} A_n) = 1$ then $\sum_n \mu(A_n) = \infty$. Thus $\sum_n \mu(A_n) = \infty$ iff $\mu(\limsup_{n=\infty} A_n) = 1$ and, in particular, if $\mu(\limsup_{n=\infty} A_n) = 0$ then $\sum_n \mu(A_n) < \infty$. Consequently the problem is reduced to showing that if $\sum_n \mu(A_n) = \infty$ then $\mu(\limsup_{n=\infty} A_n) = 1$.

Since

$$\limsup_{n=\infty} A_n = \bigcap_{n=1}^{\infty} \bigcup_{m=n}^{\infty} A_m,$$

$$\begin{aligned} 1 - \mu\Big(\limsup_{n=\infty} A_n\Big) &= \mu\Big(X \backslash \limsup_{n=\infty} A_n\Big) \\ &= \lim_{n\to\infty} \mu\Big(\bigcap_{m=n}^{\infty} (X\backslash A_m)\Big). \\ &= \lim_{n\to\infty} \lim_{M\to\infty} \mu\Big(\bigcap_{m=n}^{M} (X\backslash A_m)\Big). \end{aligned}$$

The assumption that the A_m are independent now leads to the equality: $\mu(\bigcap_{m=n}^{M}(X\backslash A_m)) = \prod_{m=n}^{M} \mu(X\backslash A_m) = \prod_{m=n}^{M}(1-\mu(A_m))$. Since the infinite product $\prod_{m=n}^{\infty}(1-\mu(A_m))$ converges to a number different from zero iff $\sum_{m=n} \mu(A_m) < \infty$ the result follows. □

477. Note first that if $\{b_n\}_{n=1}^{\infty} \subset \mathbb{C}$ and if $b_n \to b$ as $n \to \infty$ then $\sum_{n=1}^{N} b_n/N \to b$ as $N \to \infty$. Indeed,

$$\sum_{n=1}^{N} b_n/N - b = \sum_{n=1}^{N} \frac{b_n - b}{N} = \Bigg(\sum_{n=1}^{[N^{1/2}]} + \sum_{[N^{1/2}]+1}^{N}\Bigg) \frac{b_n - b}{N}.$$

For some M and all n $|b_n - b| \leqq M$ and if $a > 0$ there is an n_a such that if $n \geqq n_a$, $|b_n - b| < a/2$. Thus if $[N^{1/2}] + 1 \geqq n_a$,

$$\left| N^{-1} \sum_{n=1}^{N} b_n - b \right| \leqq M \frac{N^{1/2}}{N} + a \frac{N - ([N^{1/2}]+1)}{2N}.$$

As $N \to \infty$ the first term approaches zero and the second never exceeds $\frac{1}{2}a$ whence the (preliminary) result.

If $s_N = \sum_{n=1}^N a_n/n$ then since $\lim_{N\to\infty} s_N = s$ exists it follows that $\sum_{N=1}^{N_1} s_N/N_1 \to s$ as $N_1 \to \infty$. But $N_1^{-1} \sum_{n=1}^{N_1} a_n = N_1^{-1} \sum_{n=1}^{N_1} n \,.\, a_n/n = N_1^{-1}(\sum_{n=2}^{N_1} n(s_n - s_{n-1}) + s_1) = -(N_1-1)^{-1} \sum_{N=1}^{N_1-1} s_N \,.\, (N_1-1)/N_1 + s_{N_1} \to -s+s=0$ as $N_1 \to \infty$. □

478. The sequence $\{\sum_{n=1}^N f_n/n\}_{N=1}^\infty$ will be shown to converge in $L^2(X, \mu)$. Indeed, if $N < M$,

$$\left\| \sum_{n=1}^{N} - \sum_{n=1}^{M} f_n/n \right\|_2^2 = \left\| \sum_{N+1}^{M} f_n/n \right\|_2^2 = 2 \sum_{\substack{m,n=N+1 \\ m \neq n}}^{M} E(f_n f_m)/nm + \sum_{N+1}^{M} \|f_n\|_2^2/n^2$$

$$= 2 \sum_{\substack{m,n=N+1 \\ m \neq n}}^{M} E(f_n)E(f_m) + \sum_{N+1}^{M} \sigma_n^2/n^2$$

$$= 0 + \sum_{N+1}^{M} \sigma_n^2/n^2 \to 0 \qquad \text{as } N, M \to \infty.$$

If $\|h - \sum_{n=1}^N f_n/n\|_2 \to 0$ as $N \to \infty$ then the Schwarz inequality implies $\|h - \sum_{n=1}^N f_n/n\|_1 \to 0$ as $N \to \infty$ and $\|h\|_1 \leqq |E(\sum_{n=1}^N f_n/n)| + \|h - \sum_{n=1}^N f_n/n\|_1 = 0 + \|h - \sum_{n=1}^N f_n/n\|_1 \to 0$ as $N \to \infty$. Hence $E(h) = 0 = \lim_{n\to\infty} \int_{X_{\Gamma_n}} \tilde{h}(\tilde{x}_{\Gamma_n})\, d\mu_{\Gamma_n}(\tilde{x}_{\Gamma_n}) = \lim_{n\to\infty} \sum_{m=n+1}^\infty \tilde{f}_m/m$ (Problem 469) and so $N^{-1} \sum_{n=1}^N f_n \to 0$ as $N \to \infty$ (Problem 477). □

479. For m in $\mathbb{N}$ and $0 \leqq k \leqq n-1$ let f_{mk} be

$$I \ni \sum_k p_k n^{-k} \mapsto \begin{cases} 1, & \text{if } p_m = k \\ 0, & \text{otherwise} \end{cases}.$$

Then for k fixed $\{f_{mk}\}_{m=1}^\infty$ is an independent set of functions. Furthermore $E(f_{mk}) = n^{-1}$ and $\mathrm{var}(f_{mk}) = n^{-1} - n^{-2}$ which is independent of m and k. Hence if $g_{mk} = f_{mk} - n^{-1}$ then problem 478 applies to yield $\lim_{N\to\infty} N^{-1} \sum_{m=1}^N g_{mk} = 0$, i.e., $\lim_{N\to\infty} N^{-1} \sum_{m=1}^N f_{mk} = n^{-1}$ a.e., $0 \leqq k \leqq n-1$. Since $k(t, N) = \sum_{m=1}^N f_{mk}$ the desired result follows. □

NOTE. The conclusion is valid for each n in $\mathbb{N}$ and all associated k. Hence there is a fixed null set E such that for all n in $\mathbb{N}$ if $t \notin E$, $t = \sum_k p_k n^{-k}$, and $0 \leqq k \leqq n-1$ then $\lim_{N\to\infty} k(t, N)/N = n^{-1}$.

480. The integral is a sum of integrals over the intervals

$$\left[-\pi + k \,.\, \frac{2\pi}{n}, -\pi + (k+1) \,.\, \frac{2\pi}{n}\right],$$

$0 \leqq k \leqq n-1$. In the kth integral the successive substitutions $t = u/n$, $v = u - (k-1) \,.\, 2\pi$ lead to the formula: $n^{-1} \int_{-\pi}^{\pi} g(v) f(v/n + (k-1) \,.\, 2\pi/n)\, dv$ and so

$$\int_{\mathbb{T}} f(t) g(nt)\, dt = (2\pi)^{-1} \int_{-\pi}^{\pi} g(v) \sum_{k=1}^{N} f(v/n + (k-1) \,.\, 2\pi/n) 2\pi/n\, dv.$$

The hypotheses permit passage to the limit as $n \to \infty$ and there emerges $(2\pi)^{-1} \int_{\mathbb{T}} g(v)\,dv \int_{\mathbb{T}} f(w)\,dw = 2\pi \hat{f}(0)\hat{g}(0)$. □

481. Let A_m be $\{x: \sup_{n \leqq m} s_n(x) > 0\}$. Then $A_m \subset A_{m+1} \cdots \subset A$ and $A = \bigcup_m A_m$. If $a_n(x) = f(T^n(x))$ then $s_N = \sum_{n=0}^{N-1} a_n$ and the notation of problem 110 may be adapted, i.e., $D(x)$ is the set of distinguished indices for x and so $\sum_{n \text{ in } D(x), 0 \leqq n \leqq N-1} a_n(x) = \sum_{n \text{ in } D(x), 0 \leqq n \leqq N-1} f(T^n(x)) > 0$. Hence if $B_n = \{x: n \in D(x)\}$, $0 \leqq n \leqq N-1$, it follows that $\sum_{n=0}^{N-1} \int_{B_n} f(T^n(x))\,d\mu(x) > 0$. However

$$B_n = \left\{x: \max_{n-1 < p \leqq N-1} [f(T^n(x)) + \cdots + f(T^p(x))] > 0\right\}$$
$$= T^{-(n-1)} A_{N-(n-1)}$$

and so $\int_{B_n \cap E} f(T^n(x))\,d\mu(x) = \int_{A_{N-(n-1)} \cap E} f(x)\,d\mu(x)$ and it follows that $0 \leqq \sum_{n=0}^{N-1} \int_{A_{N-(n-1)} \cap E} f(x)\,d\mu(x) = \sum_{k=2}^{N+1} \int_{A_k \cap E} f(x)\,d\mu(x)$.

Since $\bigcup_k A_k = A$ and $A_k \subset A_{k+1}$ it follows that $\int_{A_k} f(x)\,d\mu(x) \to \int_A f(x)\,d\mu(x)$ as $k \to \infty$ and so $0 \leqq N^{+1} \sum_{k=0}^{N+1} \int_{A_k \cap E} f(x)\,d\mu(x) \to \int_{A \cap E} f(x)\,d\mu(x)$ as $N \to \infty$. □

482. If, in Problem 481, f is replaced by $f - a$ and s_n/n is replaced by $s_n/n - a$ the result of Problem 481 is applicable and yields the more general result given in Problem 482. □

483. i) Since

$$\bar{F}(T(x)) = \limsup_{N \to \infty} N^{-1}(f(T(x)) + \cdots f(T^N(x))$$
$$= \limsup_{N = \infty} \left(\frac{N+1}{N}\right) \cdot \frac{s_{N+1}(x)}{N+1} - \frac{f(x)}{N}$$
$$= \bar{F}(x),$$

the result for $\bar{F}$ and similarly the result for $\underline{F}$ follows.

ii) Since $\underline{F} \leqq \bar{F}$ the set $\{x: \underline{F}(x) < \bar{F}(x)]$ is a subset of $\bigcup_{r,s \text{ in } \mathbb{Q}} \{x: \underline{F}(x) < r < s < \bar{F}(x)\} = \bigcup_{r,s \text{ in } \mathbb{Q}} C_{rs}$. The argument in i) shows that each C_{rs} is invariant ($T(C_{rs}) = C_{rs}$). Furthermore $C_{rs} = C_{rs} \cap A_s$ since if $x \in C_{rs}$ and $\sup_n n^{-1} s_n(x) \leqq s$ then $\limsup_{n=\infty} n^{-1} s_n(x) \leqq s$, a contradiction. Hence (Problem 482)

$$\int_{C_{rs}} f(x)\,d\mu(x) = \int_{C_{rs} \cap A_s} f(x)\,d\mu(x) \geqq s\mu(C_{rs} \cap A_s) = s\mu(C_{rs}).$$

Arguing with f replaced by $-f$ above leads to: $\int_{C_{rs}} f(x)\,d\mu(x) \leqq r\mu(C_{rs})$ whence $\mu(C_{rs}) = 0$ and so $\underline{F} = \bar{F}$ $(=F)$ a.e.

iii) For any set E in S,

$$\int_E |s_n(x)/n|\,d\mu(x) \leqq \sum_{k=0}^{n-1} \int_E |f(T^k(x))/n|\,d\mu(x).$$

But $\int_E f(T^k(x))\,d\mu(x) = \int_{T^{-k}(E)} f(y)\,d\mu(y)$ and since $f \in L^1(X, \mu)$, if $\mu(T^{-k}(E))$ is small so is $|\int_{T^{-k}(E)} f(y)\,d\mu(y)|$ small and the uniform integrability of $\{s_n/n\}_{n=1}^\infty$ follows. □

484. Since $s_n/n \to F$ a.e. as $n \to \infty$ and since $\{s_n/n\}_{n=1}^\infty$ is uniformly integrable, according to Egorov's theorem for every positive a there is in S an E such that $\mu(E) < a$ and $s_n/n \to F$ uniformly on $X \backslash E$ as $n \to \infty$. Thus $\int_{X\backslash E} |s_n(x)/n - s_m(x)/m|\,d\mu(x) \to 0$ as $n, m \to \infty$ and

$$\int_E \left| \frac{s_n(x)}{n} - \frac{s_m(x)}{m} \right| d\mu(x)$$

is small for all n, m if a is small. In short, $\{s_n/n\}_{n=1}^\infty$ is a Cauchy sequence in $L^1(X, \mu)$. But then if g is the limit in $L^1(X, \mu)$ of $\{s_n/n\}_n$, $F = g$ a.e. and so $F \in L^1(X, \mu)$ and $\int_X s_n(x)/n\,d\mu(x) = \int_X f(x)\,d\mu(x) \to \int_X F(x)\,d\mu(x)$ as $n \to \infty$ and the result follows. □

485. Consider first the case $f: z \mapsto z$. Then for all z_0 in $\mathbb{T}$,

$$n^{-1} \sum_{k=0}^{n-1} f(z_0^k) = \begin{cases} n^{-1}(1 - z_0^n)/(1 - z_0), & \text{if } z_0 \neq 1 \\ 1, & \text{if } z_0 = 1 \end{cases}$$

and so $\lim_{n\to\infty} n^{-1} \sum_{k=0}^{n-1} f(z_0^k)$ exists and is 0 or 1 according as $z_0 \neq 1$ or $z_0 = 1$. If f is a polynomial it follows that

$$\lim_{n\to\infty} n^{-1} \sum_{k=0}^{n-1} f(z_0^{\pm k}) = \begin{cases} f(0), & \text{if } z_0 \neq 1 \\ f(1), & \text{if } z_0 = 1 \end{cases}.$$

If $f \in C(\mathbb{T}, \mathbb{C})$ Fejér's theorem implies there are sequences $\{p_n\}_{n=1}^\infty$ and $\{q_n\}_{n=1}^\infty$ of polynomials such that $\sup_{z\in\mathbb{T}} |f(z) - p_n(z) - q_n(z^{-1})| \to 0$ as $n \to \infty$. It follows that $\lim_{n\to\infty} n^{-1} \sum_{k=0}^{n-1} f(z_0^k)$ is $f(0)$ or $f(1)$ according as $z_0 \neq 1$ or $z_0 = 1$. Thus μ_{z_0} is the point measure concentrated at 0 resp. 1 according as $z_0 \neq 1$ resp. $z_0 = 1$ and such that $\mu_{z_0}(0)$ resp. $\mu_{z_0}(1) = 1$. □

486. i) If $s_m(x) = \sum_{n=1}^m |a_n| \cdot |f_n(x)|$ then $s_m \leqq s_{m+1}$ and $\|s_{m_1} - s_{m_2}\|_1 \to 0$ as $m_1, m_2 \to \infty$. Hence there is in $L^1(I, \lambda)$ an s such that $\|s_m - s\|_1 \to 0$ as $m \to \infty$ and in particular $\lim_{m\to\infty} s_m(x)$ is finite a.e. Hence $\sum_n a_n f_n$ converges a.e., i.e., F exists and is in $L^1(I, \lambda)$. Furthermore $\|F\|_1 \leqq \|s\|_1$.

ii) If $a_n = r_n e^{i\theta_n}$ and $f_n(x) = |f(x)|\, e^{i\varphi_n(x)}$, $0 \leqq \theta_n < 2\pi$, $0 \leqq \varphi_n(x) < 2\pi$ then (*) holds iff $\theta_n + \varphi_n(x) - (\theta_m + \varphi_m(x)) \equiv 0 (\text{mod } 2\pi)$ a.e. for all m, n in $\mathbb{N}$.

iii) The relation (*) holds for all sequences $\{f_n\}_n$ iff at most one $a_n \neq 0$.

iv) The relation (*) holds for all sequences $\{a_n\}_n$ iff at most one $f_n \neq 0$. □

487. Since the result is a standard theorem in linear algebra if either Γ or Λ is finite it will be assumed that Λ is infinite. Each x_γ is uniquely expressible as a finite sum $\sum_\lambda a_{\gamma\lambda} y_\lambda$ and each y_λ appears with a nonzero coefficient in at least one of the sums representing the various x_γ. Indeed, otherwise, if some y_{λ_0} never appears then $y_{\lambda_0} = \sum_\gamma b_{\lambda_0\gamma} x_\gamma = \sum_{\gamma,\lambda} b_{\lambda_0\gamma} a_{\gamma\lambda} y_\lambda$ and the linear independence of $\{y_\lambda\}_\lambda$ is denied. It follows that Γ is also infinite.

Thus the set Ξ of all finite subsets of Λ contains a set Ξ_1 in bijective correspondence with Γ. Since $\text{card}(\Xi_1) \leqq \text{card}(\Xi) = \text{card}(\Lambda)$ it follows that $\text{card}(\Lambda) \geqq \text{card}(\Gamma)$. Since Γ is also infinite the argument is symmetric and so $\text{card}(\Gamma) = \text{card}(\Lambda)$. □ [19]

488. As in Problem 487 it will be assumed that is infinite. The argument in the first paragraph may be repeated *mutatis mutandis* to show that if some y_{λ_0} fails to appear in all the (countable) Fourier representations of the x_γ then $\{y_\lambda\}_\lambda$ is not a maximal orthonormal set. For each γ let s_γ be the set of y_λ appearing with nonzero coefficients in the countable Fourier representation of x_γ. Then each s_γ is finite or countable and $\bigcup_\gamma s_\gamma = \Lambda$. Hence $\text{card}(\Gamma)\,\text{card}(\mathbb{N}) \geqq \text{card}(\Lambda)$. Since Λ is infinite so is Γ and hence $\text{card}(\Gamma)\,\text{card}(\mathbb{N}) = \text{card}(\Gamma)$ and so $\text{card}(\Lambda) \leqq \text{card}(\Gamma)$. The symmetry of the argument yields the result. □

A second solution is the following. It may be assumed that Λ is infinite and thus if S is the set of all finite linear combinations with rational coefficients of elements of Λ then $\text{card}(S) = \text{card}(\Lambda)$. Furthermore S is dense in $\mathfrak{H}$ and hence for each x_γ there is in S an s_γ such that $\|x_\gamma - s_\gamma\| < 2^{-1/2}$. If $\|x_{\gamma'} - s_{\gamma'}\| < 2^{-1/2}$ and $x_{\gamma'} \neq x_\gamma$ then $s_{\gamma'} \neq s_\gamma$ since otherwise $2^{1/2} = \|x_\gamma - x_{\gamma'}\| < 2^{1/2}$. Hence Γ is in bijective correspondence with a subset of S and $\text{card}(\Gamma) \leqq \text{card}(S) = \text{card}(\Lambda)$. The rest of the argument is validated by symmetry. □

489. If Γ is infinite the (Gram–Schmidt) procedure used in Solution 408 yields an infinite biorthogonal set $\{x_n, x_n^*\}_{n=1}^\infty$. Then $\{2^{-n}x_n/\|x_n\|, 2^n\|x_n\|x_n^*\}_{n=1}^\infty$ is also a biorthogonal set $\{y_n, y_n^*\}_{n=1}^\infty$. However, if $S \subset \mathbb{N}$ then $\sum_{n \text{ in } S} y_n$ converges.

Let $\{r_n\}_{n=1}^\infty$ be an enumeration of $\mathbb{Q}$ and for each t in $\mathbb{R}$ let $\{r_{n_k(t)}\}_{k=1}^\infty$ be an infinite sequence such that $r_{n_k(t)} \to t$ as $k \to \infty$. If $S_t = \{n_k(t)\}_{k=1}^\infty$ and if $t \neq t'$ then $S_t \cap S_{t'}$ is finite (the S_t are pairwise "almost disjoint"). For each t let z_t be $\sum_{n \text{ in } S_t} y_n$. It will be shown that $\{z_t\}_{t \text{ in } \mathbb{R}}$ is a linearly independent set.

Indeed, if $\sum_{m=1}^M a_m z_{t_m} = 0$ then for each m there is in S_{t_m} an n not in any $S_{t_{m'}}$, $m' \neq m$ and so $y_n^*(\sum_{m=1}^M a_m z_{t_m}) = a_m = 0$, i.e., all a_m are zero.

Thus $\{z_t\}_{t \text{ in } \mathbb{R}}$ is a subset of some Hamel basis and so any Hamel basis has a cardinality not less than $\text{card}(\mathbb{R})$. □

490. If X is a separable infinite-dimensional Banach space then $\text{card}(X) = \text{card}(\mathbb{R})$ and hence any Hamel basis must have a cardinality equal to $\text{card}(\mathbb{R})$ (Problem 489) from which the result follows since a bijection between two Hamel bases may be extended linearly to an isomorphism of the spaces for which they are the bases. □

491. The function r_n partitions I into three sets, $S_n^\pm$ and S_n^0: $S_n^\pm = r_n^{-1}(\pm 1)$, $S_n^0 = r_n^{-1}(0)$. Furthermore, $S_0^+ = (0, 1)$, $S_0^- = \varnothing$, $S_0^0 = \{0, 1\}$; if $n \geqq 1$ each of $S_n^\pm$ consists of 2^{n-1} disjoint intervals, each of length 2^{-n} and S_n^0 consists of their $2^n + 1$ endpoints. The intervals of S_n^+ alternate with those of S_n^-.

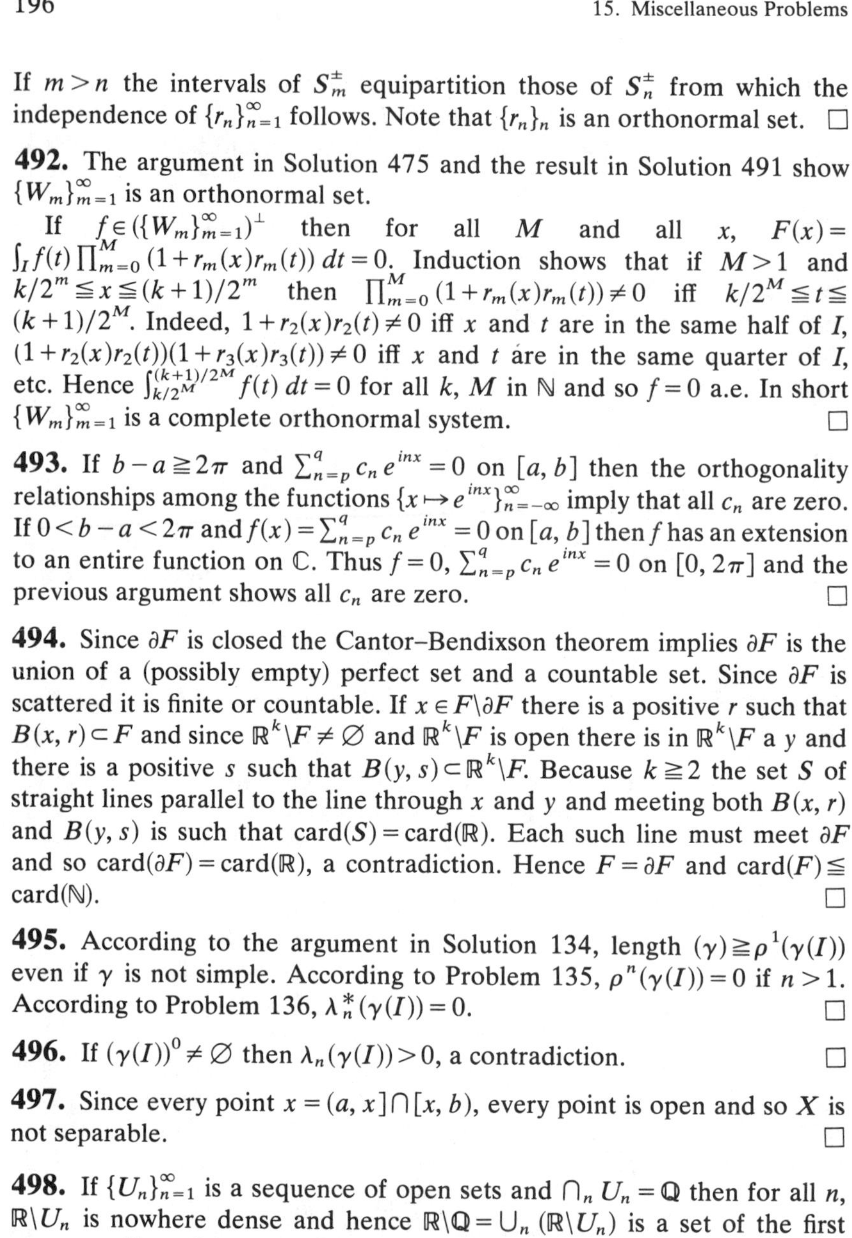

If $m > n$ the intervals of $S_m^{\pm}$ equipartition those of $S_n^{\pm}$ from which the independence of $\{r_n\}_{n=1}^{\infty}$ follows. Note that $\{r_n\}_n$ is an orthonormal set. $\square$

492. The argument in Solution 475 and the result in Solution 491 show $\{W_m\}_{m=1}^{\infty}$ is an orthonormal set.

If $f \in (\{W_m\}_{m=1}^{\infty})^{\perp}$ then for all M and all x, $F(x) = \int_I f(t) \prod_{m=0}^{M} (1 + r_m(x) r_m(t))\, dt = 0$. Induction shows that if $M > 1$ and $k/2^m \leqq x \leqq (k+1)/2^m$ then $\prod_{m=0}^{M} (1 + r_m(x) r_m(t)) \neq 0$ iff $k/2^M \leqq t \leqq (k+1)/2^M$. Indeed, $1 + r_2(x) r_2(t) \neq 0$ iff x and t are in the same half of I, $(1 + r_2(x) r_2(t))(1 + r_3(x) r_3(t)) \neq 0$ iff x and t are in the same quarter of I, etc. Hence $\int_{k/2^M}^{(k+1)/2M} f(t)\, dt = 0$ for all k, M in $\mathbb{N}$ and so $f = 0$ a.e. In short $\{W_m\}_{m=1}^{\infty}$ is a complete orthonormal system. $\square$

493. If $b - a \geqq 2\pi$ and $\sum_{n=p}^{q} c_n e^{inx} = 0$ on $[a, b]$ then the orthogonality relationships among the functions $\{x \mapsto e^{inx}\}_{n=-\infty}^{\infty}$ imply that all c_n are zero. If $0 < b - a < 2\pi$ and $f(x) = \sum_{n=p}^{q} c_n e^{inx} = 0$ on $[a, b]$ then f has an extension to an entire function on $\mathbb{C}$. Thus $f = 0$, $\sum_{n=p}^{q} c_n e^{inx} = 0$ on $[0, 2\pi]$ and the previous argument shows all c_n are zero. $\square$

494. Since ∂F is closed the Cantor–Bendixson theorem implies ∂F is the union of a (possibly empty) perfect set and a countable set. Since ∂F is scattered it is finite or countable. If $x \in F \backslash \partial F$ there is a positive r such that $B(x, r) \subset F$ and since $\mathbb{R}^k \backslash F \neq \emptyset$ and $\mathbb{R}^k \backslash F$ is open there is in $\mathbb{R}^k \backslash F$ a y and there is a positive s such that $B(y, s) \subset \mathbb{R}^k \backslash F$. Because $k \geqq 2$ the set S of straight lines parallel to the line through x and y and meeting both $B(x, r)$ and $B(y, s)$ is such that $\operatorname{card}(S) = \operatorname{card}(\mathbb{R})$. Each such line must meet ∂F and so $\operatorname{card}(\partial F) = \operatorname{card}(\mathbb{R})$, a contradiction. Hence $F = \partial F$ and $\operatorname{card}(F) \leqq \operatorname{card}(\mathbb{N})$. $\square$

495. According to the argument in Solution 134, length $(\gamma) \geqq \rho^1(\gamma(I))$ even if γ is not simple. According to Problem 135, $\rho^n(\gamma(I)) = 0$ if $n > 1$. According to Problem 136, $\lambda_n^*(\gamma(I)) = 0$. $\square$

496. If $(\gamma(I))^0 \neq \emptyset$ then $\lambda_n(\gamma(I)) > 0$, a contradiction. $\square$

497. Since every point $x = (a, x] \cap [x, b)$, every point is open and so X is not separable. $\square$

498. If $\{U_n\}_{n=1}^{\infty}$ is a sequence of open sets and $\bigcap_n U_n = \mathbb{Q}$ then for all n, $\mathbb{R} \backslash U_n$ is nowhere dense and hence $\mathbb{R} \backslash \mathbb{Q} = \bigcup_n (\mathbb{R} \backslash U_n)$ is a set of the first category. Since $\mathbb{Q}$, a countable set, is also of the first category, so is $\mathbb{R}$ and thus the Baire category theorem is contradicted. $\square$

499. Note that if $\operatorname{card}(X) = \mathscr{A}$ it is possible that $\operatorname{card}(\mathsf{O}(X)) = 2^{\mathscr{A}} > \mathscr{A}$. However, if $\{U_\gamma\}_{\gamma \in \Gamma}$ is an open cover of X then for each x in X there is a γ_x such that $x \in U_{\gamma_x}$ and so $\bigcup_{x \text{ in } X} U_{\gamma_x} = X$ and $\operatorname{card}(\{\gamma_x\}_{x \text{ in } X}) \leqq \mathscr{A}$. $\square$

500. If p is a cluster point of $\{x_\gamma\}_{\gamma \in \Gamma}$ and $p \notin \bigcap \{\bar{S} : S \in \mathscr{F}\}$ there is in $\mathscr{F}$ an S such that $p \notin \bar{S}$, i.e., there is a neighborhood $U(p)$ such that $U(p) \cap S = \emptyset$.

However x_γ is eventually in S and frequently in $U(p)$, a contradiction, whence p is a cluster point of $\mathscr{F}$.

Conversely if p is a cluster point of $\mathscr{F}$ let $\mathscr{G}$ be the set of all tails Δ of Γ. Then each $\{x_\delta\}_{\delta\in\Delta}$, Δ in $\mathscr{G}$, is in $\mathscr{F}$ and hence p is in the closure of the image of every tail, i.e., every neighborhood $U(p)$ of p meets the image of every tail and thus x_γ is frequently in $U(p)$. □

501. Let $\mathscr{F}$ be $\{A: A\subset S, A \text{ is closed}, AS\subset A\}$. Then since $S\in\mathscr{F}$, $\mathscr{F}\neq\emptyset$. If $\mathscr{F}$ is partially ordered by (reversed) inclusion ($A_1<A_2$ iff $A_2\subset A_1$) and if $\mathscr{C}=\{A_\gamma\}_{\gamma\in\Gamma}$ is a maximal chain then $\{A_\gamma\}_\gamma$ enjoys the finite intersection property and hence $\bigcap_\gamma A_\gamma = A\neq\emptyset$. Since $A_\gamma S\subset A_\gamma$, $AS\subset A_\gamma$ and so $AS\subset A$. Furthermore if $a\in A$ then $aS\subset A$ and $(aS)S\subset AS$ whence $aS=A$ since A is minimal. If $x\in S$, $ax\in A$, $axS=A=aS$ and the cancellation law shows $xS=S$. Similarly, $Sx=S$. Thus the equations $ax=b$ and $xa=b$ have unique solutions and so S is a group.

If x is close to y then xy^{-1} is close to yy^{-1} = identity of $S=e$ and so $x^{-1}(xy^{-1})=y^{-1}$ is close to $x^{-1}e=x^{-1}$. Hence $x\mapsto x^{-1}$ is continuous. [8] □

502. In S choose an A such that $(\mu\times\mu)(A\times A)>0.9M^2$. Then $(\mu\times\mu)(\theta(A\times A))=\int_A \mu(xA)\,d\mu(x)=\mu(A)^2>0.9M^2$. If π is the map $(x,y)\mapsto(y,x)$ then $(\mu\times\mu)(\pi\theta(A\times A))>0.9M^2$ and so $\pi\theta(A\times A)\cap\theta(A\times A)\neq\emptyset$. Thus there are x, y, u, and v such that $\pi\theta((x,y))=\theta((u,v))$, i.e., $(xy,x)=(u,uv)$, $x=uv$, $xy=u$, $xyv=uv=x$ and so yv serves as a right identity e_x for x, $xe_x=x$. But then for all y, $xe_xy=xy$ and the cancellation law implies $e_xy=y$, in particular $e_xx=x$, $ye_xx=yx$ and so $ye_x=y$. In sum, e_x is the unique identity of S.

If $x\in S$ there is an associated measure situation $(xS, x\mathsf{S}, \mu)$ satisfying all the conditions assumed to hold for (S, S, μ). Hence xS has an identity, x^{-1} exists and S is a group. [7] □

503. Let $F_0(X)$ be the free group generated by X: $F_0(X)=\{x_1^{a_1}x_2^{a_2}\cdots x_n^{a_n}, n \text{ in } \mathbb{N}, a_i=\pm1\}$ with the natural notion of multiplication by "juxtaposition of words" and cancellation of all terms x^ax^b in which $a+b=0$. Let $\mathscr{T}$ be the set of topologies on $F_0(X)$ and such that $F_0(X)$ is a topological group containing X topologically for each of the topologies. Then sup $\mathscr{T}$ is also a "group topology" for $F_0(X)$, denoted $F(X)$ for this topology; $F(X)$ satisfies the requirements posed as the next paragraph shows. Thus it suffices to show $\mathscr{T}\neq\emptyset$.

If G is a topological group and if $f: X\mapsto G$ is continuous let w be $x_1^{a_1}x_2^{a_2}\cdots x_n^{a_n}$. Define $f(w)$ to be $(f(x_1))^{a_1}(f(x_2))^{a_2}\cdots(f(x_n))^{a_n}$. Thus f is extended to a homomorphism of $F_0(X)$ into G. If $\{U\}$ is the topology (set of open sets) of G then $\{U\cap f(F_0(X))\}=\{V\}$ is a topology for $f(F_0(X))$ and $\{f^{-1}(V)\}=\{W\}$ is a topology for $F_0(X)$. However, although $x, y\mapsto xy^{-1}$ is continuous, $\{W\}$ is not necessarily Hausdorff and, with respect to $\{W\}$, X is not necessarily topologically embedded in $F_0(X)$. Indeed, $\{f^{-1}(U\cap f(X)\}$ is a topology for X and is not stronger than the topology originally given X. Thus $\sup(\mathcal{T}, \{W\})$ is a (Hausdorff) group topology for $F_0(X)$ and relative to $\sup(\mathcal{T}, \{W\})$ X is topologically embedded in $F_0(X)$, i.e., $\sup \mathcal{T} = \sup(\mathcal{T}, \{W\})$.

The remainder of the argument deals with the existence of $\mathcal{T}$. To this end let $\mathbb{H}^*$ denote the set of nonzero quaternions and let $C_b(X, \mathbb{H}^*)$ be the algebra of bounded continuous maps of X into $\mathbb{H}^*$. If $\mathcal{F}$ is the set of invertible elements in $C_b(X, \mathbb{H}^*)$ and if $x \in X$ let h_x be the multiplicative homomorphism $\mathcal{F} \ni f \mapsto f(x) \in \mathbb{H}^*$. If $\mathbb{H}_f = \mathbb{H}^*$ for all f in $\mathcal{F}$ and if $P = \prod_{f \text{ in } \mathcal{F}} \mathbb{H}_f$ endowed with the product topology then P is a topological group and $\theta: X \ni x \mapsto (\cdots f(x) \cdots) \in P$ is an injection since X is completely regular; θ is also a homeomorphism, by definition of the product topology.

If P_0 is the intersection of all subgroups containing $\theta(X)$ then it will be shown that $F_0(X)$ and P_0 are isomorphic.

If R is a rotation of $\mathbb{R}^3$ there is [3] a quaternion q such that for all $\mathbf{y} = (y_1, y_2, y_3)$ in $\mathbb{R}^3$ if $R(\mathbf{y}) = \mathbf{z} = (z_1, z_2, z_3)$ then $z_1 i + z_2 j + z_3 k = q(y_1 i + y_2 j + y_3 k)q'$ (q' = conjugate of q). Since the group of rotations of $\mathbb{R}^3$ contains an infinite free subset $\{R_n\}_{n=1}^{\infty}$ (i.e., no word $R_{n_1}^{a_1} R_{n_2}^{a_2} \cdots R_{n_k}^{a_k}$ = identity) [14] the set $\{q_n\}_{n=1}^{\infty}$ of corresponding quaternions is free in $\mathbb{H}^*$.

The complete regularity of X implies that if $p_1, p_2, \ldots, p_n$ are n distinct points of X there is in $C_b(X, \mathbb{H}^*)$ an f such that $f(p_i) = q_i$. This implies that $F_0(X)$ and P_0 are isomorphic. The topology inherited by P_0 (hence by $F_0(X)$) from P is in $\mathcal{T}$, i.e., $\mathcal{T} \neq \varnothing$ and the result follows. [5], [12], [14], [17] □

504. For x in $S \backslash \{0\}$ let $\bar{x}$ be the unique solution of $xux \neq 0$. Then $x\bar{x} \neq 0 \neq \bar{x}x$ and so if w is the unique solution of $\bar{x}xvx\bar{x} \neq 0$ then $xwx \neq 0$ and so $w = \bar{x}$. Since $x\bar{x}x$ and x are solutions of $\bar{x}y\bar{x} \neq 0$ it follows that $x\bar{x}x = x$ and so $x\bar{x}$ and $\bar{x}x$ are idempotents ($(x\bar{x})^2 = x\bar{x}$, $(\bar{x}x)^2 = \bar{x}x$). It will be shown that if J is the set of idempotents in $S \backslash \{0\} = S^*$ then $\mathcal{U}(J)$ and S are isomorphic.

For ease of writing, (p, q) designates the matrix unit having entry 1 at position (p, q) and 0 elsewhere. Let f be

$$S \ni x \mapsto \begin{cases} 0, & \text{if } x = 0 \\ (x\bar{x}, \bar{x}x), & \text{if } x \neq 0 \end{cases};$$

let g be

$$\mathcal{U}(J) \ni U \to \begin{cases} 0, & \text{if } U = 0 \\ \text{the unique solution of } pxq \neq 0, & \text{if } U = (p, q) \end{cases}.$$

Then f and g are semigroup homomorphisms and each is the inverse of the other, whence the result. [10] □

505. By induction it can be shown that if D is any derivation then $D(x^n) = nx^{n-1}D(x)$. Furthermore if R_x is $y \mapsto xy$ and if y' denotes $D(y)$ then $(DR_x - R_x D)(y) = D(xy) - xD(y) = D(x)y + xD(y) - xD(y) = R_{x'}(y)$.

Hence

$$R_x(DR_x - R_xD) = R_xR_{x'} = R_{xx'} = R_{x'x} = R_{x'}R_x = (DR_x - R_xD)R_x.$$

Thus if $\tilde{D} = DR_x - R_xD$ then $\tilde{D}R_x = R_x\tilde{D}$ and

$$\tilde{\tilde{D}} = (DR_x - R_xD)R_x - R_x(DR_x - R_xD) = 0.$$

If $(\ )^{(n)}$ denotes the n-fold application of $\sim$, another induction using the equation $\tilde{\tilde{D}} = 0$ shows that $(D^n)^{(n)} = n!(\tilde{D})^n$. Hence if K is the norm of the operator $\sim$ then $n!\|(\tilde{D})^n\| = \|(D^n)^{(n)}\| \leqq K^n\|D^n\| \leqq K^n\|D\|^n$, or $\|(\tilde{D})^n\|^{1/n} \leqq K\|D\|/(n!)^{1/n} \to 0$ as $n \to \infty$.

Thus the image under $\sim$ of D is a generalized nilpotent in End(A). Since $R_{x'} = \tilde{D}$ it follows that $R_{x'}$ and thus x' are generalized nilpotents whence x' is in the radical of A, i.e., $D(A) \subset \bigcap_{M \in \sigma(A)} M$. □

506. If A contains more than one point, if Z is a subset of A and card(Z)>1 then since A is closed Z may be assumed to be closed. Choose z in Z and U in $\mathscr{U}$ so that $z \in U$ and $Z \backslash U \neq \varnothing$. Thus $\varnothing \subset \{z\} \subset Z \cap (U \backslash V_U) = Z \cap U$ since $Z \subset A$. Hence $Z = (Z\backslash U) \cup (Z \cap U)$, i.e., Z is the union of two disjoint nonempty sets one, $Z\backslash U$, of which is closed. It will be shown that $Z \cap U$ is also closed, i.e., that Z is not connected whence A is totally disconnected. Indeed, $\bar{U}\backslash V_U = U \backslash V_U$ since if $x \in \bar{U}\backslash V_U$ and if $x \notin U$ then $x \in \partial U$, $x \notin V_U$, $x \in \bar{U}\backslash V_U$, a contradiction. Hence $Z \cap U = Z \cap (\bar{U}\backslash V_U)$ and is closed. □

507. If $X = I$ and B is a countable dense subset of *the* Cantor set C, then $\bar{B} = C$ which is nowhere dense and card(C) = card($\mathbb{R}$). □

508. If X is not compact there is in X a sequence $\{x_n\}_{n=1}^{\infty}$ having no cluster point. Each x_n is not isolated and so for all n there is a y_n such that $0 < d(x_n, y_n) < 1/n$. The set $S = \{x_n\}_n \cup \{y_n\}_n$ has no cluster point since any cluster point of S is also a cluster point of $\{x_n\}_n$. Thus S is closed and according to the Tietze extension theorem there is in $C(X, \mathbb{R})$ an f such that $f(x_n) = n$, $f(y_n) = 2n$. However, such an f is not uniformly continuous since for any positive a there are points x_n, y_n such that $|f(x_n) - f(y_n)| = n$ while $d(x_n, y_n) < a$. □

509. Since $\mathbb{N}^{\mathbb{N}}$ is infinite there is a bijection $F: \mathbb{N}^{\mathbb{N}} \ni \rho \mapsto F(\rho) = (\mu, \nu) \in (\mathbb{N}^{\mathbb{N}})^2$. If g is $\mathbb{N}^{\mathbb{N}} \ni \mu \to \mathscr{A}(\mathscr{M})^{\mathbb{N}}$ then $g(\mu) = \{g(\mu)_m\}_{m=1}^{\infty}$ and for all m in $\mathbb{N}$ there is an $f_{m\mu}: \mathbb{N}^{\mathbb{N}} \ni \nu \to \mathscr{M}^{\mathbb{N}}$ and $g(\mu)_m = \bigcup_{\nu \in \mathbb{N}^{\mathbb{N}}} \bigcap_{k=1}^{\infty} f_{m\mu}(\nu)_k$. For each pair μ, ν let $\{f_{m\mu}(\nu)_k\}_{m,k=1}^{\infty}$ be enumerated as $\{s(\mu, \nu)_p\}_{p=1}^{\infty} = \{s(F(\rho))_p\}_{p=1}^{\infty} = \{r(\rho)_p\}_{p=1}^{\infty}$. Then

$$\begin{aligned} N_g &= \bigcup_{\mu \in \mathbb{N}^{\mathbb{N}}} \bigcap_{m=1}^{\infty} g(\mu)_m = \bigcup_{\mu \in \mathbb{N}^{\mathbb{N}}} \bigcap_{m=1}^{\infty} \bigcup_{\nu \in \mathbb{N}^{\mathbb{N}}} \bigcap_{k=1}^{\infty} f_{m\mu}(\nu)_k \\ &= \bigcup_{(\mu,\nu) \in (\mathbb{N}^{\mathbb{N}})^2} \bigcap_{m,k=1}^{\infty} f_{m\mu}(\nu)_k = \bigcup_{(\mu,\nu) \in (\mathbb{N}^{\mathbb{N}})^2} \bigcap_{p=1}^{\infty} s(\mu, \nu)_p \\ &= \bigcup_{\rho \in \mathbb{N}^{\mathbb{N}}} \bigcap_{p=1}^{\infty} r(\rho)_p. \end{aligned}$$

Since each $f_{m\mu}(\nu)_k = s(\mu, \nu)_p = r(\rho)_p \in \mathscr{M}$ it follows that $N_g \in \mathscr{A}(\mathscr{M})$. □

510. i) For each $\nu=\{n_1,n_2,\ldots\}$ let $f(\nu)_k=M_{n_1}$, $k=1,2,\ldots$. Then $\bigcap_{k=1}^\infty f(\nu)_k=M_{n_1}$ and $M_f=\bigcup_{\nu\in\mathbb{N}^\mathbb{N}}\bigcap_{k=1}^\infty M_{n_1}=\bigcup_{\nu\in\mathbb{N}^\mathbb{N}}M_{n_1}=\bigcup_{n=1}^\infty M_n$.

ii) Let f be the constant map $f:\nu\mapsto(M_1,M_2,\ldots)$. Then $f(\nu)_k=M_k$, $\bigcap_k f(\nu)_k=\bigcap_k M_k$, and $M_f=\bigcup_{\nu\in(\mathbb{N}^\mathbb{N})}\bigcap_k M_k=\bigcap_k M_k$. □

511. For each $\nu=\{n_1,n_2,\ldots\}$ let $g(\nu)_k$ be $\bigcap_{m=1}^k f(\nu)_m\in\mathcal{M}$. Then g is regular, $\bigcap_k g(\nu)_k=\bigcap_m f(\nu)_m$, and the result follows. □

512. i) If x is in the left member, for some m in $\mathbb{N}$ and some $\nu=\{n_1,n_2,\ldots\}$, $x\in\bigcap_{k=1}^\infty M_{n_1,n_2,\ldots,n_i,m,n_{i+1},\ldots,n_{i+k}}$. If

$$\tilde{\nu}=\{n_1,n_2,\ldots,n_i,m,n_{i+1},\ldots\}=\{\tilde{n}_1,\tilde{n}_2,\ldots\}$$

then $\bigcap_{k=i+2}^\infty f(\tilde{\nu})_k=\bigcap_{k=1}^\infty M_{n_1,\ldots,n_i,m,n_{i+1},\ldots,n_{i+k}}$. Since f is regular,

$$\bigcap_{k=i+2}^\infty f(\tilde{\nu})_k=\bigcap_{k=1}^\infty f(\tilde{\nu})_k$$

and so x is in the right member.

If x is in the right member, for some $\nu=\{n_1,n_2,\ldots\}$, $x\in\bigcap_k f(\nu)_k$ and since f is regular, $x\in\bigcap_{k=i+2}^\infty f(\nu)_k$ which is $\bigcap_{k=1}^\infty M_{n_1,\ldots,n_i,m,n_{i+2},\ldots,n_{i+k}}$ if $m=n_{i+1}$ and the result follows.

ii) Since f is regular, $\bigcup_k M_{n_1,n_2,\ldots,n_k}=M_{n_1}$ and $\bigcup_{\nu\in\mathbb{N}^\mathbb{N}}\bigcup_k M_{n_1,n_2,\ldots,n_k}=\bigcup_{\nu\in\mathbb{N}^\mathbb{N}}M_{n_1}=M_1\cup M_2\cup\cdots=\bigcup_n M_n$.

iii) If x is not in the right member and $x\in M$ then, $'$ denoting complement, $x\in\bigcap_{\nu\in\mathbb{N}^\mathbb{N}}\bigcap_{k=0}^\infty(M'_{n_1,n_2,\ldots,n_k}\cup\bigcup_{m=1}^\infty M_{n_1,n_2,\ldots,n_k,m})$. Hence if $x\in M_{n_1,n_2,\ldots,n_k}$ then there is an m such that $x\in M_{n_1,n_2,\ldots,n_k,m}$. Since, by hypothesis, $x\in M$ $(=M_{n_1,n_2,\ldots,n_k}$ if $k=0)$ there is an $m=m_1$ such that $x\in M_{m_1}$. Thus there is an $m=m_2$ such that $x\in M_{m_1,m_2}$, etc., and so $x\in\bigcap_{k=1}^\infty M_{m_1,m_2,\ldots,m_k}$ and $x\in\bigcup_{\nu\in\mathbb{N}^\mathbb{N}}\bigcap_{k=1}^\infty M_{m_1,m_2,\ldots,m_k}$, i.e., x is not in the left member and the result follows. □

513. In the style of Problem 7 let E be $\mathsf{F}(I)$. If $\nu=\{n_1,n_2,\ldots\}$ and ultimately $M_{n_1,n_2,\ldots,n_k}\in\mathsf{E}$ then according to Problem 510 all $M_{n_1,n_2,\ldots,n_k}$ belong to $\mathscr{A}(\mathsf{E})$ and so $\mathsf{S}_\beta(I)\subset\mathscr{A}(\mathsf{F}(I))$. □

514. Since all members of $\mathcal{M}$ are compact so are all members of $H(\mathcal{M})=\{H(M):M\in\mathcal{M}\}$. In view of Problem 511 if $E\in\mathscr{A}(\mathcal{M})$ it may be assumed that for some regular g, $E=M_g=\bigcup_{\nu\in\mathbb{N}^\mathbb{N}}\bigcap_{k=1}^\infty g(\nu)_k$. Let $g(\nu)_k$ be K_k, a compact set. It will be shown that $H(\bigcap_{k=1}^\infty K_k)=\bigcap_k H(K_k)$. Indeed, $\bigcap_k H(K_k)\supset H(\bigcap_k K_k)$ no matter what the map H is and no matter what the sets K_k are. Conversely, if $y\in\bigcap_{k=1}^\infty H(K_k)$, then for each k, $y=H(x_k)$, $x_k\in K_k\supset K_{k+1}$. If x is a cluster point of $\{x_k\}_k$ then $x\in\bigcap_k K_k$ and if $x_{k_p}\to x$ as $p\to\infty$ then $H(x_{k_p})=y\to H(x)$ as $p\to\infty$, i.e., $y\in H(\bigcap_k K_k)$. Hence $H(E)=\bigcup_{\nu\in\mathbb{N}^\mathbb{N}}H(\bigcap_k K_k)=\bigcup_{\nu\in\mathbb{N}^\mathbb{N}}\bigcap_k H(K_k)\in\mathscr{A}(\mathsf{F}(\mathbb{R}))$ as required. □

515. If f is a map $\mathbb{N}^\mathbb{N}\to(\mathsf{S}_\lambda(I))^\mathbb{N}$ let M_f be $\bigcup_{\nu\in\mathbb{N}^\mathbb{N}}\bigcap_k f(\nu)_k=\bigcup_{\nu\in\mathbb{N}^\mathbb{N}}\bigcap_k M_{n_1,n_2,\ldots,n_k}$. Owing to Problem 511 it may be assumed that f is regular. There is in $\mathsf{S}_\beta(I)$ (which is contained in $\mathsf{S}_\lambda(I)$) an A such that

$A \supset M_f$ and $\lambda(A) = \lambda^*(M_f) =$ outer measure of M_f. More generally, for any finite sequence $\{m_1, m_2, \ldots, m_i\}$ there is in $\mathsf{S}_\beta(I)$ an $A_{m_1,m_2,\ldots,m_i}$ containing $\bigcup_{\nu\in\mathbb{N}^\mathbb{N}} \bigcap_{k=1}^\infty M_{m_1,m_2,\ldots,m_i,n_1,n_2,\ldots,n_k}$ and such that $\lambda(A_{m_1,m_2,\ldots,m_i}) = \lambda^*(\bigcup_{\nu\in\mathbb{N}^\mathbb{N}} \bigcap_{k=1}^\infty M_{m_1,m_2,\ldots,m_i,n_1,n_2,\ldots,n_k})$. Since $A_{m_1,m_2,\ldots,m_i} \cap M_{m_1,m_2,\ldots,m_i}$ serves as well as $A_{m_1,m_2,\ldots,m_i}$ it may be assumed that $M_{m_1,m_2,\ldots,m_i} \supset A_{m_1,m_2,\ldots,m_i}$. Since $M_f = A\backslash(A\backslash M_f)$ it suffices to prove that $\lambda(A\backslash M_f) = 0$. By virtue of Problem 512, iii),

$$A\backslash M_f = A\backslash \bigcup_{\nu\in\mathbb{N}^\mathbb{N}} \bigcap_{k=1}^\infty M_{n_1,n_2,\ldots,n_k} \subset A\backslash \bigcup_{\nu\in\mathbb{N}^\mathbb{N}} \bigcap_{k=1}^\infty A_{n_1,n_2,\ldots,n_k}$$

$$\subset \bigcup_{\nu\in\mathbb{N}^\mathbb{N}} \left(\bigcup_{k=0}^\infty \left(A_{n_1,\ldots,n_k} \backslash \bigcup_{m=1}^\infty A_{n_1,\ldots,n_k,m} \right)\right).$$

Since there are only countably many finite subsets of $\mathbb{N}$ there are at most countably many different sets $A_{n_1,\ldots,n_k}\backslash \bigcup_{m=1}^\infty A_{n_1,\ldots,n_k,m}$ and so if they are enumerated say as $\{B_p\}_{p=1}^\infty$ then $\bigcup_{\nu\in\mathbb{N}^\mathbb{N}} \bigcup_{k=0}^\infty (A_{n_1,\ldots,n_k}\backslash \bigcup_{m=1}^\infty A_{n_1,\ldots,n_k,m}) = \bigcup_p B_p$. Thus it suffices to prove that each B_p is a null set. Since (Problem 512, part i)

$$\bigcup_{\nu\in\mathbb{N}^\mathbb{N}} \bigcap_{k=1}^\infty M_{m_1,\ldots,m_i,n_1,\ldots,n_k} = \bigcup_{q=1}^\infty \bigcup_{\nu\in\mathbb{N}^\mathbb{N}} \bigcap_{k=1}^\infty M_{m_1,\ldots,m_i,q,n_1,\ldots,n_k} \subset \bigcup_q A_{m_1,\ldots,m_i,q}$$

it follows that

$$A_{m_1,\ldots,m_i} \backslash \bigcup_q A_{m_1,\ldots,m_i,q} \subset A_{m_1,\ldots,m_i} \backslash \bigcup_{\nu\in\mathbb{N}^\mathbb{N}} \bigcap_{k=1}^\infty M_{m_1,\ldots,m_i,n_1,\ldots,n_k}.$$

Since $A_{m_1,\ldots,m_i}\backslash\bigcup_q A_{m_1,\ldots,m_i,q}$ is a Borel set B_p contained in a set of inner measure zero, $\lambda(B_p) = 0$ and so M_f is Lebesgue measurable. □

NOTE. An examination of the proof shows that the following generalization of Szpilrajn is valid. Let $\mathcal{B}$ be a set of subsets of a set X and assume that $\mathcal{B}$ is closed with respect to the formation of countable unions and complements and furthermore if $E \subset X$ there is in $\mathcal{B}$ a set A such that $A \supset E$ and such that if $E \subset B \in \mathcal{B}$ and $F \subset A\backslash B$ then $F \in \mathcal{B}$. Then for all B in $\mathcal{B}$ and any $f: \mathbb{N}^\mathbb{N} \mapsto \mathcal{B}^\mathbb{N}$, $B_f \in \mathcal{B}$.

516. Since $\mathsf{F}(I)$ is closed with respect to the formation of (arbitrary) intersections, if $S = M_f \in \mathcal{A}(\mathsf{F}(I))$ it may be assumed that f is regular: $S = M_f = \bigcup_{\nu\in\mathbb{N}^\mathbb{N}} \bigcap_{k=1}^\infty f(\nu)_k = \bigcup_{\nu\in\mathbb{N}^\mathbb{N}} \bigcap_k M_{n_1,\ldots,n_k}$, $M_{n_1,\ldots,n_k,n_{k+1}} \subset M_{n_1,\ldots,n_k}$. Let E_i be $\{\nu: \nu = \{i, n_2, n_3, \ldots\}\}$ let E_{ij} be $\{\nu: \nu = \{i, j, n_3, n_4, \ldots\}\}$, etc., and let $T(i)$ be $\bigcup_{\nu\in E_i} \bigcap_k f(\nu)_k$, let $T(i,j)$ be $\bigcup_{\nu\in E_{ij}} \bigcap_k f(\nu)_k$, etc. Then $M_f = \bigcup_i T(i)$, $T(i) = \bigcup_j T(i,j)$, etc.

If $\operatorname{card}(S) > \operatorname{card}(\mathbb{N})$, as in Problem 458 let D be $\{x_0: x_0 \in S$, for every neighborhood $U(x_0)$ of x_0, $\operatorname{card}(U(x_0) \cap S) > \operatorname{card}(\mathbb{N})\}$. Then (Problem 458) $\operatorname{card}(D) > \operatorname{card}(\mathbb{N})$. Choose in D two distinct points d_0, d_1 and let $V_0(d_0)$ and $V_1(d_1)$ be neighborhoods with disjoint closures. Then $\operatorname{card}(V_0(d_0) \cap S)$,

$\text{card}(V_1(d_1)\cap S)>\text{card}(\mathbb{N})$ whence for some i_0, i_1, $\text{card}(T(i_0)\cap \bar{V}_0(d_0))$, $\text{card}(T(i_1)\cap \bar{V}_1(d_1))>\text{card}(\mathbb{N})$. In these sets choose distinct points d_{p0}, d_{p1}, $p=0, 1$, and neighborhoods $V_{pq}(d_{pq})$, $q=0, 1$ with disjoint closures. Continue in this way and produce for each (dyadic) sequence $\{a_m\}_{m=1}^{\infty}$ of zeros and ones a sequence $\{i_{a_1}, i_{a_1a_2}, \ldots\}$ such that for all k,

$$\text{card}(T(i_{a_1}, i_{a_1a_2}, \ldots, i_{a_1a_2\cdots a_k})\cap \bar{V}_{a_1\cdots a_k})>\text{card}(\mathbb{N}).$$

Since each $T(i_{a_1}, i_{a_1a_2}, \ldots, i_{a_1a_2\cdots a_k})\subset M_{i_{a_1}, i_{a_2},\ldots, i_{a_k}}$ it follows that for all k, $\text{card}(M_{i_{a_1}, i_{a_2},\ldots, i_{a_k}}\cap \bar{V}_{a_1a_2\cdots a_k})>\text{card}(\mathbb{N})$.

It may be assumed that $\text{diam}(\bar{V}_{a_1a_2\cdots a_k})<1/k$ and so it follows that for all $\{i_{a_1}, i_{a_2}, \ldots\}$, $\bigcap_k (M_{i_{a_1}, i_{a_2},\ldots, i_{a_k}}\cap \bar{V}_{a_1a_2\cdots a_k})$ is a single point $x_{a_1a_2\cdots}$ belonging to $K=\bigcup_{a_1} \bar{V}_{a_1}\cap \bigcup_{a_1,a_2} \bar{V}_{a_1a_2}\cap\cdots$ and so S. Since K is a Cantor-like discontinuum and $K\subset S$ it follows that $\text{card}(S)=\text{card}(\mathbb{R})$ as required. □

517. Let $\mathcal{M}$ be $\mathsf{F}(I)$. Then $\mathsf{S}_\beta(I)\subset \mathcal{A}(\mathcal{M})$ and hence $H(\mathsf{S}_\beta(I))\subset H(\mathcal{A}(\mathcal{M}))\subset \mathcal{A}(\mathsf{F}(\mathbb{R}))\subset \mathcal{A}(\mathsf{S}_\lambda(\mathbb{R}))$ (Problems 513, and 514). Since $\mathbb{R}$ is the countable union of intervals, the argument of Solution 515 may be extended to show that $\mathcal{A}(\mathsf{S}_\lambda(\mathbb{R}))=\mathsf{S}_\lambda(\mathbb{R})$ and the result follows. □

518. If there is no such x_0 then $f-g$ is always positive or always negative. Thus it may be assumed that there is a positive a such that $f>g+a$. Then $f\circ f>g\circ f+a=f\circ g+a>g\circ g+2a$ and by induction it follows that $f^n=\underbrace{f\circ f\circ\cdots\circ f}_{n}>\underbrace{g\circ g\circ\cdots\circ g}_{n}+na=g^n+na$. Since all maps f^n, g^n are in $C(I, I)$ and $a>0$ a contradiction results. □

Bibliography

[1] Banach, S. 1932. Théorie des opérations linéaires. Monografje Matematyczne, Tom I. Warsaw: Z Subwencji Funduszu Kultury Narodowej. ~ 1963. ~. 2d ed. New York: Chelsea Publishing Co.

[2] Besicovitch, A. S. 1945. On the definition and value of the area of a surface. *Quarterly Journal of Mathematics*. **16**, 86–102.

[3] Dickson, L. E. 1926. Modern algebraic theories. New York: B. H. Sanborn & Co.

[4] Gelbaum, B. R. 1958. Conditional and unconditional convergence in Banach spaces. *Anais de Academia Brasileira de Ciencias*. **30**, 21–27.

[5] —— 1961. Free topological groups. *Proceedings of the American Mathematical Society*. **12**, 737–743.

[6] —— 1976. Independence of events and of random variables. *Wahrscheinlichkeitstheorie*. **36**, 333–343.

[7] ——, Kalisch, G. K. 1952. Measure in semigroups. *Canadian Journal of Mathematics*. **4**, 396–406.

[8] ——, Kalisch, G. K., Olmsted, J. M. H. 1951. On the embedding of topological semigroups and integral domains. *Proceedings of the American Mathematical Society*. **2**, 807–821.

[9] ——, Olmsted, J. M. H. 1964. Counterexamples in analysis. San Francisco: Holden-Day, Inc.

[10] ——, Schanuel, S. 1980. A characterization of the semigroup of matrix units. *Journal of the Australian Mathematical Society* (Series A). **29**, 291–296.

[11] Graves, L. M. 1956. Theory of functions of real variables. 2d ed. New York: McGraw-Hill.

[12] de Groot, J., Dekker, T. 1954. Free subgroups of the orthogonal group. *Compositio Mathematica*. **12**, 134–136.

[13] Halmos, P. R. 1950. Measure theory. New York: Van Nostrand.

[14] Hausdorff, F. 1914. Grundzüge der Mengenlehre. Leipzig: Von Veit.

[15] —— 1927. Mengenlehre. 2. Aufl. Berlin: Gruyter. 1962. Set theory. Translation of the 1937 ed. by John R. Aumann. 2d ed. New York: Chelsea Publishing Co.

[16] Hewitt, E., Stromberg, K. 1965. Real and abstract analysis. New York: Springer-Verlag.
[17] Kakutani, S. 1944. Free topological groups and infinite direct product topological groups. *Proceedings of the Imperial Academy of Japan.* **20**, 595–598.
[18] Köthe, G. 1969. Topological vector spaces, I. New York: Springer-Verlag.
[19] Löwig, H. 1934. Über die Dimension linearer Räume. *Studia Mathematica.* **5**, 18–23.
[20] Pólya, G., Szegö, G. 1925. Aufgaben und Lehrsätze aus der Analysis, I, II. Berlin: Springer-Verlag. ~ 1972. Problems and theorems in analysis. Translation by D. Aeppli. New York: Springer-Verlag.
[21] Rosenblatt, M. 1962. Random processes. New York: Oxford University Press.
[22] Royden, H. L. 1968. Real analysis. New York: Macmillan.
[23] Rudin, W. 1974. Real and complex analysis. New York: McGraw-Hill.
[24] Saks, S. 1937. Theory of the integral. Monografie Matematyczne, Tom VII. New York: G. E. Stechert & Co.
[25] Schaeffer, H. H. 1980. Topological vector spaces. New York: Springer-Verlag.
[26] Yosida, K. 1968. Functional analysis. New York: Springer-Verlag.

Glossary of symbols

$\mathscr{A}$ — For a set $\mathfrak{M}$ of sets and the set $\mathfrak{F}$ of maps f from $\mathbb{N}^{\mathbb{N}}$ to $\mathfrak{M}^{\mathbb{N}}$ the map taking each f into $\bigcup_{\nu \text{ in } \mathbb{N}^{\mathbb{N}}} \bigcap_{k=1}^{\infty} f(\nu)_k$.

$\bar{A}$ — the closure of A

A' — the complement of A

A° — the interior of A

$AC(\mathbb{R}, \mathbb{C})$ — the set of absolutely continuously $\mathbb{C}$-valued functions on $\mathbb{R}$

a.e. — almost everywhere

$A(x)$ — the set of polynomials (in x) over the ring A

$A(E)$ — the algebra (of sets) generated by the set E (of sets)

a.u. — almost uniformly

Aut (E) — the set of automorphisms of (the algebraic structure) E

$B(a, r)$ — in a metric space (X, d) the closed ball $\{x : d(a, x) \leqq r\}$

$BV(\mathbb{R}, \mathbb{C})$ — the set of $\mathbb{C}$-valued functions of bounded variation on $\mathbb{R}$

$\mathbb{C}$ — the set of complex numbers

$C(X, \mathbb{K})$ — the set of continuous $\mathbb{K}$-valued functions on X

$C_0(X, \mathbb{K})$	the set of continuous $\mathbb{K}$-valued functions vanishing at infinity on (the locally compact space) X
$C_{00}(X, \mathbb{K})$	the set of $\mathbb{K}$-valued continuous functions having compact support
$C^{(k)}(\mathbb{R}^n, \mathbb{R})$	the set of $\mathbb{R}$-valued functions having k continuous derivatives on $\mathbb{R}^n$
$c_0(\mathbb{N})$	the set of $\mathbb{C}$-valued sequences converging to zero
$C_b(X, \mathbb{C})$	the set of bounded $\mathbb{C}$-valued continuous functions on X
card (A)	the cardinality of the set A
cont (f)	the set of points of continuity of the function f
conv (S)	the convex hull of the set S
deg (p)	the degree of the polynomial p
df	the differential (derivative) of the map f
$d(\alpha, \beta)$	the distance between the objects (points, sets) α and β
diam (S)	the diameter of the set S
dim (E)	the dimension of the vector space E
det (M)	the determinant of the matrix M
E^*	the dual of the vector space E
$E(f)$	the expected value of the function f
End (E)	the set of endomorphisms of the (algebraic structure) E
Sur (E, F)	the set of homomorphisms of (the algebraic structure) E onto (the algebraic structure) F
$\mathsf{F}(X)$	the set of closed sets in (the topological space) X
f^+ (f^-)	$f \vee 0$ $(-(f \wedge 0))$
$f_1 \vee f_2$	$\frac{1}{2}(f_1+f_2+\|f_1-f_2\|)$
$f_1 \wedge f_2$	$\frac{1}{2}(f_1+f_2-\|f_1-f_2\|)$
$\hat{f}$	for f in $L^1(\mathbb{R}, \lambda)$ the map $\hat{f}: \mathbb{R} \ni t \mapsto \int_{\mathbb{R}} f(x)\, e^{-itx}\, dx/(2\pi)^{1/2}$
$f_{(t)}$	for f in $X^{\mathbb{R}}$, the map $\mathbb{R} \ni x \mapsto f(x+t)$

$f \circ g$	for f in Z^Y and g in Y^X the map $X \ni x \mapsto f(g(x)) \in Z$
f^n	for f in X^X the map $X \ni x \mapsto \underbrace{f(f(\ldots f}_{n}(x)))$, i.e., $f^n = \underbrace{f \circ \ldots \circ f}_{n}$
f_S	for f in Y^X and S a subset of X the map $f: S \mapsto Y$.
F_σ	the union of a countable set of closed sets
G_δ	the intersection of countably many open sets
$\mathbb{H}$	the set of quaternions
$\mathfrak{H}$	Hilbert space
$H(G)$	for a connected open subset G of $\mathbb{C}$ the set of functions holomorphic in G
Hom (E, F)	the set of homorphisms of (the algebraic structure) E into (the algebraic structure) F
I	$[0, 1]$
id	the identity map
iff	if and only if
inf	infimum
$\mathbb{K}$	$\mathbb{R}$ or $\mathbb{C}$
$K(X)$	the set of compact sets in (the topological space) X
LRN	Lebesgue–Radon–Nikodým
l_γ	the length of the curve γ
l^p	$L^p(\mathbb{N}, \nu)$ (ν is counting measure)
$L^p(X, \mu)$	for the measure situation $(X, \boldsymbol{S}, \mu)$, the set of equivalence classes of $\mathbb{C}$-valued measurable functions f such that $\lVert f \rVert_p^p = \int_X \lvert f(x) \rvert^p \, d\mu(x)$ is finite
$L^p_{\mathbb{R}}(X, \mu)$	$L^p(X, \mu) \cap \mathbb{R}^X$
$L^\infty(X, \mu)$	the set of measurable $\mathbb{C}$-valued functions f such that $\lVert f \rVert_\infty = \inf \{M : \lvert f \rvert \leqq M \text{ a.e.}\} < \infty$
$L^\infty_{\mathbb{R}}(X, \mu)$	$L^\infty(X, \mu) \cap \mathbb{R}^X$
Lip (α)	$\mathbb{C}^{\mathbb{R}} \cap \{f : \lvert f(x) - f(y) \rvert \leqq K \lvert x - y \rvert^\alpha$, some K in $[0, \infty)\}$
$\lim_{x \to a} f(x)$	limit of $f(x)$ as $x \to a$
$\limsup_{n = \infty} a_n$	$\inf_n \sup_{m \geqq n} a_m$

$\liminf_{n=\infty} a_n$	$\sup_n \inf_{m\geqq n} a_m$
$\limsup_{x=a} f(x)$	$\inf_{U\ni a} \sup\{f(x): x \text{ in } U, U \text{ open}\}$
$\liminf_{x=a} f(x)$	$\sup_{U\ni a} \inf\{f(x): x \text{ in } U, U \text{ open}\}$
$\limsup_{n=\infty} A_n$	$\bigcap_{n=1}^{\infty} \bigcup_{m\geqq n} A_m$
$\liminf_{n=\infty} A_n$	$\bigcup_{n=1}^{\infty} \bigcap_{m\geqq n} A_m$
lsc	lower semicontinuous
max	maximum
min	minimum
$M(X)$	the Banach space of signed or complex Borel measures on (the topological space) X
$\mathsf{M}(\mathsf{E})$	$\bigcap\{\mathsf{M}: \mathsf{M}$ a monotone subset of 2^X, $\mathsf{M} \supset \mathsf{E}\}$
$\mathfrak{M}(J)$	the set of matrices indexed by $J \times J$
$\mathbb{N}$	the set of positive integers
$(\mathbb{N}, 2^{\mathbb{N}}, \nu)$	measure situation on $\mathbb{N}$ with counting measure ν
$\\|\ldots\\|$	norm of an element in a vector space
$\\|\ldots\\|_p$	norm in $L^p(X, \mu)$
$\\|\ldots\\|_\infty$	norm in $L^\infty(X, \mu)$
osc (f, E)	oscillation of f on E
$o(x)$	a function of x such that $o(x)/\vert x\vert \to 0$ as $\vert x\vert \to 0$
$\mathsf{O}(X)$	the set of open sets in (the topological space) X
$\vert P\vert$	norm of the partition P (of an interval $[a, b]$)
$\mathbb{Q}$	the set of rational numbers
$\mathbb{R}$	the set of real numbers
$\mathsf{R}(\mathsf{E})$	the ring (of sets) generated by the set E (of sets)
$\mathbb{R}^k$	the k-fold Cartesian product of $\mathbb{R}$ with itself
sgn (z)	signum of z
S_N	$\sum_{n=-N}^{N} c_n e^{int}$
span (S)	the span of the set S
sup	supremum
supp	support

S	a sigma ring of sets
S_β	the sigma ring of Borel sets in a topological space
S_λ	the sigma ring of Lebesgue measurable sets in $\mathbb{R}^k$
T^*	for Banach spaces E, F and a T in Hom(E, F), in Hom(F^*, E^*) the map T^* such that $T^*(y^*)(x) = y^*(T(x))$.
$\mathbb{T}$	the set of complex numbers of absolute value one
T_f	total variation (map) of f
T_{fP}	total variation of f with respect to the partition P
$\mathcal{U}(X)$	a uniform structure for X
usc	upper semicontinuous
var	variance
(X, S, μ)	measure situation
2^X	the set of all subsets of X
X^Y	the set of all maps of Y into X
X/Z	quotient algebraic structure for an algebraic structure X and an appropriate substructure Z
X/R	quotient structure of a set X with respect to an equivalence relation R
x/Z	image of x in the quotient structure X/Z
x/R	image of x in X/R
$\mathbb{Z}$	the set of integers (positive, negative, zero)
$f: A \mapsto B$	the map f of A into B
$a_n \to a$	a_n converges to a
$a_n \downarrow a$	a_n descends on a
$a_n \uparrow a$	a_n ascends to a
$x \mapsto f(x)$	the map taking x into $f(x)$
$f_n \to g$ a.e.	f_n converges to g almost everywhere
$[x]$	the greatest integer not greater than the real number x
$\oplus$	direct sum
$\oint$	line integral

$\dfrac{\partial(y_1, y_2, \ldots y_m)}{\partial(x_1, x_2, \ldots, x_n)}$	the Jacobian matrix of the map $\mathbb{R}^n \ni (x_1, x_2, \ldots, x_n) \mapsto (y_1, y_2, \ldots y_m) \ni \mathbb{R}^m$
$\{x : P(x)\}$	the set of all x such that $P(x)$
$D_{\pm}f$	the $\left\{\begin{matrix}\text{right}\\ \text{left}\end{matrix}\right.$ derivative of f
$D^{\pm}f$	the upper $\left\{\begin{matrix}\text{right}\\ \text{left}\end{matrix}\right.$ derivative of f
$M^{\perp}$	for a subset M of a Banach space E, $\{x^* : x^* \text{ in } E^* \text{ and } x^*(M) = 0\}$
$M_{\perp}$	for a subset M of the dual E^* of a Banach space E, $\{x : x \text{ in } E, M(x) = 0\}$
∂S	boundary of the set S
$\|x\|$ (x in $\mathbb{C}^k$)	$(\sum_{i=1}^k x_i\bar{x}_i)^{1/2}$
$A + B$	$\{a + b : a \text{ in } A, b \text{ in } B\}$
$A \backslash B$	$\{x : x \text{ in } A, x \text{ not in } B\}$
$A \Delta B$	$(A\backslash B) \cup (B \backslash A)$
$A \doteq B$	$A \Delta B$ is a null set
δ_{ab}	1 if $a = b$, 0 otherwise
Δ	Laplacian: $C^2(\mathbb{R}^k, \mathbb{C}) \ni \mapsto \sum_{i=1}^k \partial^2 f/\partial x_i^2$; also for a set X, in $X \times X$ the set $\{(x, x) : x \text{ in } X\}$
∇	nabla: $C^1(\mathbb{R}^k, \mathbb{C}) \ni f \mapsto (\partial f/\partial x_1, \ldots, \partial f/\partial x_k)$
$\Delta_h f$	$(f(x+h) - f(x))/h$, $h \neq 0$
λ, λ_k	Lebesgue measure, more specifically in $\mathbb{R}^k$
μ_E	$A \mapsto \mu(A \cap E)$
$\mu \ll \nu$	μ is absolutely continuous with respect to ν
$\mu \perp \nu$	μ and ν are mutually singular
μ_*, μ^*	inner, outer measure
$\mu^{\pm}$	the positive and negative parts of the signed measure μ; if P, N are a Hahn decomposition, then $\mu^+(A) = \mu(A \cap P)$, $\mu^-(A) = \mu(A \cap N)$
$\|\mu\|$	total variation of the complex measure μ
$\mu \times \nu$	the product measure corresponding to μ and ν

Π	symbol for product
ρ^p	p-dimensional Hausdorff measure
$\sigma(A)$	spectrum of the Banach algebra A
$\sigma(E, E^*)$	the weak topology for the Banach space E
$\sigma(E^*, E)$	the weak* topology for the dual space E^* of the Banach space E
$\sigma\mathsf{A}(\mathsf{E})$	the sigma algebra (of sets) generated by the set E (of sets)
$\sigma\mathsf{R}(\mathsf{E})$	the sigma ring (of sets) generated by the set E (of sets)
σ_N	the average of the first $N+1$ partial sums $S_0, S_1, \ldots, S_N$
χ_E	the characteristic function of the set E
Ω	the first uncountable ordinal number

Index/Glossary

A number in parentheses, e.g., (341), refers to both the Problem numbered 341 and to the Solution numbered 341.

Abelian Of a group G, denoting that for all a, b in G, $ab = ba$.

Abel summation The rearrangement of a sum $\sum_{n=1}^{N} a_n b_n$ into the sum $(s_N b_N - s_0 b_0) - \sum_{m=1}^{N-1} s_m(b_{m+1} - b_m)$ where $s_0 = b_0 = 0$ and $s_m = \sum_{n=1}^{m} a_n$. If ν is counting measure on $\{1, 2, \ldots, N\}$ and a_n resp. b_n is written $a(n)$ resp. $b(n)$ then $\sum_{n=1}^{N} a_n b_n = \int_0^N a(n)b(n)\,d\nu(n)$ and the formula for Abel summation is that of integration by parts.

Abel's theorem If $\{b_n\}_{n=1}^{\infty} \subset (0, \infty)$ and $\sum_{n=1}^{\infty} b_n = \infty$ then $\sum_{n=1}^{\infty} b_n/(\sum_{k=1}^{n} b_k)^a$ is finite or infinite according as $a > 1$ or $a = 1$.

absolutely continuous Of an element f in $\mathbb{C}^{\mathbb{R}}$, denoting that for each positive a there is a positive b such that if $a_1 < b_1 \leqq a_2 < b_2 \leqq \cdots \leqq a_n < b_n$ and

$$\sum_{k=1}^{n} (b_k - a_k) < b$$

then $\sum_{k=1}^{n} |f(b_k) - f(a_k)| < a$; for measure situations (X, S, μ_i), $i = 1, 2$, denoting e.g., that $\mu_1(E) = 0$ whenever $\mu_2(E) = 0$ in which case μ_1 is absolutely continuous with respect to μ_2 $(\mu_1 \ll \mu_2)$.

Alaoglu's theorem If E is a Banach space the unit ball $B(0, 1)$ of E^* is compact in the weak* topology $\sigma(E^*, E)$.

algebra A ring A that is a module over a field $\mathbb{K}$.

almost disjoint Of a set $\{A_\gamma\}_\gamma$ in 2^X, denoting that $A_\gamma \cap A_{\gamma'}$ is finite (or empty) whenever $\gamma \neq \gamma'$.

almost everywhere Of a statement, denoting that it holds off a null set (a.e.).

analytic Of an element f in $\mathbb{C}^G$, G an open connected subset of $\mathbb{C}$, denoting that f' exists at each point of G; of an element A in $2^{\mathbb{R}}$, denoting that $A \in \mathscr{A}(\mathsf{F}(\mathbb{R}))$.

approximate identity For a Banach algebra A a net $\{u_\gamma\}_\gamma$ such that for all a in A, $\lim_\gamma \|u_\gamma a - a\| = \lim_\gamma \|au_\gamma - a\| = 0$.

arithmetic vs. geometric mean theorem If $t \in [0, 1]$ and $u, v \geqq 0$ then $u^t v^{1-t} \leqq tu + (1-t)v$.

Arzelà-Ascoli theorem If K is a compact metric space and $\{f_n\}_{n=1}^\infty$ is a uniformly bounded equicontinuous sequence in $\mathbb{C}^K$ there is a subsequence $\{f_{nk}\}$ converging uniformly on K.

atom For a measure situation (X, S, μ), in S an E such that $0 < \mu(E)$ and such that if $E \supset F \in \mathsf{S}$ then $\mu(F) = \mu(E)$ or $\mu(F) = 0$.

automorphism For an algebraic structure A (e.g., a group, a ring, an algebra, a field, a vector space), a bijection $\alpha: A \mapsto A$ that respects the structure; if A is a topological algebraic structure, an automorphism is bicontinuous.

Baire category theorem In a complete metric space the intersection of a countable sequence of dense open sets is dense. In particular a complete metric space is not of the first category.

ball In a metric space (X, d), for a in X and r in $(0, \infty)$, the set $B(a, r) = \{x: d(x, a) \leqq r\}$.

Banach algebra An algebra A that is a Banach space with norm $\|\cdot\cdot\cdot\|$ and such that for a, b in A, $\|ab\| \leqq \|a\| . \|b\|$.

Banach space Over $\mathbb{R}$ or $\mathbb{C}$ a normed vector space complete with respect to the norm-induced metric.

basis For a set X, in 2^X a subset $\mathcal{N}$ such that: i) for all x in X there is in $\mathcal{N}$ at least one U_x containing x; ii) if U_x and V_x are in $\mathcal{N}$ there is in $\mathcal{N}$ a W_x contained in $U_x \cap V_x$; iii) if $y \in U_x$ there is in $\mathcal{N}$ a U_y contained in U_x. The set of all unions of elements (neighborhoods) in $\mathcal{N}$ is the set of open sets of X, the topology of X.

For a vector space without topology (discrete topology), a set $\{x_\gamma\}_\gamma$ (a Hamel basis) such that each x in X is uniquely representable as a finite sum $\sum_{n=1}^N a_{\gamma_n} x_{\gamma_n}$.

For a topological vector space X, a set $\{x_\gamma\}_\gamma$ such that if Δ is the directed set of finite subsets of $\Gamma = \{\gamma\}$ then for each x there is a unique set $\{a_\gamma\}_\gamma$ such that $x = \lim_{\sigma\in\Delta} \sum_{\gamma\in\sigma} a_\gamma x_\gamma$.

For a Banach space X, in 2^X a sequence $\{x_n\}_n$ (a Schauder basis) such that for all x in X there is a unique sequence $\{a_n\}_n$ such that $\|\sum_{n=1}^N a_n x_n - x\| \to 0$ as $N \to \infty$.

Bessel's inequality If $\{f_\gamma\}_\gamma$ is an orthonormal set in a Hilbert space $\mathfrak{H}$ and if $f \in \mathfrak{H}$ then $\sum_\gamma |(f, f_\gamma)|^2 \leqq \|f\|^2$.

bijection A map $f: A \mapsto B$ that is one-one and such that $f(A) = B$.

biorthogonal system For a vector space X and its conjugate space X^*, in $X \times X^*$ a set $\{x_\gamma, x_\gamma^*\}_\gamma$ such that $x_\gamma^*(x_{\gamma'}) = \delta_{\gamma\gamma'}$.

Bochner measurable For a measure situation (X, S, μ), a normed vector space E, and an f in E^X, denoting that there is a sequence $\{f_n = \sum_{m=1}^{M_n} a_{nm}\chi_{A_m}: a_{nm}$ in $\mathbb{C}, A_m$ in $\mathsf{S}\}_n$ such that $\|f_n - f\| \to 0$ in measure as $n \to \infty$ and $\int_X \|f_n(x) - f_m(x)\| \, d\mu(x) \to 0$ as $m, n \to \infty$.

Borel–Cantelli lemma If (X, S, μ) is a measure situation, $\{A_n\}_n \subset \mathsf{S}$, and $\sum_n \mu(A_n) < \infty$ then $\mu(\limsup_{n=\infty} A_n) = 0$ (158).

Borel set For a topological space X, in 2^X an element of the σ-ring generated by the set $\mathsf{K}(X)$ of compact sets in X.

bounded convergence theorem If $\{f_n\}_{n=0}^\infty$ is a sequence of integrable functions *re* the measure situation (X, S, μ), if $|f_n| \leqq |f_0|$ for all n, and $f_n \to g$ a.e. as $n \to \infty$ then g is integrable and $\int_X f_n(x) \, d\mu(x) \to \int_X g(x) \, d\mu(x)$ as $n \to \infty$. In particular, if

$\mu(X)<\infty$ and $|f_n|\leqq M<\infty$ for all n then $|g|\leqq M$ and $\int_X f_n(x)\,d\mu(x)\to \int_X g(x)\,d\mu(x)$ as $n\to\infty$.

bounded variation Of an f in $\mathbb{C}^{\mathbb{R}}$, denoting that there is a finite M such that if $a_1<a_2<\cdots<a_n$, n in $\mathbb{N}$, then $\sum_{k=1}^{n-1}|f(a_{k+1})-f(a_k)|\leqq M$.

boundary For a subset A of a topological space X the set $\{x$: for every neighborhood U_x, $U_x\cap A$ and $U_x\backslash A$ are nonempty$\}$.

bridging function If $a<b$, in $\mathbb{R}^{\mathbb{R}}$ a function f (usually infinitely differentiable) such that $0\leqq f\leqq 1$, $f((-\infty,a])=0=1-f([b,\infty))$.

Brouwer's fixed point theorem If f is a continuous self-map of the unit disc $\bar{U}=\{z: z\in\mathbb{C}, |z|\leqq 1\}$ there is in $\bar{U}$ a z such that $f(z)=z$.

Brouwer's invariance of domain theorem If U is an open subset of $\mathbb{R}^n$ and if $f: U\mapsto V\subset\mathbb{R}^n$ is a homeomorphism then V is also open.

Cantor–Bendixson theorem A closed set in a separable metric space is the union of a (possibly empty) perfect set and a countable set.

canonical basis for $\mathbb{C}^n$ or $\mathbb{R}^n$ The set of vectors $(1,0,\ldots,0)$, $(0,1,\ldots,0)$, $\ldots$, $(0,0,\ldots,1)$.

canonical map For a vector space E the map $E\ni x\mapsto F_x\in E^{**}$ such that for all x^* in E^*, $x^*(x)=F_x(x^*)$.

Caratheodory measurable Of a set A with respect to an outer measure μ^*, denoting that for every set B, $\mu^*(B)=\mu^*(B\cap A)+\mu^*(B\backslash A)$.

cardinal number In the class of sets an equivalence class with respect to the relation of bijectivity between sets. The class of cardinal numbers is ordered: if α and β are cardinal numbers then $\alpha=\beta$ or $\alpha\neq\beta$ and for all A in α and B in β there is an injection $A\mapsto B$ $(\alpha<\beta)$ or for all B in β and all A in α there is an injection $B\mapsto A$ $(\alpha>\beta)$.

Cartesian product For a set $\{X_\gamma\}_\gamma$ of sets the set $\prod_\gamma X_\gamma$ of all maps $f:\{\gamma\}\mapsto\bigcup_\gamma X_\gamma$ such that $f(\gamma)\in X_\gamma$. Alternatively, $\prod_\gamma X_\gamma$ is the set of all "vectors" $(\ldots,x_\gamma,\ldots)$, x_γ in X_γ.

category (first, second) Of a set E in a topological space, denoting that E is (for first) or is not (for second) the union of countably many nowhere dense sets.

Cauchy net In a uniform space a net $\{x_\gamma\}$ such that for every element U of the uniform structure there is a γ_U such that if $\gamma,\gamma'>\gamma_U$ then $(x_\gamma,x_{\gamma'})\in U$.

Cauchy sequence In a metric space (X,d) a sequence $\{x_n\}_{n=1}^\infty$ such that if $a>0$ there is in $\mathbb{N}$ an n_a such that $d(x_n,x_m)<a$ if $n,m>n_a$.

characteristic function For a subset A of a set X, in $\{0,1\}^X$ the map χ_A such that $\chi_A(x)$ is 1 or 0 according as x is or is not in A.

closed graph theorem If f is a linear map from the Banach space E to the Banach space F then f is continuous iff graph$(f)=\{(x,f(x)): x$ in $E\}$ is closed in the product topology for $E\times F$.

closed map For topological spaces X and Y a map $f: X\mapsto Y$ such that graph(f) is closed in the product topology for $X\times Y$.

closed set In a topological space X a set F such that $X\backslash F$ is open.

closure For a set A in a topological space X the set $\bar{A}$ that is the intersection of all closed sets containing A.

cluster point For a set A in a topological space X a point p such that every open set U containing p meets $A\backslash\{p\}$.

coefficient functionals For a basis $\{x_\gamma\}_\gamma$ of a vector space E the maps $x_\gamma^*: E\ni x\mapsto a_\gamma$ such that $\sum_\gamma a_\gamma x_\gamma$ is the (basis) representation for x.

cofinal Of a subset A of a partially ordered set B, denoting that if $b \in B$ there is in A an a such that $a > b$.

commutative Of a group, a ring, or an algebra, denoting that multiplication is independent of the order of the factors: $ab = ba$.

compact operator For a pair E, F of Banach spaces a map $T: E \mapsto F$ carrying bounded sets into sets having compact closures.

compact set In a topological space X a set K such that if K is covered by the union of a set of open sets then a union of finitely many of them also covers K.

complete Of a metric or a uniform space, denoting that every Cauchy sequence resp. Cauchy net converges (to a limit in the space).

completely regular Of a topological space X, denoting that if F is a closed subset and if p is a point not in F then there is in $C(X, [0, 1])$ an f such that $f(p) = 1 = 1 - f(F)$.

complex measure For a measure situation (X, S, μ) a $\mathbb{C}$-valued measure.

complexification The process of extending the validity of an argument made for an $\mathbb{R}$-situation to the analogous $\mathbb{C}$-situation.

component (in a topological space) A subset that is connected and contained properly in no connected subset.

concave (function) In $\mathbb{R}^{\mathbb{R}}$ a function f such that $-f$ is convex.

concentrated Of a measure μ, denoting the existence of a set A such that for every measurable set E, $\mu(E) = \mu(E \cap A)$; alternatively, μ lives on A.

conjugate space For a topological vector space E the vector space E^* consisting of the continuous linear maps of E into the field $\mathbb{K}$ (over which E is a module): $E^* = \mathrm{Hom}(E, \mathbb{K})$.

conjugate exponent For p in $(1, \infty)$ the number $p/(p-1)$, usually denoted q or p'; $1' = \infty$, $p'' = p$.

connected Of a subset A of a topological space X, denoting that for no two disjoint open sets U and V it is true that $A = (A \cap U) \cup (A \cap V)$ and $A \cap U$ and $A \cap V$ are nonempty.

continuous Of a map $f: X \mapsto Y$ between topological spaces, denoting that if V is open in Y then $f^{-1}(V)$ is open in X; of a map as in the preceding but at a point x in X, denoting that if V is open and contains $f(x)$ then there is an open U containing x and such that $f(U) \subset V$.

continuum hypothesis Of cardinalities, denoting that there is no set X such that $\mathrm{card}(\mathbb{N}) < \mathrm{card}(X) < \mathrm{card}(\mathbb{R})$. Paul Cohen showed that not only the continuum hypothesis but also its natural generalization are statements independent of the widely accepted Zermelo–Fraenkel axioms for set theory.

convergence Of a net $\{x_\gamma\}_\gamma$ taking values in a topological space, denoting that for some x and any neighborhood of x the net is eventually in the neighborhood; of a sequence $\{f_n : X \mapsto \mathbb{C}\}_{n=1}^{\infty}$ with respect to a measure situation (X, S, μ), denoting i) $f_n \to f$ off a null set (a.e.) or ii) for each positive a there is a measurable set E such that $\mu(E) < a$ and $f_n \to f$ uniformly off E (a.u.) or iii) for each positive a, $\mu\{x: |f_n(x) - f(x)| \geqq a\} \to 0$ (in measure) or iv) $\|f_n - f\|_p \to 0$ (in norm or in $L^p(X, \mu)$) or v) $f_n(x) \to f(x)$ for all x (everywhere) or vi) $\|f_n - f\|_\infty \to 0$ (uniformly), each for some f as $n \to \infty$. Since $\mathbb{N}$ in its natural ordering is partially ordered, convergence of a sequence in a topological space is a special case of convergence of a net.

convex Of a function f in $\mathbb{R}^{\mathbb{R}}$, denoting that for x, y in $\mathbb{R}$ and t in $[0, 1]$, $f(tx + (1-t)y) \leqq tf(x) + (1-t)f(y)$; of a set in a vector space, denoting that if x and y are in the set and $t \in [0, 1]$ then $tx + (1-t)y$ is also in the set.

convex hull For a subset A of a vector space E, the intersection of all convex sets containing A; equivalently the set $\{tx+(1-t)y: x, y \text{ in } A, t \text{ in } [0,1]\}$.
convolution The (binary) operation

$$L^1(\mathbb{R}^n, \lambda)\times L^1(\mathbb{R}^n, \lambda)\ni (f,g)\mapsto \left(x\mapsto \int_{\mathbb{R}^n} f(x-y)g(y)\,dy\right) = f*g\in L^1(\mathbb{R}^n,\lambda).$$

coset In a group G and for a subgroup H, a set aH, a in G; in a ring R or an algebra A and for an ideal J, a set $a+J$, a in R resp. A.
coset representative For a coset, any of its elements.
countable Of a set X, denoting that $\text{card}(X)=\text{card}(\mathbb{N})$.
countably additive For a field $\mathbb{K}$, a σ-ring S of sets, and a set function Φ in $\mathbb{K}^{\mathsf{S}}$, denoting that if $\{A_n\}_{n=1}^{\infty}$ is a disjoint sequence in S then $\Phi(\bigcup_n A_n)=\sum_n \Phi(A_n)$.
countably compact Of a topological space X, denoting that for every infinite set in X there is in X a cluster point of the set.
countably subadditive For a field $\mathbb{K}$, a σ-ring S of sets, and a set function Ψ in $\mathbb{K}^{\mathsf{S}}$, denoting that if $\{A_n\}_{n=1}^{\infty}$ is a sequence of sets in S then $\Psi(\bigcup_n A_n)\leqq \sum_n \Psi(A_n)$.
counting measure For the measure situation $(X, 2^X, \nu)$, if $E\subset X$ then $\nu(X)$ is $\text{card}(E)$ or ∞ according as E is finite or infinite.
cover For a topological space X a subset $\{A_\gamma\}_\gamma$ of 2^X and such that $\bigcup_\gamma A_\gamma = X$; usually a cover consists of open sets.
curve For a topological space X a continuous map $\gamma: [0,1]\mapsto X$.
cylinder set In a Cartesian product $\prod_\gamma X_\gamma$ a subset A such that for some subset Λ of $\{\gamma\}$ and a set $\{A_\lambda :, \lambda \text{ in } \Lambda, A_\lambda \text{ a subset of } X_\lambda\}$, $\{x_\gamma\}\in A$ iff for all λ in Λ $x_\lambda\in A_\lambda$.

Daniell integral construction For a set X, a linear space and sublattice L_0 of $\mathbb{R}^X$ in its natural order, and a linear functional $I: L_0\mapsto\mathbb{R}$ such that i) $I(f)\geqq 0$ if $f\geqq 0$ (I is positive) and ii) $I(f_n)\downarrow 0$ if $f_n\downarrow 0$, the extension of I to a countably additive integral on a lattice L_1 containing L_0.
Darboux's theorem If $f\in\mathbb{R}^{(a,b)}$, if f' exists on (a,b), if $a<c<d<b$, and if $f'(c)<q<f'(d)$ then there is in (c,d) a p such that $f'(p)=q$.
degree Of a polynomial $\sum_k a_k x^k$, denoting $\max\{k: a_k\neq 0\}$.
dense Of a set A in a topological set X, denoting that $\bar{A}=X$.
dense-in-itself Of a set A in a topological space X, denoting that A as a topological space with topology induced from that of X has no isolated points; alternatively, for each x in A and any open U containing x, $U\cap(A\backslash\{x\})\neq\emptyset$.
derivation For an algebra A an endomorphism D such that for a, b in A, $D(a\,.\,b)=D(a)\,.\,b+a\,.\,D(b)$.
derivative See differential.
diameter Of a set A in a metric space (X,d), $\sup\{d(x,y): x, y \text{ in } A\}$.
differential Of a map $f: E\mapsto F$ between normed vector spaces, in $(\text{Hom}(E,F))^E$ the map df such that for all x in E,

$$\lim_{\substack{h\neq 0\\ \|h\|\to 0}} \|h\|^{-1}.\|f(x+h)-f(x)-df(x)(h)\|=0.$$

dimension Of a vector space E, denoting the cardinality of any (hence every) Hamel basis of E.
Dini's theorem If $\{f_n\}_{n=1}^{\infty}\subset C(K,\mathbb{R})$, K is compact, and $f_n\downarrow 0$ then $f_n\to 0$ uniformly as $n\to\infty$.

direct sum   For a set $\{X_\gamma\}_\gamma$ of vector spaces a vector space X and a set $\{f_\gamma : X_\gamma \mapsto X\}$ of monomorphisms such that every x in X is uniquely expressible as a finite sum $\sum_\gamma f_\gamma(x_\gamma)$.

directed set   A partially ordered set Γ such that if $\gamma, \gamma' \in \Gamma$ there is in Γ a γ'' such that $\gamma'' > \gamma, \gamma'$.

discrete measure   For a measure situation $(X, 2^X, \delta)$ a map $w: X \mapsto [0, \infty)$ such that for E in 2^X, $\delta(E) = \sum_{x \text{ in } E} w(x)$.

discrete topology   For a set X the topology 2^X (every set is open).

disjoint   Of a set of subsets of a set X, denoting that any pair of different subsets have an empty intersection.

distance   In a metric space (X, d), between two points p and q, $d(p, q)$; between a point p and a set B, $\inf_{b \text{ in } B} d(p, b)$; between two sets A and B, $\sup_{a \text{ in } A} d(a, B)$.

dominated convergence theorem   See bounded convergence theorem.

dual space   See conjugate space.

dyadic construction   For a set X and for every dyadic rational number $\sum_{k=1}^{K} a_k 2^{-k}$, $a_k = 0$ or 1, K in $\mathbb{N}$, the construction of a set $A_{a_1 a_2 \ldots a_K}$ such that i) if $\{a_1, a_2, \ldots, a_K\} \neq \{a_1', a_2', \ldots, a_K'\}$ then $A_{a_1 a_2 \ldots a_K} \cap A_{a_1' a_2' \ldots a_K'} = \varnothing$ and ii) for all L in $\mathbb{N}$, $A_{a_1 a_2 \ldots a_K} \supset A_{a_1 a_2 \ldots a_K a_{K+1} \ldots a_{K+L}}$.

dyadic discontinuum   In a complete metric space (X, d), if $t = \sum_{k=1}^{\infty} a_k 2^{-k} \in [0, 1]$, $a_k = 0$ or 1, and if (see preceding) all sets A are compact and $\lim_{K \to \infty} \operatorname{diam}(A_{a_1 a_2 \cdots a_K}) = 0$, the set $\bigcup_{t \text{ in } [0,1]} \bigcap_{K=1}^{\infty} A_{a_1 a_2 \cdots a_K}$.

dyadic rational number   A number of the form $n + \sum_{k=1}^{K} a_k 2^{-k}$, $a_k = 0$ or 1, n in $\mathbb{Z}$.

Eberlein's theorem   A Banach space is reflexive iff its unit ball is weakly sequentially compact in the weak topology.

Egorov's theorem   If (X, S, μ) is a measure situation, $\mu(X) < \infty$, and $\{f_n\}_{n=1}^{\infty}$ is a sequence of measurable functions converging a.e. to f as $n \to \infty$ then for each positive a there is a measurable set E_a such that $\mu(E_a) < a$ and such that $f_n \to f$ uniformly on $X \backslash E_a$.

endomorphism   For an algebraic structure A (a group, a ring, a field, an algebra, a vector space, etc.,) a homomorphic self-map $f: A \mapsto A$; f is continuous if A is a topological algebraic structure.

equicontinuous   For a pair X, Y of uniform spaces and a set $\{f_\gamma : X \mapsto Y\}_\gamma$ of maps, denoting that if V is an element (vicinity) of the uniform structure of Y there is in the uniform structure of X a vicinity U such that for all γ if $(x, x') \in U$ then $(f_\gamma(x), f_\gamma(x')) \in V$.

equivalence class   For an equivalence relation R on a set X, a set $\{y : yRx\} = R(x)$.

equivalence relation   For a set X, in $X \times X$ a subset R such that for all x, $(x, x) \in R$, if $(x, y) \in R$ then $(y, x) \in R$, and if (x, y), (y, z) are in R then $(x, z) \in R$; usually $(x, y) \in R$ is written xRy or even $x \in R(y)$.

ergodic theorem   If (X, S, μ) is a measure situation, if $T: X \mapsto X$ is a bijection preserving together with T^{-1} the measure and measurability of all measurable sets, and if $f \in L^1(X, \mu)$ then $F(x) = \lim_{n \to \infty} (f(x) + f(T(x)) + \cdots + f(T^{n-1}(x)))/n$ exists a.e., $F \in L^1(X, \mu)$, and for every measurable set E such that $T(E) = E$, $\int_E F(x)\, d\mu(x) = \int_E f(x)\, d\mu(x)$ (481–484).

essentially bounded   Of a measurable function f, denoting the existence of a null set E and in $[0, \infty)$ an M such that off E, $|f| \leqq M$; the least such M is $\|f\|_\infty$.

essential supremum For an essentially bounded function f, the number $\|f\|_\infty$.
eventually Of a net $\{x_\gamma\}_\gamma$ and a property P, denoting that there is a γ_0 such that if $\gamma > \gamma_0$ then x_γ enjoys property P.
expected value For a measure situation (X, S, μ) such that $\mu(X) = 1$ and an f in $L^1(X, \mu)$ the number $\int_X f(x)\, d\mu(x) = E(f)$.
extended $\mathbb{R}$-valued function For a set X an element of $(\mathbb{R} \cup \{-\infty, \infty\})^X$.
extreme point For a convex set K in a vector space E, a point p in K and such that if q, r are in K, if $t \in [0, 1]$, and if $p = tq + (1-t)r$ then either $t = 0$ or 1 or $q = r = p$.

Fatou's lemma If (X, S, μ) is a measure situation and $\{f_n\}_{n=1}^\infty$ is a sequence of nonnegative measurable functions then

$$\int_X \liminf_{n=\infty} f_n(x)\, d\mu(x) \leqq \liminf_{n=\infty} \int_X f_n(x)\, d\mu(x).$$

Fejèr's theorem The averages of the partial sums of the Fourier series of an f in $C(\mathbb{T}, \mathbb{C})$ converge uniformly to f.
field A commutative ring in which each nonzero element has a unique multiplicative inverse.
Fourier series For an f in $L^1(\mathbb{T}, \lambda)$, the series $\sum_{n=-\infty}^\infty c_n e^{inx}$, $c_n = \int_{-\pi}^{\pi} f(x)\, e^{-inx}\, dx/2\pi$.
Fourier integral For an $L^1(\mathbb{R}, \lambda)$ the map $t \mapsto \int_\mathbb{R} f(x)\, e^{-itx}\, dx/(2\pi)^{1/2}$.
frequently Of a net $\{x_\gamma\}_\gamma$ and a property P, denoting that for each γ_1 there is a γ_2 such that $\gamma_2 > \gamma_1$ and such that x_{γ_2} enjoys the property P.
Fubini's theorem If $(X_i, \mathsf{S}_i, \mu_i)$, $i = 1, 2$, are σ-finite measure situations and if f is in $L^1(X_1 \times X_2, \mu_1 \times \mu_2)$ then $\int_{X_1 \times X_2} f(x_1, x_2)\, d(\mu_1 \times \mu_2)(x_1, x_2) = \int_{X_1} (\int_{X_2} f^{x_1}(x_2)\, d\mu_2(x_2))\, d\mu_1(x_1) = \int_{X_2} (\int_{X_1} f^{x_2}(x_1)\, d\mu_1(x_1))\, d\mu_2(x_2)$.
function See map.
F_σ In a topological space a union of countably many closed sets.

$\mathbf{G}_\delta$ In a topological space an intersection of countably many open sets.
generalized nilpotent In a Banach algebra an element x such that $\lim_{n\to\infty} \|x^n\|^{1/n} = 0$.
graph For a map $f: X \mapsto Y$ in $X \times Y$ the set $\{(x, f(x)) : x \text{ in } X\}$.
Gram-Schmidt process The passage from a linearly independent sequence $\{x_n\}_{n=1}^\infty$ in a normed vector space X to a sequence $\{y_n, y_n^*\}_{n=1}^\infty$ in $X \times X^*$ and such that $y_n^*(y_m) = \delta_{nm}$, viz., $y_1 = x_1/\|x_1\|$, $y_1^*(y_1) = 1$; if $\{y_k, y_k^*\}_{k=1}^K$ are constructed so that $y_k^*(y_l) = \delta_{kl}$, then

$$y_{K+1} = \frac{x_{K+1} - \sum_{k=1}^K y_k^*(x_{K+1}) y_k}{\left\| x_{K+1} - \sum_{k=1}^K y_k^*(x_{K+1}) y_k \right\|}$$

and $y_{K+1}^*(y_k) = \delta_{K+1,k}$, $k = 1, 2, \ldots, K+1$.
Green's theorem If S is open in $\mathbb{R}^2$, ∂S is the union of finitely many rectifiable Jordan curves, and P, Q are in $C^\infty(\mathbb{R}^2, \mathbb{C})$ then $\oint_{\partial S} P(x, y)\, dx + Q(x, y)\, dy = \iint_S (\partial Q(x, y)/\partial x - \partial P(x, y)/\partial y)\, dx\, dy$.
group A set G and a map $G \times G \ni (a, b) \mapsto a \,.\, b \in g$ such that $(a \,.\, b) \,.\, c = a \,.\, (b \,.\, c)$ and such that for all a, b in G the equations $ax = b$ and $xa = b$ have solutions; usually $a \,.\, b$ is written ab.

Haar measure For a locally compact group G a measure μ such that $(G, \sigma\mathsf{R}(\mathsf{K}(G)), \mu)$ is a measure situation, such that if E is a measurable set then so is xE for all x in G and $\mu(xE)=\mu(E)$, and such that if K is compact and U is open and nonempty then $0 \leqq \mu(K) < \infty$ and $0 < \mu(U) \leqq \infty$.

Hahn–Banach theorem If E is a topological vector space, if K is a convex set such that $K^0 \neq \varnothing$ and if F is a subspace such that $F \cap K^0 = \varnothing$ then there is in E a hyperplane H ($\dim(E/H)=1$) such that $H \cap K^0 = \varnothing$. Alternatively: If E is a vector space and if $p: E \mapsto [0, \infty)$ is a map satisfying i) $p(x+y) \leqq p(x)+p(y)$ and ii) for t in $\mathbb{C}$, $p(tx)=|t|p(x)$, if F is a subspace of E, if $f \in \mathrm{Hom}(F, \mathbb{C})$, and if for all y in F, $|f(y)| \leqq p(y)$ then there is in $\mathrm{Hom}(E, \mathbb{C})$ an f_1 such that for all x in E, $|f_1(x)| \leqq p(x)$ and $f_1 = f$ on F.

Hahn decomposition For a measure situation (X, S, μ), μ signed, a partition $P, N (P \cap N = \varnothing,\ P \cup N = X)$ such that for all A in S, $\mu(A \cap P) \geqq 0$, $\mu(A \cap N) \leqq 0$.

half-open rectangle In $\mathbb{R}^n$ a set $\prod_{i=1}^n [a_i, b_i)$, $a_i < b_i$.

Hamel basis See basis (of a vector space).

Hausdorff measure For a metric space (X, d) and p in $(0, \infty)$ the map $2^X \ni Y \mapsto \rho^p(Y) = \sup_{\varepsilon>0} \rho_\varepsilon^p(Y) = \sup_{\varepsilon>0} (\inf\{\sum_{n=1}^\infty (\mathrm{diam}(U_n))^p : U_n \text{ open}, \ \bigcup_n U_n \supset Y, \mathrm{diam}(U_n) < \varepsilon\})$; ρ^p is a Caratheodory outer measure.

Hausdorff maximality principle If S is a partially ordered set there is a linearly ordered subset properly contained in no other linearly ordered subset.

Hausdorff space A topological space X such that if p, q are in X and $p \neq q$ then there are disjoint open sets U, V such that $p \in U$ and $q \in V$.

Heine–Borel theorem A countably compact metric space is compact.

Helly selection principle If $\{f_n\}_{n=1}^\infty$ is a sequence in $BV(I, \mathbb{C})$ and if for all n, $\|f_n\|_\infty + T_{f_n}(I) \leqq M < \infty$ then there is a subsequence $\{f_{n_k}\}_{k=1}^\infty$ such that $f = \lim_{k\to\infty} f_{n_k}$ exists everywhere, $f \in BV(I, \mathbb{C})$, and $\|f\|_\infty + T_f \leqq M$.

Hessian For an f in $C^2(\mathbb{R}^n, \mathbb{C})$ the matrix $(\partial^2 f/\partial x_i \partial x_j)_{i,j=1}^n$; alternatively, the matrix representation with respect to the canonical basis for $\mathbb{R}^n$ of the linear map $d^2 f$.

Hilbert space The vector space $\mathfrak{H}$ over $\mathbb{C}$ and endowed with a norm $\|\cdots\|$ satisfying $\|x+y\|^2 + \|x-y\|^2 = 2(\|x\|^2 + \|y\|^2)$ (from which is derived an inner product $(x, y) = \frac{1}{4}(\|x+y\|^2 - \|x-y\|^2 + i(\|x+iy\|^2 - \|x-iy\|^2))$.

Hölder's inequality If $p \geqq 1$ and (X, S, μ) is a measure situation then for f in $L^p(X, \mu)$ and g in $L^q(X, \mu)$, $fg \in L^1(X, \mu)$ and $\|fg\|_1 \leqq \|f\|_p \cdot \|g\|_q$.

holomorphic See analytic.

homeomorphism For topological spaces X and Y a bijection $f: X \mapsto Y$ such that f and f^{-1} are continuous.

homogeneous For vector spaces E and F, of a map $f: E \mapsto F$, denoting that for t in $\mathbb{C}$ and some p in $\mathbb{R}$, $f(tx) = t^p f(x)$.

homomorphism For a pair A, B of algebraic structures (groups, rings, fields, vector spaces, etc.) a map $h: A \mapsto B$ respecting the algebraic structures of A and B; h is continuous if A and B are topological algebraic structures.

hyperplane In a vector space E a subspace F such that $\dim(E/F) = 1$; alternatively, a proper subspace F such that F is contained in no other proper subspace (F is a maximal proper subspace).

ideal In a ring R (an algebra A) a proper subset J such that for all x, y in J, z in $R(A)$, and t in $\mathbb{C}$, $x+y$, xz, zx, and tx are in J.

idempotent   In a ring an element x such that $x^2 = x$.
identity theorem   If f is analytic in a connected open subset G of $\mathbb{C}$, if A is a subset of G and there is in G a cluster point of A, and if $f(A) = 0$ then $f(G) = 0$.
independent   For a measure situation (X, S, μ) such that $\mu(X) = 1$, of a set $\{f_\gamma\}_\gamma$ of measurable functions from X to $\mathbb{R}^n$, denoting that for every finite set $\{A_k\}_{k=1}^K$ of Borel sets in $\mathbb{R}^n$ and every finite set $\{f_{\gamma k}\}_{k=1}^K$ of distinct functions, $\mu(\bigcap_{k=1}^K f_{\gamma k}^{-1}(A_k)) = \prod_{k=1}^K \mu(f_{\gamma k}^{-1}(A_k))$.
infimum   For a subset A of a partially ordered set S an x such that for no a in A is true that $a < x$ and also such that if y has the same property then $y \not> x$ (to the extent that the phrase is meaningful, "x is a greatest lower bound of A").
inner measure   For a measure situation (X, S, μ), the map $\mu_*: 2^X \ni A \mapsto \sup\{\mu(B): B \text{ in } \mathsf{S}, B \subset A\}$.
interior   For a set A in a topological space, the union of the open subsets of A.
intermediate value theorem   See Darboux's theorem.
isolated point   In a topological space a point that is also an open set.
isomorphism   A bijective homomorphism; in a topological algebraic context, also bicontinuous.

Jacobian   For the map $f: \mathbb{R}^m \mapsto \mathbb{R}^n$, $f = (f_1, f_2, \ldots, f_n)$, the matrix $(\partial f_i/\partial x_j)_{j=1,i=1}^{m\;\;\;n}$; alternatively the matrix representing df with respect to the canonical bases for $\mathbb{R}^m$ and $\mathbb{R}^n$.
Jensen's inequality   If (X, S, μ) is a measure situation, $\mu(X) = 1$, $g \in L^1(X, \mu)$, $g(X) \subset (a, b)$, and f is convex on (a, b) then $\int_X f(g(x))\, d\mu(x) \geqq f(\int_X g(x)\, d\mu(x))$ (102).
Jordan content   Lebesgue measure restricted to the ring generated by the half-open rectangles of $\mathbb{R}^n$.
Jordan curve   A homeomorphism of $\mathbb{T}$ into $\mathbb{R}^2$.
jump discontinuity   For an f in $\mathbb{R}^{\mathbb{R}}$ a point x such that $\lim_{\substack{a \to 0 \\ a > 0}} f(x + a)$ and $\lim_{\substack{b \to 0 \\ b > 0}} f(x - b)$ exist and are unequal.

Krein–Milman theorem   If K is a compact convex subset of a locally convex topological vector space and if E is the set of extreme points of K then the closure $\overline{\text{span}(E)}$ of the span of E is K.
Kronecker's lemma   If $\sum_{n=1}^\infty a_n/n$ converges then $N^{-1} \sum_{n=1}^N a_n \to 0$ as $N \to \infty$ (477).

lattice   A partially ordered set L such that if x, y are in L then $\sup(x, y) = x \vee y$ and $\inf(x, y) = x \wedge y$ in L.
Lebesgue measure   For the measure situation $(\mathbb{R}^n, \sigma\mathsf{R}(\mathsf{K}(\mathbb{R}^n)), \lambda_n)$ the measure $\lambda_n = K_n \rho^n$, ρ^n being Hausdorff measure and K_n chosen so that for a rectangle $\prod_{i=1}^n [a_i, b_i)$, $\lambda_n(\prod_{i=1}^n [a_i, b_i)) = \prod_{i=1}^n (b_i - a_i)$; alternatively, the unique extension of the Jordan content defined by the last formula to the σ-ring generated by the rectangles.
Lebesgue's theorem on derivatives   If $f \in L^1(\mathbb{R}, \lambda)$ then

$$\lim_{\substack{h \to 0 \\ h \neq 0}} h^{-1} \int_0^h |f(x+t) - f(x)|\, dt = 0 \text{ a.e.}$$

Lebesgue–Radon–Nikodým theorem If (X, S, μ_i), $i=1,2$, are measure situations such that $\mu_1(X)+\mu_2(X)<\infty$ then there are unique measures μ_a and μ_s such that $\mu_1=\mu_a+\mu_s$, $\mu_a \ll \mu_2$, $\mu_s \perp \mu_2$, $\mu_s \perp \mu_a$. There are various extensions to σ-finite, complex, signed, decomposable, etc. measures [16].

left-continuous Of an f in $\mathbb{R}^{\mathbb{R}}$, denoting that $\lim_{\substack{h\to 0\\ h>0}} |f(x-h)-f(x)|=0$.

length For a curve $\gamma: I \mapsto X$ in a metric space

$$(X, d), \quad \sup_{\substack{0=t_0<t_1<\cdots<t_n=1\\ n \text{ in } \mathbb{N}}} \sum_{k=1}^{n-1} d(\gamma(t_{k+1}), \gamma(t_k)).$$

limit point See cluster point.

Lindelöf covering theorem If $\{U_\gamma\}_\gamma$ is an open cover of a separable topological space X then there is a countable subcover $\{U_{\gamma_n}\}_{n=1}^{N\leqq\infty}$.

Lindelöf space A topological space such that every open cover admits a countable subcover.

linear Of a map f between vector spaces, denoting that for a, b in the underlying field, $f(ax+by)=af(x)+bf(y)$.

linearly independent Of a set S in a vector space, denoting that if $\{x_i\}_{i=1}^n \subset S$ and $\{a_i\}_{i=1}^n \subset \mathbb{K}$ then $\sum_i a_i x_i = 0$ iff all a_i are zero.

linearly ordered Of a partially ordered set S, denoting that if x, y are in S then $x=y$ or $x<y$ or $x>y$.

Lipschitz constant For an f in $\text{Lip}(\alpha)$ and for an x a constant K_x such that $|f(x)-f(y)| \leqq K_x |x-y|^\alpha$.

lives Of a measure, denoting that it is concentrated on some set.

locally compact Of a topological space, denoting that each point is contained in an open set having compact closure.

locally convex Of a topological vector space, denoting that it has a basis of convex neighborhoods.

lower semicontinuous Of an f in $\mathbb{R}^X$, denoting that for each a in $\mathbb{R}$, $f^{-1}((a,\infty))$ is open; of f at a point x, denoting that $\liminf_{y=x} f(y)=f(x)$.

map For two sets X and Y, in $X\times Y$ a subset f such that for each x in X there is a unique y such that (x, y) is in f; frequently denoted $f: X \mapsto Y$.

matrix For two sets A and B and a ring R, a map $M: A\times B \mapsto R$; if M and N are two matrices their sum as maps is well-defined as is their product $MN: (a, b) \mapsto \sum_c M(a,c)\,.\,N(c,b)$ if the last sum can be given some meaning, e.g., because it is finite, or because there is in R a notion of convergence and the series converges.

matrix unit For a set J a map $U: J\times J \mapsto \{0, 1\}$ such that for at most one pair (p, q) in $J\times J$, $U(p,q)\neq 0$.

maximum For a subset S of a partially ordered set, an element of $\sup(S)\cap S$; of an $\mathbb{R}$-valued function, the maximum of its range.

maximal biorthogonal For a vector space E, of a subset $\{x_\gamma, x_\gamma^*\}_\gamma = S$ in $E\times E^*$ denoting that $x_\gamma^*(x_{\gamma'})=\delta_{\gamma\gamma'}$ and that S is properly contained in no set enjoying the same property.

maximal ideal In a ring or algebra an ideal properly contained in no other ideal.

maximal orthonormal Of a set $\{x_\gamma\}_\gamma = S$ in Hilbert space, denoting that $(x_\gamma, x_{\gamma'}) = \delta_{\gamma\gamma'}$, and that S is contained properly in no set of the same kind.

mean ergodic theorem If (X, S, μ) is a measure situation, $f \in L^2(X, \mu)$ and $T: X \mapsto X$ is a self-map such that for every measurable set E, $T(E)$ and $T^{-1}(E)$ are measurable and $\mu(T(E)) = \mu(E)$ then $\{x \mapsto (N+1)^{-1} \sum_{n=0}^{N} f(T^n(x))\}_{N=0}^{\infty}$ is a norm-Cauchy sequence in $L^2(X, \mu)$ (341).

measurable Of a map $T: X_1 \mapsto X_2$ for two measure situations (X_i, S_i, μ_i), $i = 1, 2$, denoting that for all A in S_2, $T^{-1}(A) \in S_1$; of a set in, say, X_1, denoting that it is an element of S_1.

measure situation A triple consisting of a set X, a σ-ring S of subsets of X, and a countably additive set function (measure) $\mu: S \mapsto \mathbb{C}$; $\mu(\emptyset) = 0$ and unless it is further qualified, μ assumes only nonnegative values; if μ is signed it may assume at most one of the "values" $\pm\infty$; if μ is $\mathbb{C}$-valued only values in $\mathbb{C}$ are assumed.

metric density theorem If E is a measurable subset of $\mathbb{R}$ then $c(x) = \lim_{\substack{a \to 0 \\ a > 0}} \lambda(E \cap (x-a, x+a))/2a$ exists a.e. and $c(x) = \chi_E(x)$ a.e.

metric space A set X and a map $d: X \times X \mapsto [0, \infty)$ such that $d(x, y) = 0$ iff $x = y$, $d(x, y) = d(y, x)$, and $d(x, z) \leqq d(x, y) + d(y, z)$.

minimum For a subset S of a partially ordered set, an element of $\inf(S) \cap S$; of an $\mathbb{R}$-valued function, the minimum of its range.

Minkowski's inequality If $p \geqq 1$, f, $g \in L^p(X, \mu)$ then $f + g \in L^p(X, \mu)$ and $\|f+g\|_p \leqq \|f\|_p + \|g\|_p$.

module An abelian group G, such that for some ring R there is a map $R \times G \ni (r, g) \mapsto rg \in G$ such that $r(a+b) = ra + rb$, $r(sa) = (rs)a$, and $(r+s)a = ra + sa$. The preceding describes a left R-module G; analogous definitions apply for a right R-module and for an R-bimodule, usually called an R-module.

monomorphism An injective homomorphism.

monotone Of a function f in $\mathbb{R}^{\mathbb{R}}$, denoting that if $x \leqq y$ then $f(x) \leqq f(y)$ (for monotone increasing f), $f(x) \geqq f(y)$ (for monotone decreasing f); of a set M in 2^X, denoting that if $\{M_n\}_{n=1}^{\infty} \subset M$ and if $M_n \subset M_{n+1}$ then $\bigcup_n M_n \in M$, and if $M_n \supset M_{n+1}$ then $\bigcap_n M_n \in M$.

monotone convergence theorem If (X, S, μ) is a measure situation, $\{f_n\}_{n=1}^{\infty}$ is a sequence of measurable functions, $0 \leqq f_n \leqq f_{n+1}$, either $f = \lim_{n\to\infty} f_n$ exists a.e. and $f \in L^1(X, \mu)$, in which case $\int_X f_n(x)\, d\mu(x) \uparrow \int_X f(x)\, d\mu(x)$, or $\int_X f_n(x)\, d\mu(x) \uparrow \infty$.

morphism For two sets X, Y, a map $f: X \to Y$ respecting algebraic or topological or order structures of X and Y.

neighborhood In a topological space, for a point in the space, an open set containing the point.

net A map of a directed set into a set (usually a topological space).

nilpotent Of an element x of a ring, denoting that for some n in $\mathbb{N}$, $x^n = 0$.

nonatomic Of a measure situation, denoting the absence of atoms.

norm For a vector space E a map $\|\cdot\cdot\|: E \mapsto [0, \infty)$ such that for t in $\mathbb{C}$, x, y in E, $\|x\| = 0$ iff $x = 0$, $\|tx\| = |t| . \|x\|$, $\|x+y\| \leqq \|x\| + \|y\|$.

normal Of a set of functions in $\mathbb{C}^X$, denoting that the set is precompact with respect to the topology induced by the norm $\|\cdot\cdot\|_\infty$; of a topological space, denoting that any pair of disjoint closed sets are subsets of disjoint open sets; of a subgroup H of a group G; denoting that for all a in G, $aHa^{-1} = H$.

normed space A vector space endowed with a norm.

nowhere dense Of a subset of a topological space, denoting that the interior of the closure of the subset is empty.
null set A set of measure zero.

open map For topological spaces X, Y a map $f: X \mapsto Y$ such that if V is open in X then $f(V)$ is open in Y.
open mapping theorem If X and Y are Banach spaces and $f \in \mathrm{Sur}(X, Y)$ then f is open.
ordered See partially ordered.
ordinal number With respect to the equivalence relation of order-preserving bijectivity, an equivalence class of well-ordered sets.
orthogonal Of a set $\{x_\gamma\}_\gamma$ in a Hilbert space, denoting that $(x_\gamma, x_{\gamma'}) = 0$ if $\gamma \neq \gamma'$.
orthonormal Of an orthogonal set, denoting that each element has norm one.
oscillation For an f in $\mathbb{C}^X$ and a subset E of X, $\sup_{x,y \text{ in } E} |f(x) - f(y)|$.
outer measure A countably subadditive map $\mu^*: 2^X \to [0, \infty)$ such that $\mu^*(\emptyset) = 0$.

paracompact Of a topological space, denoting that each open cover admits a refinement such that some neighborhood of each point meets only finitely many elements of the refinement (each open cover admits a neighborhood-finite refinement).
Parséval's theorem If $\{x_\gamma\}_\gamma$ is a maximal orthonormal set in a Hilbert space $\mathfrak{H}$ and $x \in \mathfrak{H}$ then $\|x\|^2 = \sum_\gamma |(x, x_\gamma)|^2$.
partially ordered Of a set X, denoting in $X \times X$ a subset R such that for all x, $(x, x) \in R$, and if (x, y) and (y, z) are in R so is (x, z) in R; usually $x < y$ is written instead of $(x, y) \in R$.
partition For a set X, in 2^X a subset $\{A_\gamma\}_\gamma$ of pairwise disjoint sets such that their union is X; for an interval $[a, b]$ the intervals $[a, x_1), [x_1, x_2), \ldots, [x_{n-1}, b]$ $(x_k < x_{k+1})$.
partition of unity For a topological space X a subset $\{f_\gamma\}_\gamma$ of $C(X, \mathbb{R})$ and satisfying $0 \leqq f_\gamma$, for all x, $\{\gamma: f_\gamma(x) \neq 0\}$ is finite, and $\sum_\gamma f_\gamma = 1$; a partition of unity is subordinate to an open cover $\{U_\gamma\}_\gamma$ if each f_γ is zero off U_γ.
perfect set A closed set in which each point is a cluster point (a closed, dense-in-itself set).
Plancherel theorem If $f \in L^1(\mathbb{R}, \lambda) \cap L^2(\mathbb{R}, \lambda)$ then $\hat{f} \in L^2(\mathbb{R}, \lambda)$ and $\int_{\mathbb{R}} |\hat{f}(t)|^2 \, dt = \|f\|_2^2$.
point measure A measure μ for which there is a point p such that for every measurable set E, $\mu(E) = 1$ or 0 according as p is or is not in E.
principal ideal In a ring R an ideal J such that for some x in R, $J = xR = Rx$.
Product See Cartesian product.
product measure For a set $\{(X_\gamma, \mathsf{S}_\gamma, \mu)\}_\gamma$ of measure situations, if $\{\gamma\}$ is finite and otherwise iff $\mu_\gamma(X_\gamma) = 1$ for all γ, the measure situation $(\prod_\gamma X_\gamma, \prod_\gamma \mathsf{S}_\gamma, \prod_\gamma \mu_\gamma)$ in which $\prod_\gamma \mathsf{S}_\gamma$ is the σ-ring generated by $\{\prod_{\gamma \in \sigma} A_\gamma \times \prod_{\gamma' \notin \sigma} X_{\gamma'}: \sigma \text{ finite}, A_\gamma \in \mathsf{S}_\gamma\}$, and $\prod_\gamma \mu_\gamma(\prod_{\gamma \in \sigma} A_\gamma \times \prod_{\gamma' \notin \sigma} X_{\gamma'}) = \prod_{\gamma \in \sigma} \mu_\gamma(A_\gamma) \prod_{\gamma' \notin \sigma} \mu_{\gamma'}(X_{\gamma'})$.
product topology For a set $\{X_\gamma\}_\gamma$ of topological spaces and their product $\prod_\gamma X_\gamma$ the topology having as a basis of neighborhoods $\{\prod_{\gamma \in \sigma} U_\gamma \times \prod_{\gamma' \notin \sigma} X_{\gamma'}: \sigma \text{ finite}, U_\gamma \text{ open in } X_\gamma\}$.
projection for a vector space E an idempotent endomorphism of E.

quaternions The set $\{q: q = a + bi + cj + dk,\ a, b, c, d \text{ in } \mathbb{R},\ i^2 = j^2 = k^2 = -1, ij = k, jk = i, ki = j\}$ regarded as an algebra with basis $\{1, i, j, k\}$ over $\mathbb{R}$.

quotient For a set X and an equivalence relation R, the set of R-equivalence classes; for a group, algebra, etc., xRy iff $xy^{-1} \in H$ (a normal subgroup) or iff $x - y \in J$ (an ideal), etc.

quotient map For a set X and an equivalence relation R the map $X \ni x \mapsto x/R$.

quotient norm For a normed space E, a closed subspace F, and the quotient space $E/F = \{Q\}$ the norm $\|\cdot\cdot\cdot\|: E/F \ni Q \mapsto \inf\{\|x\|: x \text{ in } Q\}$.

quotient topology For a topological space X, a set Y, and a map $f: X \mapsto Y$, on Y the strongest topology such that f is continuous.

Rademacher functions The functions $x \mapsto \text{sgn}(\sin 2^n \pi x)$, n in $\mathbb{N} \cup \{0\}$.

radical In a commutative Banach algebra, the set of generalized nilpotent elements.

range Of map f in Y^X, the set $f(X)$.

real analytic Of an element f in $\mathbb{R}^{\mathbb{R}}$, denoting that for each a in $\mathbb{R}$ there is in $(0, \infty)$ an r_a such that for some sequence $\{a_n\}_{n=0}^{\infty}$ in $\mathbb{R}$ and all x in $(a - r_a, a + r_a)$ $f(x) = \sum_{n=0}^{\infty} a_n (x-a)^n$.

rectifiable Of a curve $\gamma: I \mapsto X$, (X, d) a metric space, denoting that the length l_γ is finite.

refinement For a set $\{U_\gamma\}_\gamma$ of sets, a set $\{V_\lambda\}_\lambda$ of sets such that each V is a subset of some U and $\cup_\gamma U_\gamma = \cup_\lambda V_\lambda$; of a partition $\{[a, x_1), [x_1, x_2), \ldots, [x_{n-1}, b]\}$ a partition $\{[a, y_1), [y_1, y_2), \ldots, [y_{m-1}, b]\}$ such that each x_p is some y_q.

reflexive Of a Banach space E, denoting that the canonical map $E \ni x \mapsto F_x = (E^* \ni y^* \mapsto y^*(x)) \in E^{**}$ is a bijection.

regular Of an ideal in a ring, denoting that the associated quotient ring is unital; of a map $f: \mathbb{N}^{\mathbb{N}} \to \mathcal{M}^{\mathbb{N}}$ ($\mathcal{M}$ a set of sets) denoting that if $f(\nu) = (f(\nu)_1, f(\nu)_2, \ldots)$ then $f(\nu)_k \supset f(\nu)_{k+1}$; of Borel a measure μ, denoting that for every Borel set E, i) $\mu(E) = \sup\{\mu(K): K \text{ compact}, K \subset E\}$ and ii) $\mu(E) = \inf\{\mu(U): U \text{ open}, U \supset E\}$; if i) obtains, μ is inner regular; if ii) obtains μ is outer regular; of a topological space X, denoting that if $x \in U$ and U is open there is an open V such that $x \in V \subset \bar{V} \subset U$.

Riemann–Lebesgue lemma If $f \in L^1(\mathbb{T}, \lambda)$ then $c_n = (2\pi)^{-1} \int_{\mathbb{T}} f(x)\, d\lambda(x) \to 0$ as $|n| \to \infty$; if $f \in L^1(\mathbb{R}, \lambda)$ then $\hat{f}(t) \to 0$ as $|t| \to \infty$ (c_n and $\hat{f}$ vanish at infinity).

Riesz representation theorems If $\mathfrak{H}$ is a Hilbert space and $x^* \in \mathfrak{H}^*$ there is in $\mathfrak{H}$ a y such that for all x in $\mathfrak{H}$, $x^*(x) = (x, y)$. If X is a locally compact topological space and $F \in C_0(X, \mathbb{C})^*$ there is a complex Borel measure μ such that for all f in $C_0(X, \mathbb{C})$, $F(f) = \int_X f(x)\, d\mu(x)$.

ring An algebraic structure consisting of a set R and two maps, $R \times R \ni (x, y) \mapsto x + y \in R$ and $R \times R \ni (x, y) \mapsto x\,.\,y \in R$; $+$ is commutative and associative; . is associative and distributive over $+$: $x(y+z) = xy + xz$; $(y+z)x = yx + zx$; $\sim$ of sets. For a set X a subset S of 2^X; S is closed under the formation of set differences and finite unions.

right continuous Of an f in $\mathbb{R}^{\mathbb{R}}$, denoting that $\lim_{\substack{h \to 0 \\ h > 0}} |f(x+h) - f(x)| = 0$.

Rolle's theorem If $f \in C([a, b], \mathbb{R})$, f is differentiable on (a, b), and $f(a) = f(b) = 0$ there is in (a, b) a c such that $f'(c) = 0$.

running water lemma (F. Riesz) If f is bounded and in $\mathbb{R}^{(0,1)}$ and if $S = \{x : x$ in $(0, 1)$, there is in $(x, 1)$ an x' such that $\limsup_{y=x} f(y) < f(x')\} \neq \emptyset$ then S is open and the countable union of pairwise disjoint intervals (a_n, b_n) and if $x \in (a_n, b_n)$ then $f(x) \leqq \limsup_{y=b_n} f(y)$. (110, 111).

scattered Of a set A in a topological space X, denoting that A contains no nonempty perfect subset.
Schauder (or S-) basis See basis.
Schwarz inequality In Hilbert space the inequality $|(x, y)| \leqq \|x\| . \|y\|$; in $L^2(X, \mu)$, the Hölder inequality when $p = 2$.
self-map For a set X an f in X^X.
semigroup A set S and a map $S \times S \ni (x, y) \mapsto x . y \in S$; . is assumed to be associative.
separable Of a topological space X, denoting the existence in X of a countable set $\{U_n\}_{n=1}^{\infty}$ of open sets such that every open set is the union of some of the U_n; equivalently for a metric space (X, d), denoting the existence of a countable dense subset.
separately continuous Of a map $f: \prod_\gamma X_\gamma \mapsto Y$, denoting that for each γ, f regarded as a function of x_γ, the other $x_{\gamma'}$, $\gamma' \neq \gamma$, held fixed, is continuous.
separating Of a set A in $\mathbb{R}^X$, denoting that if $x, y \in X$ and $x \neq y$ then there is in A an f such that $f(x) \neq f(y)$.
sequentially compact Of a topological space, denoting that every infinite set contains a convergent subsequence.
sigma algebra (σ-algebra) For a set X a subset A of 2^X; A is assumed to be closed with respect to the formation of complements and countable unions.
sigma finite (σ-finite) Of a measure situation, denoting that every measurable set is the countable union of sets of finite measure.
sigma ring (σ-ring) For a set X a subset S of 2^X; S is assumed to be closed with respect to the formation of (set) differences and countable unions.
signed Of a measure, denoting that its range is contained either in $[-\infty, \infty)$ or in $(-\infty, \infty]$.
signum function The map

$$\text{sgn}: \mathbb{C} \ni z \mapsto \begin{cases} 0, & \text{if } z = 0 \\ |z|/z, & \text{if } z \neq 0 \end{cases}; \qquad z\,\text{sgn}(z) = |z|.$$

simple curve A curve $\gamma: I \mapsto X$ such that γ is bijective.
singular Of two measures denoting that they live on disjoint sets.
span Of a set S in a vector space, denoting $\{\sum_{i=1}^n a_i x_i : a_i$ in $\mathbb{K}$, x_i in S, n in $\mathbb{N}\}$.
spectrum Of a commutative Banach algebra, denoting the set of regular maximal ideals; alternatively, if A is the Banach algebra, the spectrum of A is $\text{Sur}(A, \mathbb{C})$.
step function A linear combination of characteristic functions of intervals in $\mathbb{R}$.
Stieltjes measure For an f in $BV(\mathbb{R}, \mathbb{R})$ the Borel measure μ such that $\mu([a, b)) =$ the total variation of f on $[a, b) = T_f([a, b))$.
Stone-Weierstrass theorem If X is a compact Hausdorff space and if A is a separating algebra of continuous $\mathbb{R}$-valued functions on X then the $\|\cdot\cdot\cdot\|_\infty$-closure of A is either $C(X, \mathbb{R})$ or, for some x_0 in X, $C_0(X\backslash\{x_0\}, \mathbb{R})$.
stronger Of one of two topologies, denoting that its set of open sets includes the set of open sets of the other.
strong law of large numbers If $\{f_n\}_{n=1}^{\infty}$ is an independent sequence, $E(f_n) = 0$,

and $\sum_{n=1}^{\infty} \operatorname{var}(f_n)/n^2 < \infty$ then $N^{-1}\sum_{n=1}^{N} f_n \to 0$ as $N \to \infty$ (478).
subadditive Of an f in $\mathbb{R}^X$, X a vector space, denoting that $f(x+y) \leqq f(x) + f(y)$; of a set function Φ, denoting that $\Phi(A \cup B) \leqq \Phi(A) + \Phi(B)$.
subcover For a cover $\{V\}$ of a set X, a set $\{U\}$ that is both a subset of $\{V\}$ and a cover of X.
subspace In a vector space, a subset that is also a vector space.
support For a topological space X and a measure situation $(X, \sigma\mathsf{R}(\mathsf{K}(X)), \mu)$, $X \backslash \bigcup\{U\colon U$ open and measurable, $\mu(U) = 0\}$; for an f in $C(X, \mathbb{C})$, $\overline{X \backslash f^{-1}(0)}$.
supremum for a subset A of a partially ordered set S an x such that for no a in A is it true that $a > x$ and also such that if y has the same property then $y \not< x$ (to the extent that the phrase is meaningful, "x is a least upper bound of A").
surjection In X^Y an f such that $f(Y) = X$.
Suslin set See analytic set.
Suslin system For a set $\mathscr{M}$ of sets the set $\mathscr{A}(\mathscr{M})$.
symmetric difference Of two sets A and B, denoting $(A \backslash B) \cup (B \backslash A)$.

tail In a directed set Γ and for γ_0 in Γ, $\{\gamma\colon \gamma > \gamma_0\}$.
Taylor's formula If $f \in C^{n+1}(\mathbb{R}, \mathbb{C})$ then for all a in $\mathbb{R}$,

$$f(x) = \sum_{k=0}^{n} \frac{f^{(k)}(a)(x-a)^k}{k!} + \int_a^x \frac{f^{(n+1)}(t)(x-t)^n\, dt}{n!}$$

Tietze's (extension) theorem If X is a normal topological space, if F is a closed subset of X, and if $f \in C_b(F, \mathbb{R})$ there is in $C_b(X, \mathbb{R})$ an $\tilde{f}$ such that $\tilde{f} = f$ on F and $\|\tilde{f}\|_\infty = \|f\|_\infty$.
Tonelli's theorem If $(X_i, \mathsf{S}_i, \mu_i)$, $i = 1, 2$, are measure situations, both σ-finite, and if f is a nonnegative $\mathsf{S}_1 \times \mathsf{S}_2$-measurable function then all integrals that follow exist as extended $\mathbb{R}$-valued functions and

$$\int_{X_1} \left(\int_{X_2} f^{x_1}(x_2)\, d\mu_2(x_2) \right) d\mu_1(x_1) = \int_{X_2} \left(\int_{X_1} f^{x_2}(x_1)\, d\mu_1(x_1) \right) d\mu_2(x_2)$$

$$= \int_{X_1 \times X_2} f(x_1, x_2)\, d(\mu_1 \times \mu_2)(x_1, x_2).$$

topology For a set X, in 2^X a set $\mathscr{T}$ containing $\varnothing$, X, and closed with respect to the formation of arbitrary unions and finite intersections; the members of $\mathscr{T}$ are the open sets of X.
topological group A Hausdorff topological space G that is a group such that the map $G \times G \ni (x, y) \mapsto xy^{-1} \in G$ is continuous.
topological vector space A Hausdorff topological group and a vector space E such that $\mathbb{K} \times E \ni (a, x) \mapsto ax \in E$ is continuous.
total variation For an f in $\mathbb{C}^{\mathbb{R}}$, the function $T_f\colon \mathbb{R} \ni x \mapsto \sup\{\sum_{k=1}^{n-1} |f(x_{k+1}) - f(x_k)|\colon -\infty < x_1 < x_2 < \cdots < x_n = x,\ n$ in $\mathbb{N}\}$.
totally bounded Of a set A in a metric space (X, d), denoting that if $a > 0$ there is in A a finite subset $\{a_n\}_{n=1}^N$ such that $\bigcup_n B(x_n, a)^0 \supset A$.
totally disconnected Of a subset D of a topological space X, denoting that the only connected subsets of D are the empty set and the points of D.
translate For a group G, t in G, and an f in X^G the map $f_{(t)}\colon x \mapsto f(x+t)$.
triangle inequality In a metric space (X, d) the inequality $d(x, z) \leqq d(x, y) + d(y, z)$; in a normed space the (corresponding) inequality $\|x + y\| \leqq \|x\| + \|y\|$.

uniform boundedness principle If X and Y are Banach spaces and $\{T_\gamma\}_\gamma \subset \text{Hom}(X, Y)$ then $\sup_\gamma \|T_\gamma\| < \infty$ iff for all x in X, $\sup_\gamma \|T_\gamma(x)\| < \infty$.
uniformly integrable For a measure situation (X, S, μ), of a set $\{f_\gamma\}_\gamma$ in $L^1(X, \mu)$, denoting that if $a > 0$ there is a positive b such that whenever $\mu(E) < b$ then $|\int_E f_\gamma(x)\, d\mu(x)| < a$ for all γ; $\sim$ in the sense of Hewitt-Stromberg, denoting that if $a > 0$ there is in $\mathbb{N}$ an n such that for all γ, $\int_{\{x:|f_\gamma(x)| \geqq k\}} |f_\gamma(x)|\, d\mu(x) < a$ if $k \geqq n$.
uniform space A set X and in $2^{X \times X}$ a subset $\mathcal{U}$ such that i) if $U \in \mathcal{U}$ then $U \supset \Delta = \{(x, x): x \text{ in } X\}$, ii) if U_1, U_2 are in $\mathcal{U}$ there is in $\mathcal{U}$ a U_3 contained in $U_1 \cap U_2$, iii) if $U \in \mathcal{U}$ there is in $\mathcal{U}$ a V such that $V . V^{-1} = \{(x, y)$: there are in V an (x, z) and a $(y, z)\} \subset U$, iv) if $U \in \mathcal{U}$ and $U \subset V$ then $V \in \mathcal{U}$, if v) $\bigcap_{U \text{ in } \mathcal{U}} U = \Delta$ then X is separated; $\mathcal{U}$ is a uniform structure of X or a uniformity for X and its elements U are vicinities.
uniformly continuous Of a map f in X^Y, X, Y uniform spaces, denoting that for each X-vicinity U there is a y-vicinity V such that if $(x, y) \in V$ then $(f(x), f(y)) \in U$.
unital Of a ring (or an algebra), denoting that there is in the ring (algebra) an identity.
unitary map For two Hilbert spaces $\mathfrak{H}_1$, $\mathfrak{H}_2$, in $\text{Hom}(\mathfrak{H}_1, \mathfrak{H}_2)$ an isomorphism U such that $(U(x), U(y)) = (x, y)$.
upper semicontinuous Of an f in $X^{\mathbb{R}}$, denoting that for all a in $\mathbb{R}$, $f^{-1}((-\infty, a))$ is open; of f at a point x, denoting that $\limsup_{y \to x} f(y) = f(x)$.

vanish at infinity For a locally compact space X, of functions f in $C(X, \mathbb{C})$, denoting that for each positive a there is a compact set K_a such that $|f| < a$ off K_a.
variance For a measure situation (X, S, μ) such that $\mu(X) = 1$ and an f in $L^2(X, \mu)$ denoting $E((f - E(f))^2)$.
variation See total variation.
vicinity See uniform space.